PHYSICAL CHEMISTRY

Quantum Mechanics

PHYSICAL CHEMISTRY
Quantum Mechanics

Horia Metiu
University of California
Santa Barbara

 Taylor & Francis
Taylor & Francis Group

New York • London

Vice President	Denise Schanck
Senior Editor	Robert L. Rogers
Assistant Editor	Summers Scholl
Senior Publisher	Jackie Harbor
Production Editor	Simon Hill
Copyeditor	Ruth Callan
Cover Designer	Joan Greenfield
Indexer & Typesetter	Keyword Group
Printer	RR Donnelley

Cover image courtesy of Jeffry Madura

ISBN 0 8153 4087 7

Library of Congress Cataloging-in-Publication Data

Metiu, Horia, 1940-
  Physical chemistry : quantum mechanics / Horia Metiu.
    p. cm.
  ISBN 0-8153-4087-7 (acid-free paper)
  1. Quantum theory–Mathemtics–Textbooks. 2. Quantum theory–Problems, exercises, etc. I. Title.

QC174.17.M35M45 2006
530.12–dc22

                      2005032155

Published in 2006 by Taylor & Francis Group, LLC,
270 Madison Avenue, New York, NY 10016, USA and
4 Park Square, Milton Park, Abingdon, Oxon, OX14 4RN, UK.

Printed in the United States of America on acid-free paper.

10 9 8 7 6 5 4 3 2 1

CONTENTS

Chapter 18   The hydrogen atom   295

Chapter 20 The electronic structure of molecules: The H_2 molecule   353

Chapter 21 Nuclear magnetic resonance and electron spin resonance   **421**

Supplementary Resources Disclaimer

Additional resources were previously made available for this title on CD. However, as CD has become a less accessible format, all resources have been moved to a more convenient online download option.

You can find these resources available here: www.routledge.com/9780815340874

Please note: Where this title mentions the associated disc, please use the downloadable resources instead.

PREFACE

Most people write a textbook because they have not found one that presents the appropriate material in the proper way. I am no exception to this rule and here is why. When I started teaching Physical Chemistry, I was disappointed by the books available to me. We implement physical chemistry, in our professional work, by using extensive computations. Until very recently these were too laborious for classroom work. As a result, textbook writers developed what I call "pedagogical physics" or "physics with abacus": a version of physical chemistry watered down to allow doing homework with a calculator. The simplest calculations in quantum mechanics involve complicated integrals, making plots or solving eigenvalue problems for matrices of higher dimension than 2×2. Avoiding these calculations meant that there were very few meaningful examples that the student could work through.

After making such simplifications, it makes no sense to compare theory with experiments. Unless a sanitized version of reality is chosen (low pressure and high temperature, small temperature ranges, very simple reactions, etc.), the calculations disagree with the measurements, giving physical chemistry a bad name. As a result, the textbooks fail to show how the theory being taught adds value to the experimental work or to technology.

I believe that we can change this situation and teach a more realistic physical chemistry, than we currently do, by using software that allows the student to overcome the fear of mathematics and the tedium of the computations. I have in mind symbolic manipulation languages, such as **Mathematica**® or **Mathcad**®, which make it possible to easily perform the calculus manipulations that physical chemistry requires, and to quickly write programs that perform extensive numerical calculations on a personal computer. Since most students own a computer these days, using one in the class room no longer requires the existence of a computer laboratory in the department.

Since such computing is possible, a new textbook is needed in which the simplifications are removed (this requires rewriting much of the theory), realistic examples are used (which requires writing programs), and extensive comparisons with experiments are provided (which requires searching the library for data). Moreover, the topics omitted, because they were mathematically or computationally too complicated, need to be reinserted. This book, together with volumes on kinetics, thermodynamics, and statistical mechanics, is an attempt to provide such a textbook.

I developed these volumes while teaching physical chemistry at the University of California at Santa Barbara. It is perhaps useful to share our experience and explain how we have used computing in the classroom. When we started (about six years ago) very few students owned a computer and not a single chemistry student in the class knew how to use a computer to perform scientific calculations. This was rather appalling: it is, in my opinion, unacceptable to give a chemistry degree to people who are unable to perform calculations that require more than a hand-held calculator. It also posed severe constraints on what we could do. My colleague Alec Wodtke decided to help and he created a new "laboratory" course in which he was going to teach the students how to program with **Mathematica**. At the time Alec made this decision, he did not know **Mathematica**, but was eager to learn it for personal use. In about two weeks he became a decent **Mathematica** programmer and was a "master" by the time he finished teaching the class. Learning **Mathematica** can be a rather pleasant and entertaining learning experience. My task was to write new lecture notes and the **Mathematica** programs accompanying them and to teach the physical chemistry class.

Since we did not have enough computers for all the students enrolled in the course, we created a "**Mathematica** track" and offered it to a group of twenty to thirty students. This caused additional complications: I had to write lecture notes for all students, regardless whether they used a computer or not. In addition, I had to give two sets of homework. This additional labor was a great stroke of luck, because it prevented me from writing a book of physical chemistry dependent upon **Mathematica**. This textbook can be used to teach physical chemistry without a computer; but the experience is enriched substantially for those students who do learn how to read and write **Mathematica** or **Mathcad** programs.

The lecture notes and the **Mathematica** programs were posted on the web and were used in the classroom (no textbook was recommended), for three years in a row. A lack of classroom space made it very difficult to schedule separate lectures to teach programming. Because of this, our students started learning **Mathematica** and physical chemistry at the same time. The tutorial in **Mathematica** was structured so that in a few weeks the students were able to do the homework in the physical chemistry course. The **Mathematica** Workbooks that I posted on the web provided a "programming template" that the students could follow. After two weeks of Alec's tutoring, the students managed to use **Mathematica** to do their homework. While this arrangement worked for us, a more rational approach would be to teach programming in advance of physical chemistry. My preference would be to start immediately after the students finished learning calculus, before they take physics.

The choice of **Mathematica** was not based on "market research." I have used **Mathematica** almost daily since it appeared and I was sure that it would do the job well. As the lecture notes turned into a book my editor, Bob Rogers, wisely decided to expand our audience and recruited Professor Jeffry D. Madura (Duquesne University) to produce a **Mathcad** version of my **Mathematica** programs. His programs and mine are included in the CD-ROM included with the book. Whether one uses **Mathematica** or **Mathcad** depends mostly on whether the campus has a license for one or the other, and on the personal history of the instructor.

Changing the tools we use, changes the work we do. Comparison with experiments is now possible and it has become an integral part of the course. After all, we teach this material mainly for future use in practice, even though at times we teach to provide amusement and illumination.

Increasing the level of the material presented in the class entails some dangers. The students easily can become lost or frustrated if the pace is too fast or key details are skipped. Because of this, I paid very close attention to clarity. I have tried to determine what an average student can understand, by working closely with the students during office hours. I was also lucky that Celia Wrathall decided that typing the lecture notes provided her with an opportunity to learn physical chemistry. She peppered me with questions, admonitions, and suggestions regarding the clarity of the material and of the phrasing of the text. No paragraph in the book survived if Celia did not understand it.

Making the material clearer is usually a matter of dosage. I tried to outline the line of thought at the beginning and to break it up into smaller, logically connected parts.

I did not hesitate, as an argument developed, to review how far we had gotten and where we are going. I hope that a better understanding is a valuable compensation for increasing the length of the text.

Physical chemistry is where the undergraduate science and engineering student encounters the rigorous, quantitative methods of science. We teach certain material, but we also teach the method of using physics and mathematics carefully and precisely for solving physics problems of interest to chemistry. With this in mind, I avoided sloppy arguments and "linguistic physics," the habit of covering up ignorance with vague nomenclature. If a limitation in the background of the student made it impossible to prove or explain correctly a statement, I said so. I did not try to create the illusion of understanding by using fuzzy language.

One of the most difficult duties of a textbook writer is deciding what material to teach. Severe time constraints make this task harder. In the 1930s physical chemistry was taught for a year and the course covered only thermodynamics and kinetics. Since then, we have added quantum mechanics and statistical mechanics, without adding a single hour of lecture. This puts a tremendous pressure on us to condense, eliminate, and consolidate the content.

Introductory quantum mechanics poses special difficulties because its mathematical complexity makes it impossible to teach it properly in one quarter. While I did explain the mathematical structure and notation carefully, in the applications I had to give the results, the physics leading to them, and a careful explanation of their use in applications. The emphasis is on quantum physics rather than on quantum mechanics, on quantum phenomena rather than on quantum theory. In spectroscopy I emphasized transitions and made contact with experiment. For chemical bond I presented the hydrogen molecule with a minimal basis set, in a way that is as close as possible to that used in current applications of quantum chemistry. In other words I emphasized those aspects of the presentation that are more general and are used in current research. For example, rather than present valence bond theory, I calculated the properties of the hydrogen molecule by using a six-configurations CI. The book has more NMR than usual in textbooks, for several reasons. NMR is one of the few areas of quantum mechanics where one can do accurate calculations with a small effort. It is very appropriate for illustrating the methodology for solving quantum problems. In addition, NMR is probably the most important subject in physical chemistry: more chemists use NMR than any other physical method.

The choice of material was guided not only by its utility to the practicing chemist and engineer, but also by the fact that physical chemistry ought to teach a chemist how to use physics and how to interpret the results of such usage. Much of the material is concerned with a detailed examination of physical consequences of the theory. In other words, most calculations are followed by one question: what does this mean for the person in the laboratory?

I have never tried to teach all the material in this book in the quarter reserved for quantum mechanics, out of fear that by being taught too much the students will learn too little. It is said that Victor Weisskopf once conversed with a young assistant professor who was excited by how much material he managed to present to his class. Weisskopf remarked: "you young people like to cover a lot; I like to uncover a little." It is better to teach less material well, than to go too fast through a lot of material. After all, the purpose of this course is to introduce the students to quantum mechanics in a way that allows them to read a more advanced book when they need to. To accomplish this, one must explain clearly the principles and the manner in which they are used; there is no need to cover all possible applications.

The material in chapters 1–12 can be presented briefly, with emphasis on the basic concepts, with the bulk of material left for individual study. Chapters 13–18 involve many calculations that should be only outlined in the class. The instructor could better spend the time by emphasizing the manner in which the problem is set up and the physical meaning of the results. The remaining chapters have such a high density of new concepts that they must be covered in detail. Sometimes, I present the diatomic molecule in an abbreviated form and spend more time on NMR. In other times, I give more details on the diatomic molecule and condense the chapter on NMR. Finally, I note a few conspicuous absences. I do not discuss symmetry for a simple reason: to do it right it would take too much time, considering its diminishing importance, and doing it briefly leads to a presentation devoid of content. I have always thought that one should not learn at all something that one could not learn well. There are many books that present symmetry thoroughly and chemists who think they need it could learn it there.

Many recent textbooks give long lists of references at the end of every chapter. I have tried to avoid this, since long lists increase the entropy of information, not its quality. I give a few general texts at the end of the book, which can be consulted with benefit in case of confusion or offer a good start for further study.

Finally, it is a tradition in preface writing to thank the people who helped the author along the way. I want to thank all the students who came to office hours, since they taught me what the average student knows and what he or she can do. Various teaching assistants have also helped me stay anchored in reality. Professor Michelle Francl (Bryn Mawr) has been a priceless reviewer. Her comments did much to improve the book and her wit made me chuckle often. I am also grateful to Professor Hannes Jónsson (University of Washington), Professor Dmitrii Makarov (University of Texas), and Professor Steve Buratto (University of California at Santa Barbara) for useful advice and encouragement. Professor Flemming Hansen (Technical University of Denmark) also provided helpful comments. My editor Bob Rogers believed in this book and supported it and it has been a pleasure to work with him. Summers Scholl at Taylor & Francis provided useful comments and kept the project on track. My assistant Celia Wrathall typed and read the text, re-derived and tested all equations, asked many questions that helped clarify the text, and was an invaluable partner throughout the writing of the book. I am indebted to her for her intelligence and hard work. This book could not have been written without the love, the support, the humor, and the patience of my wife Jane.

HOW TO USE THE WORKBOOKS, EXERCISES, AND PROBLEMS

Workbooks on CD-ROM

A special feature of this book is the use of symbolic manipulation programs to perform some of the derivations and the calculations that show how to use the theory. These **Mathematica** and **Mathcad** programs are organized into Workbooks, which collect all the calculations related to a given chapter. All chapters that have a Workbook associated with them have an icon placed after their title. For easy reference, these ☁ icons also appear in the margin of the textbook page wherever a specific Workbook is first used and should be consulted.

These computer programs are auxiliary materials. Each chapter in the book explains fully how the calculation is done and gives the necessary intermediate steps and results. The student can follow them without consulting the Workbook. In addition, every program in the Workbooks contains a summary of the theory used in the calculation and can be understood by a student who has not learned programming. To help students who do not have much programming experience the programs are as simple as possible. If a more sophisticated syntax is necessary, its meaning is carefully explained.

The **Mathcad** programs were written by Professor Jeffry Madura (Duquesne University) and they follow, to the extent that the syntax allow, the flow of the

Mathematica programs. All references to **Mathematica** programs in the text are also valid for the **Mathcad** programs.

I believe that one does not understand a theory if one is not able to turn it into a flow chart which can be translated into an algorithm. Because of this I have avoided providing programs where the student types some numbers in a box, clicks on a button and gets a table or a graph. Turning a theory into a working algorithm deepens the understanding of the theory and is an art that can be learned only through supervised practice. For this reason, the students are encouraged to write programs, for solving problems, rather than use "canned" software. Students can imitate the programs, which cover all the applications of the theory taught in the text, or use them for inspiration.

One advantage of using **Mathematica** or **Mathcad** is that it allows the use of more advanced theory and of real data, to connect classroom chemistry with laboratory practice. By presenting realistic calculations and comparing them with the experimental results, the Workbooks prepare students for their future professional lives.

The Workbooks also make it possible to work through realistic examples without the tedium of hand calculations. The students can focus on the physical chemistry phenomena, rather than the routine work of computation. By being able to process a large amount of data in milliseconds, they can study how phenomena change when the relevant parameters are modified. For example, one can calculate the equilibrium composition of complex reaction for many temperatures and pressures.

The use of **Mathematica** or **Mathcad** reduces the fear of mathematics: doing integrals, derivatives, power series expansions, taking limits, solving algebraic or differential equations become trivial.

Mathematica

Instructions for installing the **Mathematica** programs are provided in the ReadMe file in the CD-ROM. The Workbooks were created with **Mathematica** 5.1. The **Mathematica** tutorial is written to cover the needs of a physical chemistry student. The first two lectures teach the knowledge required for writing a primitive, working program. Subsequent lectures go into more details and show how to write safe programs that run efficiently. I provide these lectures on **Mathematica** because there is no "minimalist" **Mathematica** book, a shortcut to **Mathematica** directed to the needs of the physical chemist. This tutorial has been used several times as a textbook for a one-quarter course at the University of California, Santa Barbara for teaching **Mathematica** to chemistry and biochemistry students. Wolfram's book is available on the web, from inside any **Mathematica** program, and the reader can consult it when trouble arises. This exceptionally well-organized online help makes learning **Mathematica** a lot easier than any other computer language.

Professor Jeffry D. Madura's **Mathcad** Workbooks are a translation of the **Mathematica** Workbooks, to the extent that the syntax of the two languages allows this to be done. For all pedagogical purposes they are identical and the choice of one over the other should depend on availability, price, and instructor's preference.

Mathcad

A short introduction to **Mathcad** is provided as a PDF file on the CD-ROM to help the reader get started. This "tutorial" is also used to illustrate **Mathcad** features that are encountered in solving many of the problems. Each Workbook follows the **Mathematica** numbering scheme. That is, **Mathematica** notebook QM3.3 would be **Mathcad** Workbook 3.3. The Workbooks were written in **Mathcad** 12. To help those who do not have this version there are Adobe Acrobat PDF files of all the **Mathcad** Workbooks to view and use. All documents have been saved so that they can be read by older versions of **Mathcad**. The older versions have been placed in appropriately labeled folders and should be accessible in the same manner as described above.

Exercises and Problems

There is a trend, in recent textbooks, to include a large number of repetitive problems. I have resisted this trend. Once the calculation for ethane has been performed, nothing is gained by performing it for CO, CO_2, etc. I give detailed, solved exercises in the text out of the belief that a principle or methodology is understood only when one sees how it is implemented. These exercises are an integral part of the text and of the learning process. Unsolved exercises are placed throughout the chapters where they are most relevant to the book's content. This strategic placement reinforces what the reader is learning, as he or she learns it. Additional homework problems are included at the end of some chapters. Some exercises remain unsolved for a reason. The physical chemistry students are juniors or seniors, and many of them will go to work as scientists and engineers after they finish the course. It is a fair guess that the person asking them to calculate something does not know the answer. The students must learn how to test whether a result is reasonable. They must take responsibility for their work now, when the consequence for failure is a lower grade, not dismissal or tragedy. Every type of problem that a physical chemist needs to solve is illustrated, with a detailed example in the text and with even more details in the appropriate Workbook. This should offer sufficient guidance to students doing homework.

A few exercises have been included for which no data are given. Finding data is an essential part of the practice of chemistry and should be learned while in school. If a student is asked to calculate the change of enthalpy with temperature and is given the heat capacity and the enthalpy at 298 K, he or she is prompted to look for a formula that contains these two quantities. This crutch is absent in the working world. When asked to calculate the change of enthalpy with temperature, the student should know to look for a formula giving the derivative on enthalpy with temperature. Then, the student can figure out what data is needed for implementing the formula. Occasionally, I have asked the students to create a suitable homework for a given chapter. Presumably, a student that understood the material in a chapter should be able to figure out what he or she can do with it.

Below is a chapter-by-chapter listing of unsolved Exercises that can be assigned for homework:

Chapter One: 1.1. *Chapter Two*: 2.1, 2.2, 2.3, 2.4, 2.5, 2.6, 2.7, 2.9. *Chapter Three*: 3.1, 3.2, 3.3, 3.4, 3.5, 3.6, 3.7, 3.8, 3.9, 3.10, 3.11, 3.12, 3.13, 3.14. *Chapter 4*: 4.1, 4.2, 4.3, 4.4. *Chapter Five*: 5.1. *Chapter Six*: 6.1, 6.2, 6.3. *Chapter 7*: 7.1, 7.2, 7.3. *Chapter Eight*: 8.1, 8.2, 8.3, 8.4, 8.5. *Chapter Nine*: 9.1, 9.2, 9.3. *Chapter 10*: 10.1, 10.2, 10.3, 10.4, 10.5, 10.6, 10.7. *Chapter 11*: 11.1. *Chapter 12*: 12.1, 12.2, 12.3, 12.4, 12.5, 12.6, 12.7. *Chapter 13*: 13.1, 13.2, 13.3, 13.4, 13.5, 13.6. *Chapter 14*: 14.1, 14.2, 14.3, 14.4, 14.5. *Chapter 16*: 16.1, 16.2, 16.3, 16.4, 16.5, 16.6, 16.7, 16.8, 16.9, 16.10, 16.11. *Chapter 17*: 17.1, 17.2, 17.3, 17.4, 17.5, 17.6, 17.7, 17.8, 17.9, 17.10, 17.11, 17.12, 17.13, 17.14. *Chapter 18*: 18.1, 18.2, 18.3, 18.4, 18.5, 18.6, 18.7, 18.8, 18.9, 18.10, 18.11, 18.12, 18.13,18.14. *Chapter 19*: 19.1, 19.2, 19.3, 19.4, 19.5, 19.6. *Chapter 20*: 20.1, 20.2, 20.3, 20.4, 20.5, 20.6, 20.7, 20.8, 20.9, 20.10, 20.11, 20.12, 20.13, 20.14, 20.15, 20.16, 20.17, 20.18, 20.19, 20.20, 20.21, 20.22, 20.23, 20.24. *Chapter 21*: 21.1, 21.2, 21.3, 21.4, 21.5, 21.6, 21.7, 21.8, 21.9, 21.10, 21.11, 21.12, 21.13, 21.14, 21.15, 21.16, 21.17, 21.18, 21.19, 21.20, 21.21, 21.22

1

WHY QUANTUM MECHANICS?

If you want to calculate the motion of the nuclei and electrons in a molecule, Newton has a prescription for you. Treat each particle as if it is a point that has a mass and an electrical charge. These particles exert forces on each other that are given by Coulomb's law. Write that the mass times the acceleration of each particle equals the sum of all forces acting on that particle. Solve the resulting system of equations and you will learn all there is to know about the motion of the electrons and nuclei in a molecule. Several centuries of testing have shown that this procedure works well for cars, planets, billiard balls, and most other things you would think of. But when applied to objects as light as electrons or atoms, Newton's equation fails miserably.

I am not talking about a 20% error in calculating the value of a certain quantity, such as energy. The failure is wholesale and is qualitative; even the language used by classical mechanics is inapplicable to electrons or nuclei in a molecule.

To understand the behavior of atoms and molecules, we must use quantum mechanics. This chapter gives a few examples that show the deep difference between quantum and classical behavior of a molecular system.

§1. *Molecular Vibrations.* A classical description of the oscillations of a diatomic molecule tells us how the positions and the velocities of the atoms change in time. Quantum mechanics says that such knowledge is impossible: we can only know the probability that these positions take a certain value at a given time. Classical mechanics says that if I gradually change the force acting on the atoms, their vibrational energy will change gradually. Quantum mechanics says that the energy of an oscillating molecule varies discretely. Unless the force is capable of giving a certain amount of energy to the oscillator, it would have no effect at all.

These are not inconsequential quibbles. The energy of the oscillating molecule determines how it absorbs or emits light. It also contributes to the specific heat and the entropy of a gas of molecules. All measurements of these quantities agree with the predictions made by quantum mechanics and are in conflict with those made by classical theory.

§2. *Radiation by a Hot Body.* If you heat a metal bar to high temperature, it will emit radiation: you can see the red–white glow of the metal (visible light) and your skin can feel the heat produced by the infrared emission. Classical electromagnetic theory allows us to calculate the energy of the light emitted by the bar. Such a calculation tells us that the energy of light of frequency between ω and $\omega + d\omega$, per unit volume, is

$$\rho(\omega, T)d\omega = \frac{kT\omega^2}{\pi^2 c^3}d\omega \tag{1.1}$$

Note a striking conclusion: the glowing bar emits more high-frequency radiation than low-frequency radiation. More ultraviolet light is emitted than infrared light, more X-rays than ultraviolet light, and more gamma rays than X-rays. If this theory is right, sitting near a fireplace for a few hours exposes you to enough X- and gamma rays to kill you. (I don't mean sitting on the fire, just watching it from a respectful distance.)

We have enough experience with hot bodies to know that this prediction is complete nonsense. Quantum mechanics shows that the classical equation (Eq. 1.1) is wrong for high frequencies (it is OK for low frequencies). The correct equation, given by quantum theory, is

$$\rho(\omega, T)d\omega = \frac{\hbar}{\pi^2 c^3} \frac{\omega^3}{\exp[\hbar\omega/k_B T] - 1}d\omega \tag{1.2}$$

Here $\hbar = 1.0545 \times 10^{-27}$ erg s is Planck's constant, $k_B = 1.3806 \times 10^{-16}$ erg/K is Boltzmann's constant, and $c = 2.9979 \times 10^{10}$ cm/s is the speed of light. This equation is in perfect agreement with the measurements and shows that your fireplace emits practically no X- or gamma rays.

Exercise 1.1

(a) Plot $\rho(\omega, T)$ given by the two formulae as a function of ω for $T = 2000$ K.
(b) Derive the conditions under which Eqs 1.1 and 1.2 give the same result.

§3. *The Stability of Atoms.* At the beginning of the past century, Rutherford performed a "simple" experiment. He shot alpha particles (nuclei of He atoms) at a very thin foil of gold. To his amazement, some particles passed through the gold as if it were empty space. Some were deflected backwards, as if they had hit a heavy particle having a positive charge. Rutherford concluded, correctly, that the atoms in the metal had a heavy, positively charged nucleus and the rest of the foil was mostly empty space. Since gold is electrically neutral, negative charges (electrons) had to be located somewhere. He proposed that inside an atom, the electrons move in orbits around the nucleus, like the planets around the sun. If the orbits were large compared to the size of the nucleus, the atom was mostly "empty," as observed in his experiment. This was a fine model, except for a fatal flaw. According to classical electrodynamics, a moving charge (the electron, in this case) emits electromagnetic radiation, which carries energy away from the atom. This energy is provided by the moving electron, which will have to slow down. If it does slow down, the centrifugal force becomes smaller and the nucleus is able to pull the electrons closer to it. In the end, the electrons will lose their kinetic energy and fall on the nucleus. The atom would be a very compact object, and gold would not be mostly empty space.

A young Danish physicist, Niels Bohr, found a courageous solution to this difficulty. He postulated several arbitrary rules that contradicted classical physics, but made the atom stable. In addition, these rules allowed Bohr to calculate the spectrum of the hydrogen atom with astonishing accuracy. This was a brilliant improvisation, not a theory. But it made a powerful point: to explain the stability of atoms, we must abandon classical mechanics. A new mechanics was needed.

After many false starts, the new mechanics was formulated by Heisenberg, Schrödinger, Born, and Dirac. This is not a "correction" to classical theory, but an entirely new construction. It denies most of the concepts used by classical mechanics, and it has been extremely successful in applications. In all cases in which we can solve its equations, the predictions made by the theory are in precise agreement with experiment.

The theory has many features that conflict with our everyday experience, which was accumulated by observing phenomena for which classical mechanics is valid. It tells us, among other surprises, that the idea of a trajectory is meaningless, that a precise determination of both position and velocity is impossible, that continuous changes of energy are not possible except in special cases, that you cannot calculate the result of a measurement but only the probability that a certain result will be obtained.

Quantum mechanics introduces new mathematical objects, new equations and a new way of thinking about mechanical events, and it has extensive applications to chemistry. None of the spectroscopies used by analytical chemists (nuclear magnetic resonance (NMR), electron spin resonance (ESR), microwave, infrared, visible, ultraviolet (UV), photoelectron spectroscopies) can be understood without quantum mechanics. The chemical bond is also an entirely quantum-mechanical phenomenon. The subject is much too rich to be studied in its fullness in an introductory course. We will focus mostly on describing some of the fundamental quantum phenomena. They will help you understand chemical bonding, spectroscopy, NMR, and photochemistry.

2

DYNAMIC VARIABLES
AND OPERATORS

Operators: the Definition

§1. *Introduction.* Quantum mechanics is not a "correction" to classical mechanics, but an entirely new way of describing the mechanical properties of very light objects (e.g. electrons, nuclei, atoms, molecules). The mathematics of quantum theory and the physical phenomena that take place in systems where quantum mechanics applies, are very different from those of classical mechanics.

In Newton's classical scheme, the fundamental variables (the positions, the momenta, the kinetic or the potential energy of the particles) are numbers. In quantum mechanics this is no longer true. The dynamical variables are represented by mathematical objects called *operators*. In this chapter, you learn what an operator is and which operators represent the dynamical variables of the system (e.g. position, momentum, kinetic energy).

§2. *The Definition.* A function is a set of rules that are applied to a given number, to produce another number. For example, $y = f(x) \equiv x^2$ is shorthand for the following rules: take the argument x of $f(x)$ and calculate its square. The number you obtain is the value of $f(x)$.

This concept can be generalized to define rules that use as input *a function* (not a number) and produce as a result another *function* (not a number). Such a set of rules is called an *operator*. An operator is to functions what a function is to numbers.

The operator notation analogous to $y = f(x)$ is

$$\phi(x) = \hat{O}\psi(x) \tag{2.1}$$

Here $\hat{O}$ is a symbol for the operator (the hat on top reminds me that $\hat{O}$ is an operator). This operator acts on or is applied to the function $\psi(x)$ and produces as a result a function called $\phi(x)$.

§3. *Examples of Operators.* Here I give several examples, which illustrate the kinds of operators that you will encounter in your study of quantum mechanics. The operator $\hat{O}_x$, defined by

$$\hat{O}_x f(x, y, z) = x f(x, y, z), \tag{2.2}$$

acts on a function $f(x, y, z)$ to produce the function $x f(x, y, z)$. This rule is easy to apply; for example:

$$\hat{O}_x \sin(x^2) = x \, \sin(x^2)$$

$$\hat{O}_x [xy \ln x] = x^2 y \ln x$$

No matter to what function I apply the operator $\hat{O}_x$, the result is that function multiplied by x.

Another example is

$$\hat{V} f(x, y, z) = V(x, y, z) f(x, y, z) \tag{2.3}$$

where $V(x, y, z)$ is a function specified in advance. The operator $\hat{V}$ turns a function $f(x, y, z)$ into the function $V(x, y, z) f(x, y, z)$. If $V(x) = kx^2/2$ then

$$\hat{V} f(x, y, z) = \frac{kx^2}{2} f(x, y, z)$$

Another important operator is defined by

$$\hat{p}_x \psi(x, y, z) = \frac{\hbar}{i} \frac{\partial}{\partial x} \psi(x, y, z) \tag{2.4}$$

$\hbar$ is Planck's constant and $i = \sqrt{-1}$. The previous operators involved multiplication with a function of coordinates; this one involves a derivative.

Here is an example that shows how this works: if

$$\psi(x, y, z) = \exp[ik_x x + ik_y y + ik_z z]$$

then

$$\hat{p}_x \psi(x, y, z) = i\hbar k_x \exp[ik_x x + ik_y y + ik_z z]$$

(Remember that $\partial \exp[ik_x x]/\partial x = ik_x \exp[ik_x x]$.)

Similarly, I can define

$$\hat{p}_y \psi(x, y, z) = \frac{\hbar}{i} \frac{\partial}{\partial y} \psi(x, y, z) \qquad (2.5)$$

Operations with Operators

It is easy to define the addition of two operators, the multiplication of an operator by a number, the multiplication of two operators, etc. This is what I do next.

§4. *Operator Addition.* The sum of two operators is defined just as you would expect. The operator

$$\hat{C} \equiv \hat{A} + \hat{B}$$

is defined by the rule

$$\hat{C}\psi = \hat{A}\psi + \hat{B}\psi$$

To determine how $\hat{C}$ acts on a function ψ, add the result of acting with $\hat{A}$ on ψ to the result of acting with $\hat{B}$ on ψ. You know how to add $\hat{A}\psi$ to $\hat{B}\psi$ because they are functions.

When α is a complex number, the operator

$$\hat{D} \equiv \alpha \hat{A}$$

is defined by

$$\hat{D}\psi = \alpha(\hat{A}\psi)$$

To calculate how $\hat{D}$ acts on ψ, you calculate the function $\phi = \hat{A}\psi$ and multiply the result by α.

We can combine these rules to create new operators. For example, the operator $\hat{C}$ defined by

$$\hat{C} = \hat{p}_x + 2\hat{O}_x,$$

(where $\hat{p}_x$ is defined by Eq. 2.4 and $\hat{O}_x$ by Eq. 2.2) acts on the function ψ as shown below:

$$\hat{C}\psi = \frac{\hbar}{i}\frac{\partial\psi}{\partial x} + 2x\psi$$

§5.   *Operator Multiplication.* I define the symbol $\hat{A}\hat{B}$ by the following rule:

$$\hat{A}\hat{B}\psi = \hat{A}(\hat{B}\psi) = \hat{A}\phi \ \ \text{where} \ \ \phi = \hat{B}\psi \tag{2.6}$$

This means: take the function ψ and calculate the effect the operator $\hat{B}$ has on it. Call the function obtained in this way ϕ. Apply the operator $\hat{A}$ to ϕ. The result is the action of the operator $\hat{A}\hat{B}$ on ψ.

Let us see how this works on a few examples. (Workbook QM2.1 shows how to use **Mathematica** to manipulate operators.) Consider the operator $\hat{p}_x\hat{O}_x$, where $\hat{O}_x$ is defined by Eq. 2.2 and $\hat{p}_x$ is defined by Eq. 2.4:

$$\hat{p}_x\hat{O}_x\psi(x,y,z) = \hat{p}_x(\hat{O}_x\psi(x,y,z)) = \hat{p}_x(x\psi(x,y,z))$$

$$= \frac{\hbar}{i}\frac{\partial}{\partial x}(x\psi(x,y,z))$$

$$= \frac{\hbar}{i}\psi(x,y,z) + \frac{\hbar}{i}x\frac{\partial\psi(x,y,z)}{\partial x} \tag{2.7}$$

Note that in the calculation performed above, $\frac{\partial}{\partial x}$ acts on $x\psi(x,y,z)$.

Let us now apply the operators in the opposite order:

$$\hat{O}_x\hat{p}_x\psi(x,y,z) = \hat{O}_x\left(\frac{\hbar}{i}\frac{\partial}{\partial x}\psi(x,y,z)\right) = x\frac{\hbar}{i}\frac{\partial\psi(x,y,z)}{\partial x} \tag{2.8}$$

$\hat{O}_x\hat{p}_x\psi$ *differs* from $\hat{p}_x\hat{O}_x\psi$: the operators $\hat{O}_x$ and $\hat{p}_x$ *do not commute*.

For practice, let us calculate how $\hat{O}_x \hat{p}_y$ and $\hat{p}_y \hat{O}_x$ act on a function $\psi(x,y,z)$:

$$\hat{p}_y \hat{O}_x \psi(x,y,z) = \hat{p}_y(x\psi) = \frac{\hbar}{i}\frac{\partial}{\partial y}x\psi(x,y,z) = \frac{\hbar}{i}x\frac{\partial \psi(x,y,z)}{\partial y} \qquad (2.9)$$

$$\hat{O}_x \hat{p}_y \psi(x,y,z) = \hat{O}_x\left(\frac{\hbar}{i}\frac{\partial \psi(x,y,z)}{\partial y}\right) = \frac{\hbar}{i}x\frac{\partial \psi(x,y,z)}{\partial y} \qquad (2.10)$$

Note that $\hat{O}_x \hat{p}_y$ and $\hat{p}_y \hat{O}_x$ applied to a function $\psi(x,y,z)$ give *the same result, no matter what* $\psi(x,y,z)$ is. When this happens, we say that the operators $\hat{O}_x \hat{p}_y$ and $\hat{p}_y \hat{O}_x$ are equal:

$$\hat{O}_x \hat{p}_y = \hat{p}_y \hat{O}_x$$

When two operators satisfy $\hat{A}\hat{B} = \hat{B}\hat{A}$, we say that the operators *commute*.

§6. *Powers of Operators.* We can use this multiplication rule to define powers of operators. For example,

$$\hat{p}_x^2 \psi(x,y,z) \equiv \hat{p}_x \hat{p}_x \psi(x,y,z) = \frac{\hbar}{i}\frac{\partial}{\partial x}\left(\frac{\hbar}{i}\frac{\partial}{\partial x}\psi(x,y,z)\right) = -\hbar^2 \frac{\partial^2}{\partial x^2}\psi(x,y,z)$$

I have used above the fact that $i^2 = -1$. Another example is

$$\hat{O}_x^3 \psi(x,y,z) = \hat{O}_x^2(\hat{O}_x \psi(x,y,z)) = \hat{O}_x^2(x\psi) = \hat{O}_x(x^2\psi) = x^3 \psi(x,y,z)$$

§7. *The Commutator of Two Operators.* Operator multiplication is not necessarily commutative. We say that the operators $\hat{A}$ and $\hat{B}$ commute if $\hat{A}\hat{B}\psi = \hat{B}\hat{A}\psi$ for every function ψ. The calculations performed in Eqs 2.7–2.10 show that $\hat{p}_x$ and $\hat{O}_x$ do not commute but $\hat{p}_y$ and $\hat{O}_x$ do.

Whether two operators commute or not has important physical consequences in quantum mechanics. For this reason we often use an operation called *the commutator* of two operators:

$$[\hat{A}, \hat{B}] \equiv \hat{A}\hat{B} - \hat{B}\hat{A} \qquad (2.11)$$

The commutator $[\hat{A}, \hat{B}]$ is an operator.

An operator $\hat{O}$ is said to be equal to 0 if $\hat{O}\psi$ is identically 0 for every function ψ. Two operators $\hat{A}$ and $\hat{B}$ commute if and only if their commutator is equal to 0. This means that

$$[\hat{A}, \hat{B}]\psi = (\hat{A}\hat{B} - \hat{B}\hat{A})\psi = \hat{A}\hat{B}\psi - \hat{B}\hat{A}\psi = 0$$

for any function ψ.

The commutator $[\hat{p}_x, \hat{O}_x]$ can be easily calculated from Eqs 2.7 and 2.8:

$$[\hat{p}_x, \hat{O}_x]\psi = \hat{p}_x\hat{O}_x\psi - \hat{O}_x\hat{p}_x\psi = \frac{\hbar}{i}\psi \tag{2.12}$$

Because this equality is valid for any function ψ, we can write that

$$[\hat{p}_x, \hat{O}_x] = \frac{\hbar}{i} \tag{2.13}$$

Exercise 2.1

In parts (c) and (d), $V(x)$ is a function of x and $\hat{V}\psi = V(x)\psi$.

(a)   Calculate $[\hat{p}_x, \hat{O}_x^2]$.

(b)   Calculate $[\hat{p}_x, \hat{O}_x]$.

(c)   Calculate $[\hat{p}_x, \hat{V}]$.

(d)   Calculate $[\hat{p}_y, \hat{V}]$.

(e)   Calculate $[\hat{p}_x, \hat{p}_y]$.

(f)   Calculate $[\hat{p}_x^6, \hat{p}_x]$.

Exercise 2.2

Suppose $\kappa(z)$ is a function.

(a)   Is $\hat{K}$ defined by

$$\hat{K}\psi(x) \equiv \int_{-\infty}^{+\infty} \kappa(x - y)\psi(y)dy$$

an operator?

(b)   Calculate $[\hat{p}_x, \hat{K}]$.

(c)   Take $\kappa(z) = \exp[-(z/\sigma)^2]$ in part (a) and calculate $\hat{K}\psi(x)$ for $\psi(x) = x$ and for $\psi(x) = x^2$.

§8.   *Linear Operators.* In quantum mechanics we are interested only in operators that satisfy the condition

$$\hat{A}(a\psi + b\phi) = a\hat{A}\psi + b\hat{A}\phi \tag{2.14}$$

Here a and b are complex numbers and ψ and ϕ are arbitrary functions. An operator that satisfies this condition is called *linear.*

The operators $\hat{O}_x$, $\hat{V}$, $\hat{p}_x$, $\hat{p}_y$ introduced so far are all linear operators (check that this is true). An example of a nonlinear operator is $\hat{S}\eta \equiv \eta^2$. Indeed, if a and b are numbers and ψ and ϕ are functions, then

$$\hat{S}(a\psi + b\phi) = (a\psi + b\phi)^2 = a^2\psi^2 + b^2\phi^2 + 2ab\psi\phi$$

If $\hat{S}$ were a linear operator, we would have had (see Eq. 2.14)

$$\hat{S}(a\psi + b\phi) = a\hat{S}\psi + b\hat{S}\phi = a\psi^2 + b\phi^2$$

Because the two expressions differ, $\hat{S}$ is not a linear operator.

Exercise 2.3

Test whether the following operators are linear.

$$\hat{A}\psi(x) \equiv \int_a^b K(y - x)\psi(x)dx$$

$$\hat{B}\psi(x) \equiv a\frac{d^2}{dx^2}\psi(x) + b\frac{d}{dx}\psi(x)$$

$$\hat{C}\psi(x) \equiv \sin(\psi(x))$$

Dynamical Variables as Operators

§9.   *Introduction.* We can now introduce one of the fundamental rules of quantum mechanics: all classical dynamical variables are represented in quantum mechanics by operators. You will see as we proceed what the word "represented" means.

In classical mechanics, the fundamental dynamical variables of a particle are: the position vector $\mathbf{r} = \{x, y, z\}$ (in Cartesian coordinates) and the momentum vector $\mathbf{p} = \{p_x, p_y, p_z\}$, where

$$p_x = m\frac{dx}{dt}$$

$$p_y = m\frac{dy}{dt}$$

and

$$p_z = m\frac{dz}{dt}$$

The derivatives with respect to time t are the components of the velocity vector. m is the mass of the particle.

All other dynamical variables are functions of positions and momenta. The potential energy $V(\mathbf{r}) = V(x, y, z)$ is a function of coordinates. The kinetic energy

$$\frac{\mathbf{p} \cdot \mathbf{p}}{2m} = \frac{p_x^2 + p_y^2 + p_z^2}{2m}$$

is a function of momenta. The angular momentum vector $\mathbf{L} \equiv \mathbf{r} \times \mathbf{p}$ (where $\mathbf{a} \times \mathbf{b}$ denotes the cross-product of vectors $\mathbf{a}$ and $\mathbf{b}$) and the total energy $H = (\mathbf{p} \cdot \mathbf{p}/2m) + V(\mathbf{r})$ depend on both momentum and coordinate.

If the system has N particles, the fundamental dynamical variables are the positions of all particles and the momenta of all particles. The interaction energy U is a function of the positions of the particles. The kinetic energy K is the sum of the kinetic energies of each particle. The total energy is $K + U$. The total angular momentum is the sum of the angular momenta of each particle. We see that the dynamic variables of an N-particle system are also functions of positions and momenta.

If you know how the positions and the momenta of a classical system change in time, you know all there is to know about the mechanics of that system.

§10. *Position and Potential Energy Operators.* In quantum mechanics the position operator $\hat{\mathbf{r}}$ is defined by

$$\hat{\mathbf{r}}\psi(\mathbf{r}) = \mathbf{r}\psi(\mathbf{r}) = \mathbf{i}\,x\psi(\mathbf{r}) + \mathbf{j}\,y\psi(\mathbf{r}) + \mathbf{k}\,z\psi(\mathbf{r}) \tag{2.15}$$

Here $\mathbf{i}, \mathbf{j}, \mathbf{k}$ are unit vectors along the orthogonal, Cartesian, coordinate axes, and x, y, z are the Cartesian coordinates of the particle located at $\mathbf{r}$. $\psi(\mathbf{r}) \equiv \psi(x, y, z)$ is an arbitrary function used to indicate how the operator $\hat{\mathbf{r}}$ acts on a function. The components of the vector operator $\hat{\mathbf{r}}$ are the operators $\hat{O}_x$, $\hat{O}_y$ and $\hat{O}_z$ defined earlier.

The position operator can be easily defined for a system with many particles. For example, if a system has three particles with classical position vectors $\mathbf{r}(1)$, $\mathbf{r}(2)$, and $\mathbf{r}(3)$, the position operator for particle 1, with coordinates $x(1), y(1), z(1)$, is

$$\hat{\mathbf{r}}(1)\psi(\mathbf{r}(1), \mathbf{r}(2), \mathbf{r}(3)) = \mathbf{i}\,x(1)\psi(\mathbf{r}(1), \mathbf{r}(2), \mathbf{r}(3)) + \mathbf{j}\,y(1)\psi(\mathbf{r}(1), \mathbf{r}(2), \mathbf{r}(3))$$

$$+ \mathbf{k}\,z(1)\psi(\mathbf{r}(1), \mathbf{r}(2), \mathbf{r}(3))$$

$$\equiv \mathbf{r}(1)\psi(\mathbf{r}(1), \mathbf{r}(2), \mathbf{r}(3)) \tag{2.16}$$

Here $\psi(\mathbf{r}(1), \mathbf{r}(2), \mathbf{r}(3)) \equiv \psi(x(1), y(1), z(1), x(2), y(2), z(2), x(3), y(3), z(3))$ is an arbitrary function. The operators $\hat{\mathbf{r}}(2)$ and $\hat{\mathbf{r}}(3)$ that represent the positions of the other two particles are defined similarly.

The operator corresponding to the x-coordinate of particle 1 is

$$\hat{x}(1)\psi(\mathbf{r}(1), \mathbf{r}(2), \mathbf{r}(3)) = x(1)\psi(\mathbf{r}(1), \mathbf{r}(2), \mathbf{r}(3)) \tag{2.17}$$

Exercise 2.4

Show that in a system with two particles, the operators $\hat{\mathbf{r}}(1)$ and $\hat{\mathbf{r}}(2)$ commute.

§11. *Potential Energy Operator.* The classical potential energy is a function of the coordinates of all the particles. If, for a one-particle system, the classical potential energy is $V(\mathbf{r}) = V(x, y, z)$, then the operator $\hat{V}$ representing this quantity in quantum mechanics is defined by

$$\hat{V}\psi(\mathbf{r}) \equiv V(\mathbf{r})\psi(\mathbf{r}) \tag{2.18}$$

We can generalize this to any quantity that depends only on coordinates. The rule is very simple: if a classical variable $O(\mathbf{r}(1), \mathbf{r}(2), \ldots, \mathbf{r}(N))$ is a function of coordinates (but not of momenta), then the operator $\hat{O}$ corresponding to it is

$$\hat{O}\psi(\mathbf{r}(1), \mathbf{r}(2), \ldots, \mathbf{r}(N)) = O(\mathbf{r}(1), \mathbf{r}(2), \ldots, \mathbf{r}(N))\psi(\mathbf{r}(1), \mathbf{r}(2), \ldots, \mathbf{r}(N)) \quad (2.19)$$

Exercise 2.5

Show that $[\hat{x}, \hat{V}] = 0$ and $[\hat{\mathbf{r}}, \hat{V}] = 0$.

---

§12. *The Momentum Operator.* In classical mechanics, the momentum of a particle is a vector $\mathbf{p} = \mathbf{i}\,m\,v_x + \mathbf{j}\,m\,v_y + \mathbf{k}\,m\,v_z$. Here m is the mass of the particle and v_x, v_y, and v_z are the components of the velocity vector: $v_x = dx/dt$, $v_y = dy/dt$, and $v_z = dz/dt$. This definition of momentum is valid only if you use Cartesian coordinates.

The quantum mechanical operators $\hat{p}_x$, $\hat{p}_y$, $\hat{p}_z$ corresponding to $p_x = mv_x$, $p_y = mv_y$, $p_z = mv_z$ are

$$\hat{p}_x \psi(x,y,z) = \frac{\hbar}{i} \frac{\partial}{\partial x} \psi(x,y,z) \tag{2.20}$$

$$\hat{p}_y \psi(x,y,z) = \frac{\hbar}{i} \frac{\partial}{\partial y} \psi(x,y,z) \tag{2.21}$$

$$\hat{p}_z \psi(x,y,z) = \frac{\hbar}{i} \frac{\partial}{\partial z} \psi(x,y,z) \tag{2.22}$$

Here $\hbar$ is Planck's constant, whose value is given in Appendix 1, and $i = \sqrt{-1}$ (for a review of complex numbers, see Supplement 2.1).

In a system having N particles, each particle ℓ has a momentum $\mathbf{p}(\ell)$ with components $\{p_x(\ell), p_y(\ell), p_z(\ell)\}$ and a position $\mathbf{r}(\ell)$ with components $\{x(\ell), y(\ell), z(\ell)\}$. For example, the operator corresponding to the y-component of the momentum of the third particle (denoted $\hat{p}_y(3)$) is

$$\hat{p}_y(3)\psi(\mathbf{r}(1),\ldots,\mathbf{r}(N)) = \frac{\hbar}{i} \frac{\partial}{\partial y(3)} \psi(\mathbf{r}(1),\ldots,\mathbf{r}(N))$$

§13. *The Kinetic Energy Operator.* Classical kinetic energy in a one-particle system is

$$K = \frac{\mathbf{p} \cdot \mathbf{p}}{2m} = \frac{p_x^2 + p_y^2 + p_z^2}{2m} \tag{2.23}$$

where m is the mass of the particle.

The kinetic energy operator in quantum mechanics is

$$\hat{K} = \frac{\hat{\mathbf{p}} \cdot \hat{\mathbf{p}}}{2m} = \frac{\hat{p}_x^2 + \hat{p}_y^2 + \hat{p}_z^2}{2m} \qquad (2.24)$$

where $\hat{p}_x$, $\hat{p}_y$, and $\hat{p}_z$ are defined by Eqs 2.20–2.22.

To see what this operator does when applied to a function $\psi(\mathbf{r}) = \psi(x, y, z)$, I look at $\hat{p}_x^2/2m$. According to the rule of operator multiplication,

$$\frac{\hat{p}_x^2}{2m}\psi = \frac{1}{2m}\hat{p}_x\left(\hat{p}_x\psi\right)$$

The rule says: act first with $\hat{p}_x$ and then act with $\hat{p}_x$ on the result. Using Eq. 2.20 for $\hat{p}_x$ leads to

$$\frac{\hat{p}_x^2}{2m}\psi = \frac{1}{2m}\hat{p}_x\left(\hat{p}_x\psi\right) = \frac{1}{2m}\hat{p}_x\left(\frac{\hbar}{i}\frac{\partial\psi}{\partial x}\right)$$

$$= \frac{1}{2m}\frac{\hbar}{i}\frac{\partial}{\partial x}\left(\frac{\hbar}{i}\frac{\partial\psi}{\partial x}\right) = \frac{1}{2m}\left(\frac{\hbar}{i}\right)^2\frac{\partial^2\psi(x,y,z)}{\partial x^2} \qquad (2.25)$$

Similarly

$$\frac{\hat{p}_y^2}{2m}\psi = \frac{1}{2m}\left(\frac{\hbar}{i}\right)^2\frac{\partial^2\psi(x,y,z)}{\partial y^2} \qquad (2.26)$$

and

$$\frac{\hat{p}_z^2}{2m}\psi = \frac{1}{2m}\left(\frac{\hbar}{i}\right)^2\frac{\partial^2\psi(x,y,z)}{\partial z^2} \qquad (2.27)$$

Using Eqs 2.25–2.27 in Eq. 2.24 gives

$$\hat{K}\psi = -\frac{\hbar^2}{2m}\left(\frac{\partial^2\psi}{\partial x^2} + \frac{\partial^2\psi}{\partial y^2} + \frac{\partial^2\psi}{\partial z^2}\right) \qquad (2.28)$$

The negative sign comes from the relation $i^2 = -1$. I can also use the definition

$$\Delta \equiv \frac{\partial^2}{\partial x^2} + \frac{\partial^2}{\partial y^2} + \frac{\partial^2}{\partial z^2} \qquad (2.29)$$

of the Lapacian (or Laplace operator) and write

$$\hat{K} = -\frac{\hbar^2}{2m}\Delta = -\frac{\hbar^2}{2m}\left(\frac{\partial^2}{\partial x^2} + \frac{\partial^2}{\partial y^2} + \frac{\partial^2}{\partial z^2}\right) \tag{2.30}$$

with the understanding that such equations are meaningful only when both sides act on the same function ψ.

How about systems having many particles? The classical kinetic energy of a system with N particles is

$$K = \sum_{\ell=1}^{N} \frac{\mathbf{p}(\ell) \cdot \mathbf{p}(\ell)}{2m(\ell)} = \frac{\mathbf{p}(1) \cdot \mathbf{p}(1)}{2m(1)} + \cdots + \frac{\mathbf{p}(N) \cdot \mathbf{p}(N)}{2m(N)}$$

The vector $\mathbf{p}(\ell)$ is the momentum of particle ℓ. The corresponding operator is obtained by using Eq. 2.30 for the kinetic energy of each particle:

$$\hat{K} = \sum_{\ell=1}^{N} \frac{\hat{\mathbf{p}}(\ell) \cdot \hat{\mathbf{p}}(\ell)}{2m(\ell)}$$

It is easy to see that

$$\hat{K}\psi(\mathbf{r}(1), \ldots, \mathbf{r}(N)) = -\frac{\hbar^2}{2}\sum_{\ell=1}^{N}\frac{1}{m(\ell)}\left(\frac{\partial^2\psi}{\partial x(\ell)^2} + \frac{\partial^2\psi}{\partial y(\ell)^2} + \frac{\partial^2\psi}{\partial z(\ell)^2}\right) \tag{2.31}$$

§14. *The Total Energy (the Hamiltonian).* In classical mechanics, the total energy (or the Hamiltonian)

$$H = K + V$$

is the sum of the kinetic energy K and the potential energy V. The operator $\hat{H}$ is

$$\hat{H} = \hat{K} + \hat{V} \tag{2.32}$$

For an N-particle system,

$$\hat{H}\psi(\mathbf{r}(1),\ldots,\mathbf{r}(N)) = \hat{K}\psi(\mathbf{r}(1),\ldots,\mathbf{r}(N)) + \hat{V}\psi(\mathbf{r}(1),\ldots,\mathbf{r}(N))$$

$$= -\sum_{\ell=1}^{N} \frac{\hbar^2}{2m(\ell)} \left(\frac{\partial^2\psi}{\partial x(\ell)^2} + \frac{\partial^2\psi}{\partial y(\ell)^2} + \frac{\partial^2\psi}{\partial z(\ell)^2} \right)$$

$$+ V(\mathbf{r}(1),\ldots,\mathbf{r}(N))\psi(\mathbf{r}(1),\ldots,\mathbf{r}(N)) \qquad (2.33)$$

Here $V(\mathbf{r}(1),\ldots,\mathbf{r}(N))$ is the classical potential energy of the system.

§15. *Angular Momentum.* I come now to the last operator considered here, the angular momentum. In classical mechanics this is defined by

$$\mathbf{L} = \mathbf{r} \times \mathbf{p}$$

I remind you of the rules for calculating the cross-product, which can be written as a determinant:

$$\mathbf{L} = \mathbf{r} \times \mathbf{p} = \begin{vmatrix} \mathbf{i} & \mathbf{j} & \mathbf{k} \\ x & y & z \\ p_x & p_y & p_z \end{vmatrix} \qquad (2.34)$$

Expanding the determinant in the right-hand side of Eq. 2.34 by cofactors gives

$$\mathbf{L} = \mathbf{i} \begin{vmatrix} y & z \\ p_y & p_z \end{vmatrix} - \mathbf{j} \begin{vmatrix} x & z \\ p_x & p_z \end{vmatrix} + \mathbf{k} \begin{vmatrix} x & y \\ p_x & p_y \end{vmatrix}$$

$$= \mathbf{i}(yp_z - zp_y) - \mathbf{j}(xp_z - zp_x) + \mathbf{k}(xp_y - yp_x)$$

The x-component of $\mathbf{L}$ is therefore

$$L_x = yp_z - zp_y \qquad (2.35)$$

For the other components, we have:

$$L_y = zp_x - xp_z \qquad (2.36)$$

and

$$L_z = xp_y - yp_x \tag{2.37}$$

There are all kinds of rules for remembering these equations. I give you Metiu's rule: remember a book where the equations are written down and remember where you put the book.

The operators corresponding to L_x, L_y, L_z are easily found.

$$\hat{L}_x \psi = (\hat{y}\hat{p}_z - \hat{z}\hat{p}_y)\psi = y\frac{\hbar}{i}\frac{\partial \psi}{\partial z} - z\frac{\hbar}{i}\frac{\partial \psi}{\partial y} = \frac{\hbar}{i}\left(y\frac{\partial}{\partial z} - z\frac{\partial}{\partial y}\right)\psi \tag{2.38}$$

Another way of writing this is

$$\hat{L}_x = \frac{\hbar}{i}\left(y\frac{\partial}{\partial z} - z\frac{\partial}{\partial y}\right) \tag{2.39}$$

The other components are

$$\hat{L}_y = \frac{\hbar}{i}\left(z\frac{\partial}{\partial x} - x\frac{\partial}{\partial z}\right) \tag{2.40}$$

and

$$\hat{L}_z = \frac{\hbar}{i}\left(x\frac{\partial}{\partial y} - y\frac{\partial}{\partial x}\right) \tag{2.41}$$

The square of the angular momentum

$$\mathbf{L}^2 = \mathbf{L}\cdot\mathbf{L} = L_x^2 + L_y^2 + L_z^2 \tag{2.42}$$

is important in classical mechanics because it enters into the rotational energy of a particle. The corresponding operator in quantum mechanics is

$$\hat{L}^2 = \hat{L}_x^2 + \hat{L}_y^2 + \hat{L}_z^2 \tag{2.43}$$

Exercise 2.6

Calculate $\hat{L}^2 \psi(x, y, z)$ by using Eq. 2.43 and Eqs 2.39–2.41.

Exercise 2.7

For $r = \sqrt{x^2 + y^2 + z^2}$ and a a real number, show that (see Workbook QM2.2)

$$\hat{L}_z e^{-ar} = 0$$

$$\hat{L}^2 e^{-ar} = 0$$

$$\hat{L}_z \left(e^{-ar} z \right) = 0$$

$$\hat{L}^2 \left(e^{-ar} z \right) = 2\hbar^2 e^{-ar} z$$

Exercise 2.8

Show that (see Workbook QM2.2) if

$$\psi(x, y, z) = \frac{z(x + iy)}{x^2 + y^2 + z^2}$$

then

$$\hat{L}_z \psi = \hbar \psi$$

and

$$\hat{L}^2 \psi = 6\hbar^2 \psi$$

§16. *Summary.* In quantum mechanics, the dynamical variables are represented by operators. Here is a list of the important ones.

The momentum operator of particle ℓ with Cartesian coordinates $x(\ell), y(\ell), z(\ell)$ is a vector with the Cartesian components

$$\hat{p}_x(\ell) = \frac{\hbar}{i} \frac{\partial}{\partial x(\ell)} \tag{2.44}$$

$$\hat{p}_y(\ell) = \frac{\hbar}{i} \frac{\partial}{\partial y(\ell)} \tag{2.45}$$

$$\hat{p}_z(\ell) = \frac{\hbar}{i} \frac{\partial}{\partial z(\ell)} \tag{2.46}$$

In a system for which $V(\mathbf{r}(1), \dots, \mathbf{r}(N))$ is the classical potential energy, the potential energy operator $\hat{V}$ is defined by

$$\hat{V}\psi(\mathbf{r}(1), \dots, \mathbf{r}(N)) = V(\mathbf{r}(1), \dots, \mathbf{r}(N))\psi(\mathbf{r}(1), \dots, \mathbf{r}(N)) \tag{2.47}$$

The rule is: multiply ψ by the classical formula $V(\mathbf{r}(1), \dots, \mathbf{r}(N))$ for the potential energy.

The total energy (the Hamiltonian) operator is

$$\hat{H} = \sum_{\ell=1}^{N} \frac{\hat{\mathbf{p}} \cdot \hat{\mathbf{p}}}{2m(\ell)} + \hat{V}$$

$$= -\sum_{\ell=1}^{N} \frac{\hbar^2}{2m(\ell)} \left(\frac{\partial^2}{\partial x(\ell)^2} + \frac{\partial^2}{\partial y(\ell)^2} + \frac{\partial^2}{\partial z(\ell)^2} \right)$$

$$+ V(\mathbf{r}(1), \dots, \mathbf{r}(N)) \tag{2.48}$$

The angular momentum operator of particle ℓ is a vector $\mathbf{L}$ with Cartesian components

$$\hat{L}_x(\ell) = \frac{\hbar}{i} \left(y(\ell)\frac{\partial}{\partial z(\ell)} - z(\ell)\frac{\partial}{\partial y(\ell)} \right) \tag{2.49}$$

$$\hat{L}_y(\ell) = \frac{\hbar}{i} \left(z(\ell)\frac{\partial}{\partial x(\ell)} - x(\ell)\frac{\partial}{\partial z(\ell)} \right) \tag{2.50}$$

$$\hat{L}_z(\ell) = \frac{\hbar}{i} \left(x(\ell)\frac{\partial}{\partial y(\ell)} - y(\ell)\frac{\partial}{\partial x(\ell)} \right) \tag{2.51}$$

Angular momentum squared is

$$\hat{\mathbf{L}}^2 = \hat{L}_x^2 + \hat{L}_y^2 + \hat{L}_z^2 \tag{2.52}$$

In later chapters, I will explain how these operators are connected to measurable physical quantities.

Supplement 2.1 Review of Complex Numbers

§17. *Complex Numbers.* In many problems in quantum mechanics, the values of the wave function turn out to be complex numbers. We need to remind ourselves of a few of their properties.

A complex number is defined by

$$z = a + ib \tag{2.53}$$

where a and b are two real numbers called the real and imaginary parts of z, respectively, and

$$i = \sqrt{-1} \tag{2.54}$$

The notations $\text{Re}(z)$ and $\text{Im}(z)$ are used for the real and imaginary parts of z. Obviously, i has the property

$$i^2 = -1 \tag{2.55}$$

Here are the operations you can perform with two complex numbers $z_1 = a_1 + ib_1$ and $z_2 = a_2 + ib_2$:

- Addition of two complex numbers

$$z_1 + z_2 = (a_1 + a_2) + i(b_1 + b_2) \tag{2.56}$$

- Multiplication of a complex number by a real number c

$$cz_1 = ca_1 + i(cb_1) \tag{2.57}$$

- Multiplication of two complex numbers

$$z_1 z_2 = (a_1 + ib_1)(a_2 + ib_2)$$

$$= a_1 a_2 + ib_1 a_2 + ia_1 b_2 + i^2 b_1 b_2$$

$$= (a_1 a_2 - b_1 b_2) + i(a_1 b_2 + a_2 b_1) \tag{2.58}$$

The last equality was obtained by using $i^2 = -1$.

- Division of two complex numbers

$$\frac{z_1}{z_2} = \frac{a_1 + ib_1}{a_2 + ib_2} = \frac{a_1 a_2 + b_1 b_2}{a_2^2 + b_2^2} + i \frac{b_1 a_2 - a_1 b_2}{a_2^2 + b_2^2} \qquad (2.59)$$

- Complex conjugation: if $z = a + ib$, then

$$z^* \equiv a - ib \qquad (2.60)$$

is called the complex conjugate of z. In general, to obtain the complex conjugate of a complicated expression, replace i throughout by $-i$.

- Polar representation: a complex number $z = a + ib$ can be written in the form

$$z = \rho \exp[i\theta] \qquad (2.61)$$

where ρ and θ are real numbers, $\rho > 0$, and $0 \le \theta < 2\pi$. To connect ρ and θ to a and b, we can use Euler's relation

$$\exp[i\theta] = \cos\theta + i \sin\theta \qquad (2.62)$$

Inserting this in Eq. 2.61 gives

$$z = \rho[\cos\theta + i \sin\theta] = \rho \cos\theta + i\rho \sin\theta$$

Comparison with $z = a + ib$ gives (see Fig. 2.1)

$$a = \rho \cos\theta$$

$$b = \rho \sin\theta$$

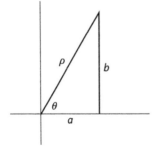

Figure 2.1 Polar representation of a complex number: $a + ib = \rho e^{i\theta}$.

From these two equations, we easily obtain

$$\rho = +\sqrt{a^2 + b^2} \tag{2.63}$$

$$\theta = \arctan\left(\frac{b}{a}\right) \quad \text{with} \quad 0 \le \theta < 2\pi \tag{2.64}$$

For a pure imaginary number $z = ib$, $\theta = 0$ and $\rho = b$.

- Absolute value: the quantity

$$|z| = \sqrt{z^*z} \tag{2.65}$$

is called the absolute value of the complex number z. If we use the polar form $z = \rho \exp[i\theta]$, then we have

$$|z| = \sqrt{(\rho \exp[-i\theta])(\rho \exp[i\theta])} = \sqrt{\rho^2} = |\rho| \tag{2.66}$$

where $|\rho|$ is the absolute value of the real number ρ. If we use the form $z = a + ib$, then

$$|z| = \sqrt{(a - ib)(a + ib)} = \sqrt{a^2 + iab - iab + b^2} = \sqrt{a^2 + b^2} \tag{2.67}$$

Exercise 2.9

(a) Calculate ρ and θ for $z = 2 + 3.1i$.

(b) Calculate the real and imaginary parts of

$$x = (2.1 + 0.3i)(3.13 - 2.1i)$$

$$y = 3(1.7 - 2.1i)$$

$$m = \frac{2 + 1.3i}{3 + 1.1i}$$

(*Hint.* Multiply the numerator and denominator of m by $3 - 1.1i$.)

(c)   Write x, y, and m from part (b) in polar form.

(d)   Use the relation $e^{i\theta} = \cos\theta + i\sin\theta$ to show that

$$\cos\theta = \frac{e^{i\theta} + e^{-i\theta}}{2}$$

and

$$\sin\theta = \frac{i}{2}\left[e^{i\theta} - e^{-i\theta}\right]$$

(e)   Use **Mathematica** to perform these exercises.

§18. *Complex Functions of Real Variables.* Now that we know how to perform operations (addition, multiplication, etc.) with complex numbers, we can define functions whose argument is a real number and whose value is a complex number. An example of such a function is

$$z = \phi(\theta) = \exp[i\theta]$$

with θ a real number. Since the values of $\phi(\theta)$ are complex numbers, you can handle it according to the rules in §17. For example, the complex conjugate of $\phi(\theta) = e^{i\theta}$ is written $\phi(\theta)^*$ or $\phi^*(\theta)$ and has the value $e^{-i\theta}$.

3

THE EIGENVALUE PROBLEM

The Eigenvalue Problem: Definition and Examples

§1. *Definition.* In the previous chapter I told you that in quantum mechanics the dynamical variables of a system (particle positions, momenta, energies, etc.) are represented by operators. The word "represented" is deliberately vague and will become more clear as we proceed. To characterize the motion of a mechanical system we measure how its dynamical variables change in time. This means that these variables must be real numbers. If quantum mechanics represents these variables by operators, how are we to measure their values?

The connection between operators and numerical values that we can measure is provided by the *eigenvalue equation*

$$\hat{O}\psi = \sigma\psi \tag{3.1}$$

Such an equation can be written for any operator $\hat{O}$. The function ψ that satisfies this equation is called an *eigenfunction* or *eigenstate* of the operator $\hat{O}$. "Eigen" is German for "own," as in "the operator's own function." The number σ is called an *eigenvalue* of $\hat{O}$. In this equation $\hat{O}$ is known, and ψ and σ are to be found.

The connection between the eigenvalues of an operator and the results of a measurement will be explained in Chapter 4. In Chapter 5 you will learn how to extract physical information from the eigenfunctions. In this chapter I explain what an eigenvalue problem is and state some of the properties of eigenvalues and eigenfunctions. A thorough understanding of the eigenvalue equation requires more time and more mathematics than is available to us. Because of this, we will learn by examining a few simple examples.

§2. *The Eigenvalue Problem for $\hat{p}_x$.* Following the general definition (Eq. 3.1), the eigenvalue equation for the operator $\hat{p}_x$ is

$$\hat{p}_x \psi(x,y,z) = p\psi(x,y,z) \tag{3.2}$$

Here $\psi(x,y,z)$ is the eigenfunction of $\hat{p}_x$ and p is the eigenvalue. The operator $\hat{p}_x$ was defined in Chapter 2, §12, and is given by

$$\hat{p}_x = \frac{\hbar}{i} \frac{\partial}{\partial x} \tag{3.3}$$

Using this in Eq. 3.2 leads to the eigenvalue equation

$$\frac{\hbar}{i} \frac{\partial \psi}{\partial x} = p\psi \tag{3.4}$$

Here p and ψ are unknown and we need to find them by solving this equation. You can check easily that

$$\psi = \exp[ikx]$$

is a solution of Eq. 3.4. Indeed, inserting $\exp[ikx]$ in Eq. 3.4 gives

$$\hat{p}_x e^{ikx} = \frac{\hbar}{i} \frac{\partial}{\partial x} e^{ikx} = \frac{\hbar}{i}(ik)e^{ikx} = \hbar k \, e^{ikx} \tag{3.5}$$

If you compare Eq. 3.2 with Eq. 3.5, you see that $\psi(x) \equiv \exp[ikx]$ is an eigenfunction of $\hat{p}_x$ with the eigenvalue $p = \hbar k$.

§3. *$\hat{p}_x$ has an Infinite Number of Eigenvalues.* The eigenvalue equation (Eq. 3.4) puts no restriction on the value of $p = \hbar k$: $\exp[ikx]$ is an eigenfunction and $\hbar k$ is the corresponding eigenvalue for any number k, be it real or complex. There are as many eigenvalues and eigenfunctions of $\hat{p}_x$ as there are complex numbers.

Moreover, if ψ is an eigenfunction of a linear operator $\hat{O}$ (i.e. if $\hat{O}\psi = o\psi$, where o is a number), then $\alpha\psi$ (where α can be any complex number) is also an eigenfunction of $\hat{O}$, corresponding to the same eigenvalue o. To see that this is true, introduce $\alpha\psi$ into the eigenvalue equation for $\hat{O}$ and use the fact that $\hat{O}\psi = o\psi$ and $\hat{O}\alpha\psi = \alpha\hat{O}\psi$.

It is not unusual that a mathematical equation describing a physical phenomenon has more than one solution and that not all of them have a physical meaning. As you will soon see, physics provides additional conditions that a physically relevant eigenfunction must satisfy.

Exercise 3.1

Show that $\exp[ikx]$, $\exp[-ikx]$, $\exp[kx]$, $\exp[-kx]$ (where k is a real number) are eigenstates of $\hat{p}_x$. Calculate the corresponding eigenvalues.

Exercise 3.2

Show that $\psi(x,y,z) = \exp[-i(k_x x + k_y y + k_z z)] \equiv e^{-i\mathbf{k}\cdot\mathbf{r}}$ (where the vector $\mathbf{k}$ has the Cartesian components $\{k_x, k_y, k_z\}$ and $\mathbf{r}$ has the Cartesian components $\{x, y, z\}$) is an eigenstate of $\hat{p}_x$ with the eigenvalue $-\hbar k_x$ *and* an eigenstate of $\hat{p}_y$ with the eigenvalue $-\hbar k_y$ *and* an eigenstate of $\hat{p}_z$ with the eigenvalue $-\hbar k_z$.

Exercise 3.3

Show that $\sin(k_x x)\sin(k_y y)\sin(k_z z)$ is an eigenstate of $\hat{p}_x^2 + \hat{p}_y^2 + \hat{p}_z^2$ with eigenvalue $\hbar^2(k_x^2 + k_y^2 + k_z^2)$.

Exercise 3.4

Show that $\exp[i\mathbf{k}\cdot\mathbf{r}]$ is an eigenstate of $\hat{p}_x^2 + \hat{p}_y^2 + \hat{p}_z^2$.

§4. *The Eigenvalue Problem for Angular Momentum.* I examine the eigenvalue problem for angular momentum to give another simple example, which has interesting features that are encountered in other problems in quantum mechanics.

In Chapter 2, §15, I postulated that the operators corresponding to the components of the angular momentum vector are:

$$\hat{L}_x = \frac{\hbar}{i}\left(y\frac{\partial}{\partial z} - z\frac{\partial}{\partial y}\right) \tag{3.6}$$

$$\hat{L}_y = -\frac{\hbar}{i}\left(x\frac{\partial}{\partial z} - z\frac{\partial}{\partial x} \right) \tag{3.7}$$

$$\hat{L}_z = \frac{\hbar}{i}\left(x\frac{\partial}{\partial y} - y\frac{\partial}{\partial x} \right) \tag{3.8}$$

The physical interpretation of these operators will be discussed later. For now I examine briefly some properties of their eigenvalues. According to the general definition (Eq. 3.1), the eigenvalue problem for $\hat{L}_z$ is

$$\hat{L}_z\psi(x,y,z) = \ell_z\psi(x,y,z) \tag{3.9}$$

Using the definition of $\hat{L}_z$ (see Eq. 3.8) turns Eq. 3.9 into

$$\frac{\hbar}{i}\left(x\frac{\partial\psi}{\partial y} - y\frac{\partial\psi}{\partial x} \right) = \ell_z\psi \tag{3.10}$$

The eigenfunction $\psi(x,y,z)$ and the eigenvalue ℓ_z are calculated by solving this equation.

The eigenvalue equation for the square of the angular momentum is

$$\hat{\mathbf{L}}^2\psi(x,y,z) = \lambda\psi(x,y,z) \tag{3.11}$$

You can calculate $\hat{\mathbf{L}}^2\psi$ from

$$\hat{\mathbf{L}}^2\psi = \hat{L}_x\hat{L}_x\psi + \hat{L}_y\hat{L}_y\psi + \hat{L}_z\hat{L}_z\psi$$

and Eqs 3.6–3.8, which give expressions for $\hat{L}_x$, $\hat{L}_y$, and $\hat{L}_z$.

Exercise 3.5

Show that (see Workbook QM.3, Cell 2)

$$\hat{\mathbf{L}}^2\psi \equiv \hat{L}\hat{L}\psi = \lambda\psi$$

leads to

$$-\hbar^2\left[-2z\frac{\partial\psi}{\partial z} + (x^2 + y^2)\frac{\partial^2\psi}{\partial z^2} - 2y\frac{\partial\psi}{\partial y} - 2yz\frac{\partial^2\psi}{\partial y\partial z} + (x^2 + z^2)\frac{\partial^2\psi}{\partial y^2}\right.$$

$$\left.-2x\frac{\partial\psi}{\partial x} - 2xz\frac{\partial^2\psi}{\partial x\partial z} + 2xy\frac{\partial^2\psi}{\partial x\partial y} + (y^2 + z^2)\frac{\partial^2\psi}{\partial x^2}\right] = \lambda\psi$$

§5. *If Two Operators Commute, They Have the Same Eigenfunctions.* Let us examine the following examples of eigenfunctions (see Workbook QM.3, Cell 3):

$$\psi_1(x,y,z) = \frac{x + iy}{\sqrt{x^2 + y^2 + z^2}} \tag{3.12}$$

$$\psi_2(x,y,z) = \frac{x - iy}{\sqrt{x^2 + y^2 + z^2}} \tag{3.13}$$

and

$$\psi_3(x,y,z) = \frac{z(x + iy)}{x^2 + y^2 + z^2} \tag{3.14}$$

They satisfy the equations

$$\hat{L}_z\psi_1(x,y,z) = \hbar\psi_1(x,y,z)$$

$$\hat{L}_z\psi_2(x,y,z) = -\hbar\psi_2(x,y,z)$$

$$\hat{L}_z\psi_3(x,y,z) = \hbar\psi_3(x,y,z)$$

and

$$\hat{\mathbf{L}}^2\psi_\alpha(x,y,z) = \hbar^2\psi_\alpha(x,y,z) \text{ for } \alpha = 1,2,3 \tag{3.15}$$

This means that ψ_1, ψ_2, and ψ_3 are eigenfunctions of both $\hat{L}_z$ and $\hat{\mathbf{L}}^2$. While the two operators have *common eigenfunctions* they have *different eigenvalues.*

One can prove the following general theorem: If two linear operators $\hat{A}$ and $\hat{B}$ commute, we can always construct joint eigenfunctions of $\hat{A}$ and $\hat{B}$. Conversely if two linear operators have the same eigenfunctions then they commute.

By definition (see Chapter 2, §7), two operators commute if

$$(AB - BA)\psi = 0$$

for any function ψ that is acceptable in quantum mechanics (see §8 through §10 on p. 32–34 for the meaning of "acceptable").

As you will see later, this rule is very important because if two dynamical variables (e.g. $\hat{p}_x$ and $\hat{V}$) do not commute, their values cannot be measured simultaneously in an experiment, and measuring one of them changes the value of the other. An example of this behavior is the Heisenberg uncertainty principle, which says that position and momentum cannot be measured simultaneously. It can be shown that this happens because the coordinate and the momentum operators do not commute.

Exercise 3.6

Show that if two operators have the same eigenvalues, then their commutator is zero when it is applied to one of the common eigenvalues.

Exercise 3.7

Show that if $\hat{A}$ and $\hat{B}$ are two linear operators and $\hat{A}\psi = a\psi$ and $[\hat{A}, \hat{B}] = 0$, then $\hat{B}\psi$ is also an eigenfunction of $\hat{A}$, with the eigenvalue a.

Exercise 3.8

Verify that $\psi_1(x,y,z) = (x + iy)/r$, $\psi_2(x,y,z) = (x - iy)/r$, and $\psi_3(x,y,z) = z(x + iy)/r^2$ are eigenfunctions of both $\hat{L}_z$ and $\hat{\mathbf{L}}^2$ where $r = \sqrt{x^2 + y^2 + z^2}$. Calculate the corresponding eigenvalues (see Workbook QM.3, Cell 2).

§6. *Degenerate Eigenvalues.* The functions

$$\psi_1(x,y,z) = \frac{x + iy}{\sqrt{x^2 + y^2 + z^2}} \tag{3.16}$$

$$\psi_2(x,y,z) = \frac{x - iy}{\sqrt{x^2 + y^2 + z^2}} \tag{3.17}$$

satisfy the equations

$$\hat{L}^2\psi_1(x,y,z) = \hbar^2\psi_1(x,y,z) \tag{3.18}$$

and

$$\hat{L}^2\psi_2(x,y,z) = \hbar^2\psi_2(x,y,z) \tag{3.19}$$

You can test that this is true by inserting Eqs 3.16 and 3.17 into Eqs 3.18 and 3.19 (see Cell 3 of Workbook QM.3).

Eqs 3.18 and 3.19 indicate that ψ_1, ψ_2 are eigenfunctions of the operator $\hat{L}^2$ that correspond to *the same eigenvalue* $\hbar^2$. When several eigenfunctions have the same eigenvalue, we say that they are *degenerate*. Degenerate eigenfunctions appear frequently in quantum mechanics and, as you will see later, degeneracy has a physical interpretation.

You can also show easily (see Cell 3 of Workbook QM.3) that ψ_1 and ψ_2 satisfy the equations

$$\hat{L}_z\psi_1(x,y,z) = \hbar\psi_1(x,y,z) \tag{3.20}$$

and

$$\hat{L}_z\psi_2(x,y,z) = -\hbar\psi_2(x,y,z) \tag{3.21}$$

which indicate that ψ_1 and ψ_2 are also eigenfunctions of $\hat{L}_z$, corresponding to the eigenvalues $\hbar$ and $-\hbar$, respectively.

We can now understand why the eigenvalues of $\hat{L}^2$ are degenerate. The functions $\psi_1(x,y,z)$ and $\psi_2(x,y,z)$ are eigenvalues of both $\hat{L}^2$ and $\hat{L}_z$. These two operators are related: one gives the square of the length of the angular momentum vector and the other gives the projection of the angular momentum vector on the OZ axis. From classical mechanics we know that the length of the angular momentum squared is proportional to the rotational energy of the particle; the projection of the angular momentum vector on the OZ axis tells us the orientation of the axis of rotation. Since the eigenstates $\psi_1(x,y,z)$ and $\psi_2(x,y,z)$ correspond to the same eigenvalue of $\hat{L}^2$, they describe states in which the particle has *the same rotational energy*. If the particle in the states $\psi_1(x,y,z)$ has the same rotational energy as a particle in the state $\psi_2(x,y,z)$, why are these states different? The answer is given by Eqs 3.20 and 3.21, which tell us that the two states correspond to particles having different eigenvalues of $\hat{L}_z$, which means that their axis of rotation has a different orientation.

This behavior is general. The degenerate states of an operator $\hat{O}$ are usually non-degenerate states of another operator. In the example given above, the degenerate states of $\hat{L}^2$ are nondegenerate states of $\hat{L}$.

§7. *A Summary of Some of the Observations Made so Far.* The particular cases studied so far have revealed behavior that is general and must be remembered. An eigenvalue problem does not have a unique solution. As you have seen in §3 and §5 the operators $\hat{p}_x$, $\hat{L}_z$, and $\hat{L}^2$ all have more than one eigenfunction. This is true for all operators encountered in quantum mechanics. In many cases the number of eigenfunctions is infinite.

Two different operators can have common eigenfunctions, but with different eigen-values. This means that given two linear operators $\hat{A}$ and $\hat{B}$, it is sometimes possible to find a function ψ such that $\hat{A}\psi = a\psi$ and $\hat{B}\psi = b\psi$. For example, $\exp[-ikx]$ is an eigenfunction of $\hat{p}_x$ with the eigenvalue $-\hbar k$ and an eigenfunction of $\hat{p}_x^2/2m$ with the eigenvalue $\hbar^2 k^2/2m$.

Finally, we have seen that sometimes an operator has several eigenfunctions that correspond to the same eigenvalue. For example, $\exp[-ikx]$ and $\exp[ikx]$ are eigenfunctions of $\hat{p}_x^2/2m$; they both correspond to the eigenvalue $\hbar^2 k^2/2m$ (show that this is true). When this happens, the eigenvalue is called *degenerate*.

Which Eigenfunctions Have a Meaning in Physics?

§8. *Not all Eigenfunctions are Physically Meaningful.* So far we have looked at the eigenvalue problem as a mathematical problem. It often happens that some of the solutions of an equation that represents a physical process have no physical meaning and need to be discarded. This is true in the case of the eigenvalue problem.

§9. *Normalization.* To select those eigenfunctions that describe the physical properties of a system, we must impose additional conditions. One of them is written in terms of the symbol

$$\langle \phi | \psi \rangle \equiv \int_{-\infty}^{+\infty} d\mathbf{r}(1) \cdots \int_{-\infty}^{+\infty} d\mathbf{r}(N)\, \phi^*(\mathbf{r}(1), \ldots, \mathbf{r}(N)) \psi(\mathbf{r}(1), \ldots, \mathbf{r}(N)) \quad (3.22)$$

which is called the *scalar product* of ϕ with ψ. Here $d\mathbf{r}(k) \equiv dx(k)\,dy(k)\,dz(k)$ and $x(k), y(k), z(k)$ are the coordinates of particle k, located at $\mathbf{r}(k)$. $\phi^*(\mathbf{r}(1), \ldots, \mathbf{r}(N))$ is the complex conjugate of $\phi(\mathbf{r}(1), \ldots, \mathbf{r}(N))$.

The physical interpretation of the eigenfunctions ψ of an operator representing a dynamical variable (momentum, kinetic energy, total energy, ...) requires that they must satisfy the *normalization condition*

$$\langle \psi | \psi \rangle \equiv \int_{-\infty}^{+\infty} d\mathbf{r}(1) \cdots \int_{-\infty}^{+\infty} d\mathbf{r}(N) \psi^*(\mathbf{r}(1), \ldots, \mathbf{r}(N))$$

$$\times \, \psi(\mathbf{r}(1), \ldots, \mathbf{r}(N)) = 1 \tag{3.23}$$

A function that satisfies Eq. 3.23 is called *normalized.*

The condition in Eq. 3.23 follows from the physical interpretation of eigenfunctions, which postulates that

$$\psi^*(\mathbf{r}(1), \ldots, \mathbf{r}(N))\psi(\mathbf{r}(1), \ldots, \mathbf{r}(N))d\mathbf{r}(1) \cdots d\mathbf{r}(N)$$

is the probability that the particle labeled 1 is in the volume element $d\mathbf{r}(1)$, centered around the point $\mathbf{r}(1)$, the particle labeled 2 is in the volume element $d\mathbf{r}(2)$, centered around the point $\mathbf{r}(2)$, ... The integral in Eq. 3.23 is the probability that the particles are located somewhere in space, and this must be equal to 1.

§10. *A Relaxed Condition.* This seems to be a very stringent mathematical condition. However, we can relax it considerably based on the following observation. Suppose that ψ satisfies the eigenvalue equation of a *linear* operator $\hat{O}$:

$$\hat{O}\psi = \lambda \psi$$

Define $\eta = C\psi$ where C is an arbitrary complex number. Then the function η is also an eigenfunction of $\hat{O}$ with the eigenvalue λ:

$$\hat{O}\eta = \hat{O}(C\psi) = C(\hat{O}\psi) = C(\lambda\psi) = \lambda(C\psi) = \lambda\eta$$

Why is this of any help to us? Let us assume that we have found an eigenfunction ϕ of $\hat{O}$ with the eigenvalue λ, so that $\hat{O}\phi = \lambda\phi$, but ϕ does not satisfy the normalization condition (Eq. 3.23). This does not mean that we need to throw this function away and look for one that is normalized. We can use the procedure described below to turn ϕ into a normalized eigenfunction.

Assume that ϕ satisfies the weaker condition

$$\langle\phi|\phi\rangle = \int_{-\infty}^{+\infty} d\mathbf{r}(1)\cdots\int_{-\infty}^{+\infty} d\mathbf{r}(N)\, \phi^*(\mathbf{r}(1),\ldots,\mathbf{r}(N))$$

$$\times\, \phi(\mathbf{r}(1),\ldots,\mathbf{r}(N)) = M \tag{3.24}$$

where M is some positive real number. The specific value of M is irrelevant for what follows; what counts is that $\langle\phi|\phi\rangle$ is not infinite (it is easy to show that it is real and positive). The function

$$\psi \equiv \frac{\phi}{\sqrt{M}} \tag{3.25}$$

is also an eigenfunction of $\hat{O}$, with the eigenvalue λ, and it *is normalized*. This statement is very easy to check by evaluating $\hat{O}\psi$ and $\langle\psi|\psi\rangle$.

By this procedure, we turn the eigenfunction ϕ into the physically acceptable eigenfunction ψ. We say that we have *normalized* ϕ, or that ψ is the normalized version of ϕ.

We can now replace the normalization condition (Eq. 3.23) with the weaker condition (Eq. 3.24), with the proviso that if we obtain an eigenfunction ϕ that satisfies Eq. 3.24 then the physically meaningful eigenfunction is given by Eq. 3.25. You will see soon an example using this idea.

An Example: the Eigenfunctions of Kinetic Energy

§11. The fact that mathematics provides more solutions than are needed by physics is not unusual. This is not a great nuisance, as long as we have a criterion that allows us to pick the solutions we want. In quantum mechanics, the condition in Eq. 3.24 is often sufficient.

There are cases in which additional conditions are necessary. These are not general, but follow from an examination of the physics of the specific system. The eigenfunctions of the kinetic energy operator provide an example of this sort.

§12. *One-Dimensional Motion in the Force-Free, Unbounded Space.* We study the case of a particle moving in one dimension to examine why we need such additional conditions, to show what these conditions are, to explain how they are used, and to indicate how they affect the eigenvalues and the eigenfunctions.

We assume that no force acts on this particle and it moves in an unbounded space. In other words, this particle is alone in the universe. Since no force acts on the particle, its total energy is equal to its kinetic energy.

§13. *The Eigenfunctions of the Kinetic Energy Operator.* The eigenvalue problem for the kinetic energy operator is

$$\hat{K}\psi(x) = K\psi(x) \tag{3.26}$$

Since, for a one-dimensional system, $\hat{K} = \hat{p}^2/2m = -(\hbar^2/2m)(\partial^2/\partial x^2)$, Eq. 3.26 becomes

$$-\frac{\hbar^2}{2m}\frac{\partial^2\psi(x)}{\partial x^2} = K\psi(x) \tag{3.27}$$

This is a well-known differential equation. The following functions satisfy it: $\exp[kx], \exp[-kx], \sin(kx), \cos(kx), \exp[ikx], \exp[-ikx]$. I will verify this statement for $\psi(x) = \exp[kx]$. I have $\partial\psi/\partial x = k\exp[kx]$ and $\partial^2\psi/\partial x^2 = k^2\exp[kx]$. Inserting this in Eq. 3.27 leads to

$$-\frac{\hbar^2 k^2}{2m}e^{kx} = Ke^{kx}$$

Dividing by $\exp[kx]$ gives $K = -\hbar^2 k^2/2m$. Therefore $\psi(x) = \exp[kx]$ is an eigenfunction of $\hat{K}$ with eigenvalue $K = -\hbar^2 k^2/2m$. In Workbook QM.3, Cell 4, I show that all the functions listed above are eigenfunctions. Their eigenvalues are given in Table 3.1.

Workbook

Table 3.1 Eigenfunctions and corresponding eigenvalues (for simplicity, I constrain k to be a real number) for the kinetic energy operator $\hat{K}$.

$\psi(x)$	$\sin(kx)$	$\cos(kx)$	e^{-ikx}	e^{ikx}	e^{kx}	e^{-kx}
$\hat{K}\psi/\psi$	$\dfrac{k^2\hbar^2}{2m}$	$\dfrac{k^2\hbar^2}{2m}$	$\dfrac{k^2\hbar^2}{2m}$	$\dfrac{k^2\hbar^2}{2m}$	$-\dfrac{k^2\hbar^2}{2m}$	$-\dfrac{k^2\hbar^2}{2m}$

§14. *More general eigenfunctions.* The eigenfunctions $\sin(kx)$ and $\cos(kx)$ are degenerate: they correspond to the same eigenvalue $\hbar^2 k^2/2m$. If two functions ψ_1 and ψ_2 are eigenfunctions of a linear operator $\hat{O}$ having the same eigenvalue σ, then for any complex (or real) numbers A and B, the function $\psi(x) \equiv A\psi_1(x) + B\psi_2(x)$ is also an eigenfunction of $\hat{O}$, with the eigenvalue σ. To put this in symbols, if

$$\hat{O}\psi_1 = \sigma\psi_1$$

and

$$\hat{O}\psi_2 = \sigma\psi_2,$$

and

$$\psi(x) = A\psi_1(x) + B\psi_2(x),$$

then

$$\hat{O}\psi = \sigma\psi$$

You can easily see that this follows from the linearity of $\hat{O}$:

$$\hat{O}\psi = \hat{O}(A\psi_1 + B\psi_2) = A\hat{O}\psi_1 + B\hat{O}\psi_2$$

$$= A\sigma\psi_1 + B\sigma\psi_2 = \sigma(A\psi_1 + B\psi_2) = \sigma\psi$$

This "combining" property for degenerate eigenvalues is general. Let us use it for the eigenfunctions $\sin(kx)$ and $\cos(kx)$ of the kinetic energy operator. The conclusion is that

$$\psi_1(x) = A_1 \sin(kx) + B_1 \cos(kx) \tag{3.28}$$

is an eigenfunction of $\hat{K}$, with the eigenvalue $K = k^2\hbar^2/2m$.
Similarly

$$\psi_2(x) = A_2 e^{ikx} + B_2 e^{-ikx} \tag{3.29}$$

is an eigenfunction of $\hat{K}$, with the eigenvalue $K = k^2\hbar^2/2m$. And

$$\psi_3(x) = A_3 e^{kx} + B_3 e^{-kx} \tag{3.30}$$

is an eigenfunction of $\hat{K}$, with the eigenvalue $K = -k^2\hbar^2/2m$.

Exercise 3.9

Show that ψ_1 and ψ_2 defined by Eqs 3.28 and 3.29 are not independent: for a proper choice of C and D, $\psi_1(x) = A_1 \sin(kx) + B_1 \cos(kx)$ is the same function as $C \exp[ikx] + D \exp[-ikx]$.

§15. *Which of These Eigenfunctions can be Normalized?* We need now to establish which of these eigenfunctions and eigenvalues are physically acceptable. This means that we will test which one satisfies the condition in Eq. 3.24.

For the eigenfunction $\psi(x) = \sin(kx)$, the normalization condition requires that

$$\langle \psi | \psi \rangle = \int_{-\infty}^{+\infty} dx \, \psi^*(x) \phi(x) = \int_{-\infty}^{+\infty} dx \, \sin^2(kx) = M < \infty$$

It is easy to see that this integral is not finite. Indeed (see your calculus book or use **Mathematica**)

$$\int_{-z}^{+z} \sin^2(kx) dx = z - \frac{\sin(2kz)}{2k}$$

and this expression is not defined as z becomes infinite.

The function $\eta(x) \equiv \exp[ikx]$ does not fare better:

$$\langle \eta | \eta \rangle = \int_{-\infty}^{+\infty} dx \, \exp[-ikx] \exp[ikx] = \int_{-\infty}^{+\infty} dx$$

This integral is not finite, and this eigenfunction must be thrown away.

We come at last to $\lambda(x) \equiv \exp[kx]$:

$$\langle \lambda | \lambda \rangle = \int_{-\infty}^{+\infty} dx \, \exp[kx] \exp[kx] = \int_{-\infty}^{+\infty} e^{2kx} dx \frac{\exp[2k\infty]}{2k} - \frac{\exp[-2k\infty]}{2k} = \infty$$

This integral is not finite, so it does not satisfy Eq. 3.24.

If $\sin(kx)$, $\exp[ikx]$, and $\exp[kx]$ do not satisfy Eq. 3.24, neither will ψ_1, ψ_2, or ψ_3 from Eqs. 3.28–3.30. None of the eigenfunctions I found for $\hat{K}$ is physically meaningful. This is distressing and mathematics gives us no relief, since we have exhausted all eigenfunctions. The solution to this dilemma must come from physics.

§16. *Boundary Conditions: the Particle in a Box.* We have run into trouble because our formulation of the problem misses an important aspect of the physics of the problem. Any particle on which we perform an experiment is confined inside an apparatus. This means that the probability of finding the particle outside the apparatus is zero.

Let us assume that in our one-dimensional problem, the walls of the apparatus are at $x = 0$ and $x = L$. As long as the coordinate of the particle satisfies

$$0 \leq x \leq L$$

the particle is inside the apparatus. Since the particle cannot escape from it, the probability $\psi^*(x)\psi(x)$ that the particle is outside the "box" is zero. This means that we must look for a solution of the eigenvalue equation for the kinetic energy which satisfies the condition

$$|\psi(x)|^2 = 0 \ \text{ if } \ x < 0 \ \text{ or } \ x > L \tag{3.31}$$

This condition must be imposed in addition to the normalization condition (Eq. 3.24).

Furthermore, we impose the additional condition that *the wave function must be continuous.* You can see that this is a reasonable condition: the momentum operator cannot be applied to a discontinuous function, because such a function does not have a derivative.

Because of the continuity requirement, the function $\psi(x)$ must satisfy

$$\psi(x = 0) = 0 \tag{3.32}$$

and

$$\psi(x = L) = 0 \tag{3.33}$$

In the theory of differential equations these are called *boundary conditions*, because they force the eigenfunction to take certain values at the boundaries $x = 0$ and $x = L$.

§17. *Force the Eigenfunctions to Satisfy the Boundary Conditions.* Let us see what the boundary conditions do to the eigenfunction ψ_1 defined in Eq. 3.28.

For $x = 0$, I must have

$$\psi_1(0) = A \sin(k \cdot 0) + B \cos(k \cdot 0) = 0 \tag{3.34}$$

and for $x = L$,

$$\psi_1(L) = A \sin(kL) + B \cos(kL) = 0 \tag{3.35}$$

Because $\sin(k \cdot 0) = 0$ and $\cos(k \cdot 0) = 1$, Eq. 3.34 gives $B = 0$. Using this in Eq. 3.35 leads to

$$A \sin(kL) = 0 \tag{3.36}$$

One of the solutions of this equation is $A = 0$. This is unacceptable: if $A = 0$ and $B = 0$ then $\psi_1 \equiv 0$ which makes no sense; it would imply that the probability of finding the particle anywhere in the box is zero.

If $A \neq 0$, Eq. 3.36 becomes $\sin(kL) = 0$, which has the solutions

$$k_n = \frac{\pi}{L}n, \ \ n = 0, \pm 1, \pm 2, \ldots \tag{3.37}$$

This is very interesting. Before we imposed the boundary conditions (Eqs 3.32 and 3.33), $\sin(kx)$ satisfied the eigenvalue equation and the eigenvalue was $\hbar^2 k^2/2m$. There was no constraint on the values of k: $\sin(kx)$ and $\hbar^2 k^2/2m$ were eigenfunctions and eigenvalues of $\hat{K}$ for any value of k. When we impose the boundary conditions, k can no longer take arbitrary values; the only acceptable values of k are given by Eq. 3.37. As a consequence, the eigenvalues of the kinetic energy operator $\hat{K}$ can only take the discrete values

$$K_n = \frac{\hbar^2 k_n^2}{2m} = \frac{\hbar^2 \pi^2}{2mL^2}n^2, \ \ n = 0, \pm 1, \pm 2, \ldots \tag{3.38}$$

The corresponding eigenfunctions are

$$\psi_n(x) = \begin{cases} A \sin\left(\dfrac{\pi n}{L}x\right) & \text{if } 0 \leq x \leq L \\ 0 & \text{if } x \leq 0 \text{ or } x \geq L \end{cases} \tag{3.39}$$

Note that we use the subscript n to label the eigenfunctions and the eigenvalues.

§18. *Quantization.* As you will learn soon, one of the rules of quantum mechanics says that if I measure the kinetic energy of a particle I will obtain one of the eigenvalues of the kinetic energy operator. For the example of the particle in a one-dimensional box this means that the result of the measurement can be $(\hbar^2\pi^2/2mL^2)\times 1^2$ or $(\hbar^2\pi^2/2mL^2)\times 2^2$ or any other discrete value given by Eq. 3.38. No other value is allowed by quantum mechanics. This is the famous quantization postulated by Planck and Bohr in the early days of quantum theory.

This quantization appears in our calculation because we confined the particle in a box. If the box size L or the mass m is large, the allowed values of the kinetic energy are so close to each other that our instruments are not good enough to resolve the difference. In such a case it appears to us that the kinetic energy can take continuous values, as specified by classical mechanics. We can play billiards without a conflict with quantum mechanics.

Exercise 3.10

Calculate K_n for $n = 1, 2$ when $m = 1$ gram and $L = 1$ mm. Is any experiment sufficiently accurate to measure the difference $K_2 - K_1$?

---

§19. *The Particle Cannot Have Zero Kinetic Energy.* Another curiosity is provided by the case $n = 0$. The eigenfunction corresponding to $n = 0$ is

$$\psi_0(x) = A\sin\left(\frac{\pi\times 0}{L}x\right) = 0$$

We must discard this because $|\psi_0(x)|^2$ is identically zero; for this eigenfunction, the probability that the particle is in the box is zero and this conclusion is not acceptable, unless you believe in ghosts.

Since we discard $\psi_0(x)$ as unacceptable, we also discard the energy $K_0 = 0$ as impossible. The smallest kinetic energy the particle can have is K_1. If you remember that in classical physics the kinetic energy is $mv^2/2$, the fact that the lowest kinetic energy is not zero means that the particle cannot have zero velocity. This is very strange: the particle in a box can never be at rest! On the other hand we have all seen that any large, classical object put in a box does not move if we do not act on it with a force. It appears that quantum mechanics is wrong if the object is large. How can we have two mechanics, one for small objects and one for the large ones? And what happens to the objects in between, that are neither very large nor very small?

This apparent paradox is easily resolved. If the energy $\hbar^2/mL^2$ is much smaller the accuracy ε of the instrument that measures energy, the lowest kinetic energy $K_1 = \hbar^2 1^2 \pi^2/2mL^2$ is so small that our instruments cannot determine that it is different from zero. This is why classical physics does not appear to be in error.

Exercise 3.11

Calculate the value of K_1 for $m = 1$ gram and $L = 1$ cm. Try to explain why none of our instruments could measure such a small kinetic energy. Assume that you can define a velocity v by equating K_1 with $mv^2/2$. Calculate how large this velocity is and explain why it could not be measured.

Exercise 3.12

Show that the function $\psi_2(x)$ defined by Eq. 3.29 leads to the same eigenfunctions and eigenvalues as does the function $\psi_1(x)$ defined by Eq. 3.28.

Exercise 3.13

Assume that the particle is in a box whose boundaries are located at

$$-\frac{L}{2} \leq x \leq \frac{L}{2}$$

Show that the eigenfunctions look different from those derived above (where the position of the box was defined by $0 \leq x \leq L$) but the eigenvalues are the same. The two boxes are the same except that the second box is the first box translated a distance $L/2$ to the left. Should any of the physically measurable results for the two boxes be different?

Exercise 3.14

The function $\psi_3(x) \equiv A_3 \exp[-kx] + B_3 \exp[kx]$ is an eigenstate of $\hat{K}$. Determine A_3 and B_3 so that ψ_3 satisfies the boundary conditions (Eqs 3.32 and 3.33). (*Hint*. The boundary conditions can be satisfied only if k is imaginary. If k is imaginary then ψ_3 is the same as $A \exp[-ikx] + B \exp[ikx]$.)

§20. *Imposing the Boundary Conditions Removes the Trouble we had with Normalization.* It is easy to see that the eigenfunctions of $\hat{K}$ obtained by imposing the boundary conditions can be normalized. Indeed (see Workbook QM.3, Cell 6)

$$\int_{-\infty}^{+\infty} |\psi_n(x)|^2 dx = \int_0^L A^2 \sin^2\left(\frac{n\pi x}{L}\right) dx$$

$$= A^2 \left[\frac{L}{2} - \frac{L\sin(2n\pi)}{4n\pi}\right]$$

The first equality follows because $\psi_n(x) = 0$ for $x < 0$ or $x > L$ and $\psi_n(x) = A\sin(\pi n x/L)$ for $x \in [0, L]$ (see Eq. 3.39). Because $\sin(2\pi n) = 0$, the integral is equal to $A^2 L/2$. This is finite and therefore the normalization condition (Eq. 3.24) is satisfied with $M = A^2 L/2$.

We can normalize the eigenfunction in Eq. 3.39 by using Eq. 3.25. This leads to the normalized eigenfunctions

$$\phi_n(x) = \begin{cases} \left(\frac{2}{L}\right)^{1/2} \sin\left(\frac{\pi n x}{L}\right) & \text{if } 0 \leq x \leq L \\ 0 & \text{if } x \leq 0 \text{ or } x \geq L \end{cases} \tag{3.40}$$

The corresponding eigenvalues are

$$K_n = \frac{\hbar^2 k_n^2}{2m} = \frac{\hbar^2}{2m}\left(\frac{\pi n}{L}\right)^2, \quad n = \pm 1, \pm 2, \dots \tag{3.41}$$

The eigenfunctions corresponding to n and $-n$ differ by the constant factor -1 (i.e. $\phi_n(x) = -\phi_{-n}(x)$) and therefore they correspond to the same physical state. For this reason, we consider only positive values of the index n.

In a future chapter we will return to this example, to bring out the physics contained in this result. Its purpose, so far, has been to illustrate how an eigenvalue problem can be solved and to point out how important the boundary conditions are.

§21. *Some Properties of Eigenvalues and Eigenfunctions.* I conclude this section by pointing out two general properties of the eigenvalue problems that appear in quantum mechanics.

The eigenvalues of the operators appearing in quantum mechanics *are always real numbers.* This is important because, as you will learn in the next chapter, the eigenvalues of these operators correspond to physical, measurable quantities. Therefore, they must be real numbers.

Another important property is that if $\psi_n(\mathbf{r}(1), \ldots, \mathbf{r}(N))$ and $\psi_m(\mathbf{r}(1), \ldots, \mathbf{r}(N))$ are eigenfunctions of an operator $\hat{O}$ representing a physical quantity (e.g. energy, momentum, angular momentum) that correspond to *two different eigenvalues*, then their scalar product is zero:

$$\langle \phi | \psi \rangle \equiv \int_{-\infty}^{+\infty} \psi_n^*(\mathbf{r}(1), \ldots, \mathbf{r}(N)) \psi_m(\mathbf{r}(1), \ldots, \mathbf{r}(N)) d\mathbf{r}(1) \cdots d\mathbf{r}(N) = 0 \quad (3.42)$$

Two functions satisfying this condition are said to be *orthogonal*. A set $\psi_1(\mathbf{r}(1), \ldots, \mathbf{r}(N))$, $\psi_2(\mathbf{r}(1), \ldots, \mathbf{r}(N))$, ... of eigenfunctions of an operator $\hat{O}$ satisfying Eqs 3.23 and 3.42 are said to form *an orthonormal set*. Unlike the normalization condition, which we must impose, the orthogonality condition comes free: the operators representing dynamical variables are such that their eigenfunctions corresponding to different eigenvalues are automatically orthogonal.

Let us check whether this is true for the eigenfunctions (Eq. 3.40) of the kinetic energy operator. We have $\psi_n(x) = \sqrt{2/L} \sin(\pi n x/L)$ for $0 \le x \le L$ and $\psi_n(x) = 0$ otherwise. For $m \ne n$ we have (see Workbook QM.3, Cell 7)

Workbook

$$\langle \psi_n | \psi_m \rangle = \int_{-\infty}^{+\infty} \psi_n^*(x) \psi_m(x) dx$$

$$= \int_0^L dx \, \frac{2}{L} \, \sin\left(\frac{\pi n x}{L}\right) \sin\left(\frac{\pi m x}{L}\right)$$

$$= \frac{\sin((m-n)\pi)}{m-n} - \frac{\sin((m+n)\pi)}{m+n}$$

Since m and n are positive integers and $m \ne n$, $\frac{\sin((m-n)\pi)}{m-n} = \frac{\sin((m+n)\pi)}{m+n} = 0$ and the orthogonality condition is satisfied.

§22. *Summary.* If $\hat{O}$ is an operator, the equation

$$\hat{O}\psi = \lambda\psi,$$

where λ is a number and ψ is a function, is called the eigenvalue equation for the operator $\hat{O}$. The function ψ satisfying this condition is called an eigenfunction or eigenstate of the operator $\hat{O}$, and the number λ is the eigenvalue associated with that eigenfunction.

For the operators that appear in quantum mechanics, this equation has a large number of solutions, labeled $\psi_n(\mathbf{r}(1),\dots,\mathbf{r}(N))$ and λ_n. If $\hat{O}$ represents a dynamic variable, then λ_n is a real number.

Only normalized eigenfunctions, satisfying

$$\langle\psi|\psi\rangle = \int_{-\infty}^{+\infty} d\mathbf{r}(1)\cdots d\mathbf{r}(N)\psi_n^*(\mathbf{r}(1),\dots,\mathbf{r}(N))\psi_n(\mathbf{r}(1),\dots,\mathbf{r}(N)) = 1,$$

have physical meaning. However, we can impose the less stringent condition

$$\langle\psi|\psi\rangle = \int_{-\infty}^{+\infty} d\mathbf{r}(1)\cdots d\mathbf{r}(N)\psi_n^*(\mathbf{r}(1),\dots,\mathbf{r}(N))\psi_n(\mathbf{r}(1),\dots,\mathbf{r}(N)) = M < \infty$$

If this condition is satisfied for an eigenfunction ψ, then function ϕ defined by

$$\phi = \frac{\psi}{\sqrt{M}} = \frac{\psi}{\sqrt{\langle\psi|\psi\rangle}}$$

is a normalized eigenfunction of the same operator, corresponding to the same eigenvalue as ψ.

In addition, we often must impose boundary conditions, to select physically meaningful solutions.

4

WHAT DO WE MEASURE WHEN WE STUDY QUANTUM SYSTEMS?

Introduction

§1. Understanding fully how the properties of a quantum system are measured requires a detailed analysis of each specific measurement. We do not have the time and you do not have the knowledge required for this analysis. There are, however, features that are common to many measurements and we study them here by looking at a few examples.

I will discuss a class of measurements that detect energy change. They begin by "preparing" the molecules on which measurements are performed in an *initial state* Ψ_i in which their energy has a well-defined value E_i. After this preparation, the measurement exposes the system to an "external" agent, such as light or a stream of electrons of known energy or a stream of molecules, etc. The interaction between the system and the external agent causes both of them to change state. For example, a molecule in the state Ψ_i that is exposed to light may absorb a photon and change its state to Ψ_f. We say that photon absorption caused a *transition* from Ψ_i to Ψ_f. During this change of state the energy changes from E_i to E_f. Energy conservation requires that the energy $\hbar\Omega$ (Ω is the frequency of the light) of the lost photon matches the energy increase $E_f - E_i$ in the molecule. When we determine the frequencies of the absorbed photons we measure the magnitude of the *energy difference $E_f - E_i$*.

To connect theory to such measurements, quantum mechanics *postulates* that the energies that can be measured in an experiment must be eigenvalues of the energy operator. This rule is valid in general: *the result of a measurement of a dynamical variable must be one of the eigenvalues of the operator representing that dynamical variable.* This postulate places the eigenvalue problem at the center of quantum mechanics. If theory can calculate the eigenvalues of various operators, it can predict the results of various measurements.

In this chapter I discuss the "preparation" of the initial state, give three examples of energy eigenvalues and use them to discuss electron energy-loss spectroscopy and emission spectroscopy as two examples of measurements of energy differences. The purpose of the chapter is to exemplify how quantum mechanics connects measurements to operator eigenvalues.

People performing measurements on quantum systems are very ingenious and imaginative and have developed an astonishing number of clever methods for obtaining information about molecules and for testing quantum theory. The ones discussed here are the simplest.

The Preparation of the Initial State

§2. *Not all Quantum Systems have "Well-Defined" Energy.* I will discuss here only measurements in which the system was prepared in an initial state Ψ_i that has a well-defined energy E_i. Here "well-defined" means that we are certain that the energy of the molecule is equal to a specific eigenstate E_i of the energy operator.

Preparing a system in a state of well-defined energy is not always easy. There are practical situations in which we know only the probability P_1 that the system is in the state Ψ_1 having the energy E_1, the probability P_2 that the system is in the state Ψ_2 having the energy $E_2, \ldots$. Here E_1, E_2, etc., are eigenvalues of the energy operator. This situation occurs whenever our system consists of many molecules interacting with each other (i.e. they form a gas or a liquid).

Why is there such an uncertainty? The molecules in a gas or a liquid collide with each other and the energy of any given molecule is changed by these collisions. At a given moment a given molecule might be in the lowest energy state E_1. But, a short time later it might collide with a very energetic molecule and its energy may increase to E_2. Then it may lose energy in a subsequent collision and return to E_1. Thus, the energy of any given molecule will change in time in a very chaotic manner.

We call the lowest energy state the *ground* state and we say that the molecule is *excited*, or that it is in an *excited state*, when its energy is greater than that of the ground state. While a given molecule will change its state in a random manner, the number of excited molecules, at any given time, in a gas in thermodynamic equilibrium, is constant. The higher the temperature, the higher the mean energy of collision between the molecules and the higher the number of excited molecules in the ensemble.

Statistical mechanics provides equations that allow us to calculate the probability P_n that we find, in the gas, a molecule having the energy E_n. I will not give this formula here, but inform you of a simple criterion. You know, from your physics studies, that the average translational energy of a molecule or an atom in an ensemble (gas, liquid, solid, or a polymer) is equal to $3k_BT/2$. Here $k_B = 1.3807 \times 10^{-23}$ J/K and T is the temperature in kelvin. This is the mean energy of two colliding molecules. Energy conservation tells us that it is unlikely that a molecule with energy E_1 will be excited by a collision to an energy E_n, if

$$E_n - E_1 \gg k_BT \tag{4.1}$$

The word "unlikely" is not accidental: the energy of a few molecules exceeds this mean value (a student's grade can be higher than the average) and they can excite their collision partner to a state of higher energy than the one allowed by Eq. 4.1. But highly energetic molecules are as rare as perfect students, and few molecules reach the high-energy states that violate the rule given above.

It so happens that the energy required to excite the electrons in a molecule is much larger than k_BT, which is of order of 0.025 eV at room temperature. This means that the electrons in molecules are in the ground state, at most of the temperatures we are likely to use in the laboratory.

The energy to excite the vibrational motion is not as high. Nevertheless, at room temperature most molecules are in the ground vibrational state, and only a small fraction is excited.

The energy needed to excite the rotational motion is much smaller than k_BT and many molecules are rotationally excited at room temperature. It is in such situations that we say that the energy is not well defined even though we know from statistical mechanics how to calculate the probabilities P_1, P_2, ... that the molecules have the rotational energies E_1, E_2, ... When we say that the energy is not well defined, we mean that we do not know what the energy of a specific molecule in the ensemble is.

A similar (but not identical) situation is encountered when the molecules are exposed to an extremely short pulse of light, before a measurement. Such pulses contain photons of a variety of energies and we do not know which one is absorbed by the molecule. Because of this, the molecule can have one of several energies and we have no way of knowing *for sure* which one it is.

We do not discuss in this chapter measurements on systems in which the quantity being measured does not have a well-defined initial value. We analyze such a situation when we discuss the "hot bands" in the absorption and emission spectra of diatomic molecules (see Chapter 17).

Three Examples of Energy Eigenstates

§3. *Introduction.* To make the discussion of measurements in quantum mechanics as concrete as possible (at this stage), I will give three examples of energy eigenstates, for three simple systems.

In general, the energy eigenstates are *discrete*, as long as the system is confined to a finite volume. Sometimes the volume is so large that the difference between two consecutive energy eigenvalues is so small that we cannot detect it experimentally. In such cases it appears that the energy of the system takes continuous values.

We call the discrete energy eigenvalues E_n, $n = 1, 2, \ldots$ the *energy spectrum* of the system. Often we represent the energy spectrum by an *energy level diagram* (see Fig. 4.1). This figure implies the existence of a vertical energy axis (not shown) and indicates the energy eigenvalues by horizontal lines. The arrows between these lines show possible transitions in the molecule when it interacts with an external agent (e.g. light, a passing electron). For example, A indicates a transition from an initial state of energy E_1 to the state of energy E_2. Since the molecule gains

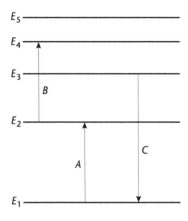

Figure 4.1 In this graph the vertical axis (which is not shown explicitly) represents the energy. The horizontal lines indicate the energy eigenvalues of the system (e.g. a molecule). The vertical arrows show possible transitions in the system, as a result of its interaction with an external agent (e.g. light, a passing electron).

the energy $E_2 - E_1$, the external agent must lose the same amount of energy. If we measure the energy lost by the external agent we know the energy gained by the molecule. C shows a transition in which the molecule gives the external agent the energy $E_3 - E_1$ (for example, the molecule emits a photon, and the emitted light is the external agent). We can measure the energy of the emitted photon and determine the magnitude of $E_3 - E_1$.

Below I give energy spectra for three systems and discuss in more detail how such a measurement can be performed.

§4. *A Particle in a One-Dimensional Box.* The energy eigenvalues of a particle (e.g. an electron) in a one-dimensional box,

$$E_n = \frac{\hbar^2 \pi^2 n^2}{2mL^2}, \quad n = \pm 1, \pm 2 \ldots, \tag{4.2}$$

were derived in Chapter 3 (see §20, Eq. 3.41). Here m is the mass of the particle, L is the length of the one-dimensional box in which the particle is trapped, and $\hbar$ is Planck's constant.

Quantum mechanics postulates that it is impossible for such a particle to have an energy ϵ that is not one of the values given by Eq. 4.2

This is in stark contrast to classical mechanics, which has no principle that imposes such a restriction. Classical theory allows the energy (or any other dynamic variable) to change continuously as we act on the system.

Let us try to reach a better understanding of these statements. To change the energy of the particle I have to act on it with a force. By using the laws of classical mechanics it can be shown that if I change the force infinitesimally and act with it for a short time, the energy of the particle changes by an infinitesimal amount. We do not need Newton to tell us that this is true: if I change the force with which I hit a tennis ball by a small amount, its velocity (hence energy) will change very little. I can thus give the ball any energy I want, by manipulating the force. This is what we mean when we say that, in classical mechanics, the energy can be changed continuously.

This statement is not correct in a system that obeys quantum mechanics. If the system is in the state E_1 and the force is too small to place the particle in the state E_2 or E_3, etc., it would have no effect on the energy of the particle.

§5. *Isn't This a Contradiction?* At first sight it seems that classical and quantum mechanics are mutually exclusive: a measurement can either satisfy the restrictions imposed by Eq. 4.3 or not. This puts us in a tight spot: we know that both theories agree very well with a very large number of different experiments. Why then are we in a situation that seems to say that if one theory is correct, the other must be wrong?

This apparent mystery can be easily solved by examining the quantum-mechanical condition Eq. 4.2. If mL is sufficiently large, the energy difference between two successive energy levels is smaller than the sensitivity of our energy measurements. Because of this it appears to us that the energy can be changed continuously.

Exercise 4.1

Use Eq. 4.2 to calculate the energy difference $E_{n+1} - E_n$ for $n = 1$ or 10, for two cases: (a) $L = 1$ cm and $m = 1$ gr, and (b) $L = 20$ Å and m equal to the mass of the electron.

This kind of behavior is general. Whenever the predictions of the quantum and classical theory are in conflict, one finds that under certain conditions (e.g. large mass) our instruments are not fine enough to detect the restrictions imposed by quantum mechanics and the world behaves as if the classical theory is correct. Classical mechanics is correct when our measurements are not sensitive enough to reveal its inadequacy.

§6. *The Vibrational Energy of a Diatomic Molecule.* The properties of a diatomic molecule will be studied in detail in Chapters 16 and 17. Here I give a minimum of information needed for understanding how the vibrational energy is connected to the observations made by certain experiments.

The motion of a diatomic molecule is a composite of its translational, rotational, and vibrational motions. To avoid being overwhelmed by details, let us assume that somehow we can create an ensemble of diatomic molecules that do not translate and rotate, but are allowed to vibrate. We can come close to this by trapping the molecules in an argon ice.

The energy eigenvalues of a molecule that vibrates, but does not rotate or translate, are given by

$$E_v = \left(v + \frac{1}{2}\right)\hbar\omega - \left(v + \frac{1}{2}\right)^2 x_e \hbar\omega, \quad v = 0, 1, 2, \ldots \tag{4.3}$$

Here $\hbar$ is Planck's constant, and ω and x_e are numbers that change from molecule to molecule. They are determined experimentally or are calculated by using quantum mechanics. ω is the frequency of the vibration and has units of s^{-1}. Because $\hbar$ has units of energy × time, $\hbar\omega$ has units of energy and so does $x_e\hbar\omega$. v is an integer called *the vibrational quantum number.*

Spectroscopists have a habit of inventing strange energy units and they use cm^{-1} as a unit for vibrational energy. Do not ask why, the story is too long and this is not the place to tell it. You can use the table given in Appendix 2 to convert cm^{-1} to other energy units.

For the CO molecule, the constants have the values:

$$\hbar\omega = 2170.21 \ cm^{-1} \tag{4.4}$$

$$x_e\hbar\omega = 13.46 \ cm^{-1} \tag{4.5}$$

These values allow us to calculate the vibrational energies given by Eq. 4.3 (see Workbook QM4.1).

The diagram of the vibrational energy levels of CO is shown in Fig. 4.2. The energies were calculated in Workbook QM4.1, from Eq. 4.3 and the parameter values given by Eqs 4.4 and 4.5.

§7. *The Energy Eigenstates of the Hydrogen Atom.* I will also use the hydrogen atom, along with the vibrational energy of a diatomic molecule, to explain how energy measurements are made. The energy eigenvalues of the electron in the hydrogen atom are

$$E_n = -\frac{\mu e^4}{2(4\pi\varepsilon_0)^2\hbar^2 n^2} , \quad n = 1, 2, \ldots \tag{4.6}$$

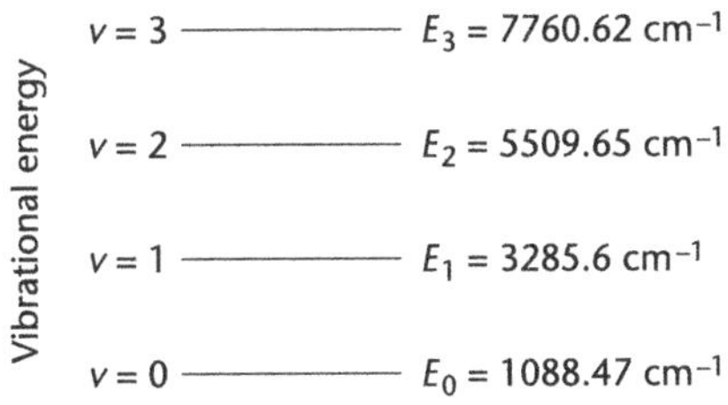

Figure 4.2 Energy level diagram for the vibrational energies of CO. v is the vibrational quantum number and E_v is the corresponding vibrational energy eigenvalue.

$v = 3$ ———— $E_3 = 7760.62 \ cm^{-1}$

$v = 2$ ———— $E_2 = 5509.65 \ cm^{-1}$

$v = 1$ ———— $E_1 = 3285.6 \ cm^{-1}$

$v = 0$ ———— $E_0 = 1088.47 \ cm^{-1}$

Vibrational energy

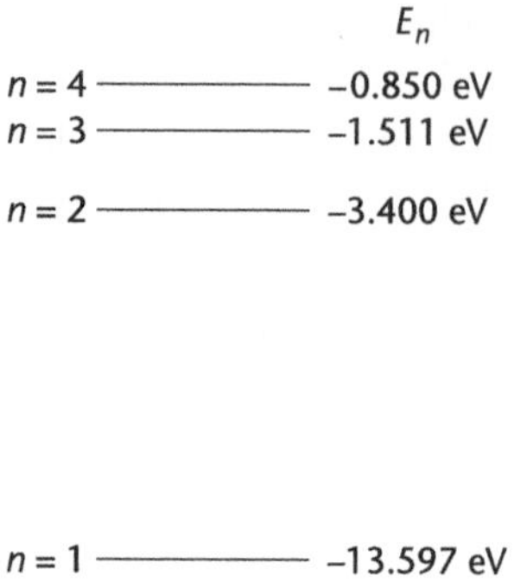

E_n

$n = 4$ ———————— –0.850 eV
$n = 3$ ———————— –1.511 eV

$n = 2$ ———————— –3.400 eV

Figure 4.3 Energy level diagram for the hydrogen atom. The energies were calculated from Eq. 4.6, in

$n = 1$ ———————— –13.597 eV   Workbook QM4.2.

Here $e = 1.6022 \times 10^{-19}$ C is the electron charge in coulomb, $\hbar = 1.0546 \times 10^{-34}$ is Planck's constant in joule seconds, $\varepsilon_0 = 8.854 \times 10^{-12} \, C^2 \, N^{-1} \, m^{-2}$ (N = newton, m = meter) is the permittivity of vacuum, and $\mu = (m_e \times m_p)(m_e + m_p)^{-1}$ kg is the reduced mass of the atom. In this expression, $m_e = 9.109 \times 10^{-31}$ kg is the electron mass and $m_p = 1.673 \times 10^{-27}$ kg is the proton mass. Eq. 4.6 gives the energies in joules. To transform joule into a more convenient unit, such as eV, we use the table given in Appendix 2. Eq. 4.6 ignores the translational energy of the atom; the errors introduced by this are very small.

There are an infinite number of energy eigenvalues, and the lowest are given in the energy level diagram in Fig. 4.3. The values of E_n were calculated in Workbook QM4.2.

Energy Measurements by Electron Scattering

§8. *Introduction.* We can now get back to our main purpose: to test the statement that the energy of a system can only be one of the eigenvalues of the total energy operator $\hat{H}$ (the Hamiltonian). To do this, we have to figure out how to measure the energy. It will turn out that we can only measure energy differences.

A standard method for measuring energy in quantum physics is to bombard the system with a stream of simple particles (e.g. electrons, neutrons or photons), which I will call "projectiles." The molecules whose energy I want to study are the "targets." I will make the sample containing the molecules so thin that a projectile either collides with one target molecule, or goes through the ensemble without a collision. I will send projectiles having a known energy e_i towards the target and measure their energy e_f after the collision.

The collision must conserve energy, so if E_i is the initial energy of the molecule and E_f is its final energy (after a collision) then

$$E_i + e_i = E_f + e_f \qquad (4.7)$$

The total energy $E_i + e_i$ *before* the collision is equal to the total energy $E_f + e_f$ *after* the collision.

Since in this experiment e_i and e_f are measured $E_f - E_i$ can be calculated from the energy conservation condition Eq. 4.7. In most practical situations, the target (the diatomic molecule or the hydrogen atom) is in the ground state (the state of lowest possible energy). To keep things simple I will analyze here the experiments in which this is the case.

Classical mechanics says that in such collisions the energy of the target (and therefore the energy of the projectile) can take any value compatible with energy conservation. In particular, in an experiment in which E_i and e_i are fixed, the energy of the projectile after the collision can vary *continuously* between e_i (the projectile gave no energy to the target) and zero (the projectile gave all its energy to the target).

However, quantum mechanics says that this is not true: E_i and E_f cannot take any good old value; they must be eigenvalues of the energy operator $\hat{H}$. Since the energy eigenvalues take *discrete* values, the difference

$$e_i - e_f = E_f - E_i \qquad (4.8)$$

(which we measure in this experiment) must also take discrete values. This equation says that if we perform the experiment at a fixed incident projectile energy e_i, the energy e_f of the projectile after the collision must take discrete values.

The predictions made by the classical and the quantum mechanics are dramatically different, and they can be tested by scattering experiments. Such experiments were first performed by Franck and Hertz, who used electron scattering by hydrogen atoms, and showed that the energy lost by electrons was discrete and could be calculated from Eq. 4.6. This equation was derived by Bohr, before the experiments were performed.

§9. *Electron Scattering from a Gas of Diatomic Molecules.* Let us look now at the results of an experiment in which the projectiles are electrons and the target is a gas of diatomic molecules that can vibrate but cannot rotate or translate.

According to Eq. 4.8, the difference $e_f - e_i$ must be equal to the difference between two energy eigenvalues of the vibrating molecule. These eigenvalues are given by Eq. 4.3. Combining the two equations leads to

$$e_i - e_f = E_{v_f} - E_{v_i}$$

$$= \left[\left(v_f + \frac{1}{2} \right) \hbar\omega - \left(v_f + \frac{1}{2} \right)^2 x_e \hbar\omega \right]$$

$$- \left[\left(v_i + \frac{1}{2} \right) \hbar\omega - \left(v_i + \frac{1}{2} \right)^2 x_e \hbar\omega \right]$$

$$= (v_f - v_i) \hbar\omega - \left[\left(v_f + \frac{1}{2} \right)^2 - \left(v_i + \frac{1}{2} \right)^2 \right] x_e \hbar\omega \qquad (4.9)$$

Let us apply this formula to a gas of CO molecules. When you learn statistical mechanics, you will find out that at a sufficiently low temperature, most CO molecules in the gas have the lowest possible vibrational energy

$$E_0 = \frac{\hbar\omega}{2} - \frac{x_e \hbar\omega}{4}$$

which corresponds to $v_i = 0$. This simplifies Eq. 4.9 to

$$e_i - e_f = \hbar\omega v_f - x_e \hbar\omega \left[\left(v_f + \frac{1}{2} \right)^2 - \frac{1}{4} \right]$$

$$= \hbar\omega v_f - x_e \hbar\omega \left[v_f^2 + v_f \right] \qquad (4.10)$$

Since v_f can be 0, 1, 2, ..., quantum mechanics makes a very peculiar prediction: the difference $e_f - e_i$ can only take the values

$$e_f - e_i = \quad 0 \qquad\qquad\qquad \text{if } v_f = 0$$
$$e_f - e_i = \quad \hbar\omega - 2x_e \hbar\omega \quad \text{if } v_f = 1$$

Table 4.1 The energies that an electron can lose when colliding with a vibrating CO molecule.

v_f	$e_f - e_i \, (cm^{-1})$
1	2197.131
2	4421.180
3	6672.150
4	8950.040
5	11254.900

$$e_f - e_i = \quad 2\hbar\omega - 6x_e\hbar\omega \quad \text{if } v_f = 2$$
$$\vdots$$

For a CO molecule, $\hbar\omega = 2170.21 \text{ cm}^{-1}$ and $x_e\hbar\omega = 13.46 \text{ cm}^{-1}$, and the allowed energy losses for the electrons are given in Table 4.1 for the first few values of v_f (see Workbook QM4.3).

Workbook

This is a remarkable prediction. Classical physics proclaims that the projectile may lose any amount of energy. The discreteness of the energy lost by the electron, which reflects the discreteness of the values of the vibrational energy, is a purely quantum phenomenon.

Countless such experiments have been performed, and the energy lost by electrons is that predicted by quantum mechanics. These and many other experiments are a direct test of the statement that the energy of a system (the molecule) must be an eigenvalue of its Hamiltonian.

Exercise 4.2

Calculate the possible energy losses in an electron-scattering experiment in which the target is a gas of hydrogen atoms. Give the results in eV.

Exercise 4.3

In a gas of CO molecules, a small minority have vibrational energy corresponding to $v = 1$. Calculate the possible energies of the electrons colliding with these molecules, if the initial energy of the electrons is $e_i = 1.5$ eV.

Exercise 4.4

What do you think: is the analysis performed for electrons also valid if the projectiles are neutrons? What kind of differences between electron and neutron scattering do you expect? Consider both theoretical and practical differences.

---

§10. *A Practical Application of this Experiment.* The observation that the change in energy of the electron beam is given by Eq. 4.10, can be used to measure the vibrational frequency ω and the constant x_e for a diatomic molecule. Normally we use optical spectroscopy (photon absorption) for such measurements, but there is one situation where electron scattering is advantageous. Many people are interested in the behavior of molecules adsorbed (this means stuck) on a solid surface. If I expose a clean metal surface (e.g. Ni) to a gas (e.g. CO) and the surface is not very hot, the surface will be covered with a layer of CO molecules. By estimating the size of a molecule, I can calculate that there will be about 10^{14} molecules on a cm^2 of surface. This is an awfully small amount of material (calculate how many moles). To make the situation worse, these molecules are spread out in a very thin layer (about 3×10^{-8} cm thick). If I send a photon to bounce off this surface, the chance that the photon is absorbed is very small.

However, the interaction of an electron with a molecule is much stronger than the interaction of the photon with the molecule. Moreover, electron detection is more efficient than photon detection. If I send a beam of electrons to collide with the surface, a sufficient number of them will lose energy to the diatomic molecule. This number is big enough to allow me to detect how much energy was lost. By measuring how much energy the electrons have lost, I measure the frequency of the diatomics stuck on the surface. Since different diatomic molecules have different vibrational frequencies, I can tell (by measuring the frequency) what kind of molecules are present on the surface. In this way, I can perform chemical analysis of very small numbers of molecules adsorbed on a surface.

I can also use this tool to study chemistry. Let us assume that I suspect that CO dissociates when adsorbed on Ni. How can I tell if this happens? If the molecule dissociates, the CO frequency is absent in the measurement and I will observe the frequency of the O and C atoms vibrating in the direction perpendicular to the surface. If CO does not dissociate, I can do some interesting experiments. I can heat the surface and monitor the departure of CO from the surface by observing that the number of electrons losing energy goes down with temperature (less CO on the surface, fewer electrons lose energy). Or I can put H_2 on the surface and monitor if it reacts with CO. If it does, the signal from CO decreases in time. These procedures can be used not only for diatomics but for any kind of molecule; they all have specific vibrational frequencies which can be used to identify them.

This technique, called *electron energy-loss spectroscopy*, is routinely used in surface science laboratories all over the world.[a] It is more sensitive than photon absorption experiments, but it has lower resolution (it does not measure frequency as accurately as photon absorption does). Other disadvantages are: the experiment can be performed only at very low pressure (to avoid collisions of the electrons with gas molecules) and sometimes the electrons destroy the molecule we want to study.

Energy Measurements by Photon Emission

§11. *Photon Emission.* Now that you know the story of electron energy-loss experiments, you should wonder what happens to the system (e.g. a vibrating molecule) after it takes energy from the electron. What is the excited molecule going to do next? It turns out that quantum systems "do not like" to be in an excited state. In most cases (but not always), they will lower their energy by emitting photons.

In this emission process, energy must be conserved. If the energy of the system (e.g. a molecule) after photon emission is E_f, then

$$E_f - E_i = \hbar\Omega \qquad (4.11)$$

where Ω is the frequency of the emitted photon and $\hbar\Omega$ is its energy.

Since Ω must be positive, E_f must be smaller than E_i: the energy of the final state of the molecule must be lower than the energy of its initial state. This is not surprising, since to create the photon (which has energy) the molecule must lower its own energy.

Because the energy of the molecule is discrete, a molecule that is initially in a state having the energy E_3 can only emit photons of energy $E_3 - E_2$, or $E_3 - E_1$. I assume here, as is usual, that $E_1 < E_2 < E_3 < \dots$.

The emission process is often described by using an energy-level diagram with the transitions indicated by arrows between the initial and the final state of the molecule. For the vibrating CO molecule, such a diagram is shown in Fig. 4.4. The arrow pointing down indicates that a molecule in the state with $v = 3$ and energy $E_3 = 7760.62$ cm^{-1} emitted a photon of frequency $\hbar\Omega = 7760.62 - 3285.6$ cm^{-1}, leaving the molecule in the state with $v = 1$ and the energy $e_1 = 3285.6$ cm^{-1}.

[a] H. Ibach and D. L. Mills, *Electron Energy Loss Spectroscopy*, Academic Press, New York, 1982; G. A. Somorjai, *Chemistry in Two Dimensions: Surfaces*, Cornell University Press, Ithaca NY, 1981.

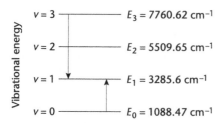

Figure 4.4 The energy level diagram for the vibrational motion of CO. The downward-pointing arrow indicates that the molecule in a state with $v = 3$ emitted a photon and ended in the state with $v = 1$. The photon energy is $E_3 - E_1$. The upward-pointing arrow indicates an excitation of the molecule (by photon absorption or by collision with an electron) from a state with $v = 0$ to one with $v = 1$.

The arrow pointing up indicates that the molecule was excited from the state with $v = 0$ to the state with $v = 1$. If the excitation was caused by collision with an electron or a neutron, the projectile lost the energy $E_1 - E_0 = 3285.6 - 1088.47$ cm^{-1}. Because the energies of the molecule take discrete values, the energy of the emitted photons must also be discrete.

§12. *Applications.* Photon emission by excited molecules has many interesting implications. Each excited atom or molecule emits light of a *specific frequency* and this can be used, like a fingerprint, in analytical chemistry. If your doctor wants to know if you have mercury poisoning, he could take a sample of hair (Hg accumulates in hair) and zap it with a powerful laser. This heats the hair and breaks it up into atoms, some of which are excited. The excited atoms will emit light. If the emitted light has a frequency typical to the mercury atoms, then you have some degree of mercury poisoning. The doctor can tell, from the number of photons emitted with that frequency, how bad the poisoning is. This method, when pushed to extremes, can detect a single atom (under favorable conditions).

Another obvious application is in lighting. If you take a neon gas at low pressure and pass an electric discharge through it, the atoms become excited and emit light. Since they emit light typical of neon, a source of colored light has been produced. If you want "white" light, the tube containing the gas should be surrounded by a ceramic material that converts light of one color into light of many colors. Neon lights, white lights, halogen lights, all function by using the same principle: use a current to produce fast, charged particles, which hit the atoms, excite them, and make them emit light.

Lasers are based on the same idea but use additional tricks that make the emitted light have special properties. The scanner at the supermarket, your CD player, and other laser-containing devices work because excited systems (in this case electrons in a solid) feel an urge to lose energy by emitting light.

While this urge to be dull (unexcited) is widespread in the physical world it is not always satisfied by photon emission. Many molecules find relief from excitation by doing chemistry. A whole field of physical chemistry, namely photochemistry, is based on the fact that some excited molecules lose excess energy by breaking into fragments. We benefit from that. Your eye contains molecules that absorb light and perform chemical reactions that send electric signals to your brain. The brain turns these signals into an image. The food chain is based on the photo-chemistry taking place in the leaves of plants, which absorb photons and use their energy for chemical synthesis. The chemical reaction caused by photon absorp-tion can take place only if it is faster than photon emission. There would hardly be any life on Earth if the molecules in the leaves lost energy by re-emitting the absorbed photons.

§13. I conclude this chapter by warning you that our analysis of the measurements performed in quantum mechanics is incomplete. As far as energy conservation is concerned, the projectile can lose the energy $(E_f - E_i)/\hbar$ if this quantity is smaller than its initial energy. However, Nature is subtle and there are other rules in this game, besides energy conservation. When a collision with a projectile changes the energy of a quantum system from E_i to E_f, we say that a transition from E_i to E_f has taken place. A transition allowed by energy conservation takes place with a certain probability: some transitions are very probable, others take place rarely, and still others do not take place at all. To see how this works, let us assume that we perform an energy-loss experiment in which the projectile (e.g. an electron) has enough energy to excite the system from E_0 to E_1 or to E_2 or to E_3. It may happen that the transition $E_0 \rightarrow E_1$ is very probable, that $E_0 \rightarrow E_2$ has low probability, and that the probability of the transition $E_0 \rightarrow E_3$ is so low that we might not be able to detect it. For some transitions the transition probability is equal to zero: when this happens we say that the transition *is forbidden*. Given these facts, what will we observe in an electron energy-loss experiment?

The energy of most electrons after the collision will be the same as the initial energy (before the collision) since there is a fair probability that the electrons pass through the sample without losing energy. A large number will have the final energy

$e_f = e_i - (E_1 - E_0)/\hbar$, a few will have energy $e_f = e_i - (E_2 - E_0)/\hbar$, and the number of those emerging with the energy $e_i - (E_3 - E_0)/\hbar$ is so low that it is below our detection limit. These results could cause some embarrassment. If calculations predict that the eigenvalue E_3 exists and you do not see it in the experiment, you might conclude that the theory is wrong. Not so. The same theory, when used properly, will predict that the transition $E_0 \rightarrow E_3$ is very improbable and your experiment is unable to detect the existence of the state with energy E_3. Other experiments, using other projectiles will obey different rules and are able to detect the transition $E_0 \rightarrow E_3$, confirming the existence of the state of energy E_3.

I will tell you more about the probability of observing a given transition in future chapters.

5

SOME RESULTS ARE CERTAIN, MOST ARE JUST PROBABLE

Introduction

§1. You have learned in your science courses that the world is predictable. This assumption is at the foundation of the scientific method: if you repeat an experiment under identical conditions, you get the same result. Moreover, when you studied classical mechanics, or thermodynamics, or electricity, you have learned that you can calculate very precisely the outcome of any experiment. There are many situations when such a calculation is too difficult to carry out. But, you have never encountered a situation where the theory says that it is impossible, in principle, to calculate or to know precisely the outcome of an experiment. Some measurements in the classical world are impossible because we are technologically inept, not because of some limitation imposed by the laws of Nature.

Before we go on, I have to hedge a bit. There are many situations in classical physics where we are content to calculate probabilities either because our instruments are not good enough to control precisely the experimental conditions or because we are not interested in all the details. A good example is coin flipping. This kind of experiment is usually described by probabilities. However, if we were to construct a mechanical device that flips the coin and has perfect control of the force, the place where the coin is hit, the condition of the air, etc., we could predict accurately the outcome. There is no law of Nature that prevents us from attaining perfect accuracy in such an experiment.

The situation is different in the world of quantum mechanics. Quantum theory says that whenever the conservation laws (e.g. energy or momentum conservation) allow an experiment to have several outcomes, it is impossible to know with certainty which outcome occurs in the experiment. If a collision between an electron and a molecule can leave the molecule in several states, we cannot predict accurately the state of the molecule after the collision. Nor can we be certain that repeating the experiment under identical conditions will produce an identical result.

The best we can do is to determine the probability of each of the possible states. Knowing the probability that an event occurs means that we can predict the results of measurements in which we repeat, many times, the same experiment, or in which we perform simultaneously a large number of identical experiments. This situation is similar to that encountered by gamblers, who cannot predict whether flipping a coin will result in heads or tail, but can tell you that if you flip the same coin a hundred thousand times, roughly half of the flips produce heads and the other half produce tails.

Because the objects governed by quantum mechanics are very small and we are very large, in most experiments we act identically on a very large number of molecules. In such a case knowing the probability that something will happen to one molecule leads to rather precise predictions of the number of molecules having that behavior. Because of this, we were led to think that predictability is a law of Nature.

Furthermore, when we deal with objects that are very heavy, various alternative outcomes are so close to each other that our instruments are unable to distinguish them, and so they are lumped together into one event. In such cases the probability that all the alternative events occur is 1, and we can again make accurate predictions of the outcome of the experiment.

In this chapter, I will examine some examples that will clarify the meaning of the statements made above. I will use three experiments: electron scattering by a molecule, adsorption of photons by molecules, and emission of light following the excitation of a molecule.

What is the Outcome of an Electron-Scattering Experiment?

§2. *The Question We Ask.* Let us take another look at the collision experiment described in the previous chapter. There I examined what happens when electrons are passed through an ensemble of diatomic molecules that can vibrate but are not allowed to have a rotational or a translational motion. My concern was to determine the possible energies of the electron after it collided with the molecule. Because of energy conservation, the energy lost by an electron must be equal to a difference between two energy eigenvalues of the molecule.

Let us examine this experiment further. Imagine that I perform the experiment with one electron and one molecule at a time and that I can repeat it under identical conditions as many times as I want. Since, after the collision, the electron can have one of several final energies (see Chapter 4), I would like to know which one occurs in a given experiment. Quantum mechanics answers this question very differently than classical physics.

§3. *Classical Interpretation of the Experiment.* We have already established in the previous chapter that classical mechanics will erroneously predict that the electron could transfer any amount of energy to the molecule.

Quantum mechanics says that this statement is wrong. No matter how I vary the impact conditions, the energy lost by the electron (and gained by the molecule) must be a difference between the energy eigenvalues of the molecule. Since these energies are discrete, the energy lost by the electron is discrete.

What else is wrong with classical mechanics? Let us examine what classical physics has to say about the question posed here: given a particular set of conditions (initial energies, impact conditions), what is the final energy of the electron after the collision? Classical mechanics claims that it can answer this question precisely and without ambiguity.

Let me explain what that means. At a time that I will call zero, I will set my apparatus to shoot an electron from a known position, with a known velocity. At the same time I will start the molecule vibrating by specifying the positions of the atoms and their velocity. Remember that to keep things simple it is assumed that the molecule does not move around as a whole and does not rotate.

Note that I do not worry about practical limitations that may not allow me to set the positions and the velocities of these particles precisely. This should not disturb you. I want to examine only those limitations that are *intrinsic* to Nature, not those caused by my lack of technological skill. Classical mechanics has no principle that forbids me from giving, at a specified time, precise values to the positions and the velocities of all particles involved in the experiment.

Once these initial conditions are set, I can solve Newton's equation for the motion of the electron and the molecule. This will describe how the electron approaches the molecule, interacts with it, exchanges some energy with it, and then moves away and reaches the detector. This calculation will tell me exactly what the final energy of the electron will be and how the molecule vibrates after the collision.

Moreover, classical mechanics tells me that if I repeat the experiment with the same initial conditions, the same result will always be obtained . This statement is true not just for this particular experiment, but for any experiment I care to make.

§4. *The Quantum Description of the Experiment.* Quantum mechanics tells us that, in most experiments, the predictions described above are not possible. Here is what happens in reality when I perform the experiment described above. I assume that before the collision, the molecule is in the ground vibrational state and the electron approaches it with a velocity $\mathbf{v}_i$. The initial energy of the electron is

$$e_i = \frac{m\mathbf{v}_i^2}{2}$$

To describe further what can happen I need to remind you of a few things you have learned in the previous chapter. The possible energies of the vibrating diatomic molecule are discrete and are denoted E_v. Here v is the vibrational quantum number, $v = 0, 1, 2, \ldots$. The energies E_v can be calculated from Eq. 4.4.

Energy conservation told us that after the collision the final energy e_f of the electron can only take values given by the equation

$$e_i - e_f = E_f - E_0 \tag{5.1}$$

E_0 appears in the above equation because I am performing an experiment in which the molecule starts in the state $v = 0$, with the energy E_0. E_f is the energy of the molecule after the collision.

The final energy of the electron, after the collision, is

$$e_f = e_i - (E_f - E_0) \tag{5.2}$$

After the collision, the electron moves with velocity $\mathbf{v}_f$ and no force acts on it. Its energy, $e_f = m\mathbf{v}_f/2$, must be positive. This condition limits the energies E_f to values satisfying

$$e_i - (E_f - E_0) \geq 0 \tag{5.3}$$

Let us further assume that I pick the initial electron energy so that

$$E_3 < e_i < E_4 \tag{5.4}$$

Eq. 5.3 tells me that the final energy E_f could be E_0 or E_1 or E_2 or E_3. Therefore the final energy of the electron (which is what I measure) could be e_i, or $e_i - (E_1 - E_0)$, or $e_i - (E_2 - E_0)$, or $e_i - (E_3 - E_0)$.

In this chapter, I am asking the following question: if I perform the experiment under the conditions specified above, what is the energy of the electron after the collision?

Quantum theory says that it is impossible to tell, with certainty, what the final state of the molecule (or electron) will be.

What does this mean in the laboratory? Assume that the graduate student Mifune performs five identical electron-scattering experiments, with a single electron and a single molecule. In the first one, the final energy of the molecule is E_1. If he believes in classical mechanics, he expects that the remaining four experiments will all excite the molecule in the state with energy E_1. But they don't. The next four results are $\{E_2, E_0, E_0, E_3\}$, even though the conditions were identical in each experiment.

Puzzled by this, he talks with his pal DeNiro. A devout classicist, DeNiro decides that there is something wrong with the experiment and repeats it. Under the same conditions as Mifune he gets E_3, E_3, E_1, E_1, E_0. The results of the five identical experiments are different, and there is no relationship between his results and Mifune's.

§5. *Probabilities.* This is unacceptable! How can we do science if the same experiment leads to different results at different times? Besides this philosophical question, there is a practical one: if the physical world is so unpredictable, how come nobody noticed until 1927?

I will answer these questions next. Mifune performs 5000 identical experiments. He denotes by n_0 the number of experiments in which the final energy of the molecule is E_0, by n_1 the number of experiments in which the final energy of the molecule is E_1, and so on. Then he calculates the ratio

$$\nu_v = \frac{n_v}{5000}, \quad v = 0, 1, 2, \ldots$$

ν_v is called *the frequency* with which the molecule was excited to the state E_v, or the frequency with which the electron emerged with the energy $e_f = e_i - (E_v - E_0)$. (These two statements are saying the same thing.)

Now here comes a surprise. DeNiro does a new set of 5000 experiments and calculates the frequencies ν_v. He finds that the values of ν_v in his set of experiments are very close to those obtained by Mifune. After a series of laborious experiments, Mifune and DeNiro find that whenever they perform a sufficiently large number of experiments they get the same values for the frequencies ν_v. Finally, they have found something reproducible in their experiments!

The frequencies ν_v depend on the number N of identical experiments performed. As N increases, ν_v no longer changes and approaches a value that we denote by p_v. This is called *the probability* that after the collision the molecule has the state v with the energy E_v. The larger the value of N, the closer ν_v is to p_v. A mathematician would write this definition of the probability as

$$p_v = \lim_{N \to \infty} \nu_v$$

Quantum mechanics tells us that it cannot predict the outcome of one experiment but it can calculate the probability of its occurrence.

§6. *A Summary.* Quantum mechanics states that whenever an experiment can have several outcomes (i.e. the molecule could emerge, after a collision, with several different energies), we are not sure of the outcome, but we can calculate the probability that each outcome will occur. If we do the experiment only a few times, it is unpredictable which outcome will be observed, and the results of such an experiment are not reproducible. Quantum measurement behaves like throwing a die. The only predictable result is that if you throw the die N times, and N is very large, each face appears $N/6$ times.

Classical physics is *deterministic*: it states that if you reproduce the conditions of an experiment you will reproduce its result. The initial conditions of the experiment determine precisely its outcome. Quantum mechanics is *probabilistic*: it states that we can only know the probability that an experiment leads to a certain result. However, it is deterministic in a more narrow sense: a given set of initial conditions always results in the same *probabilities.*

Great scientists such as Einstein and Schrödinger believed that this behavior indicates that there is something wrong with quantum theory and that sooner or later this error will be fixed to make the world deterministic again. Einstein used to say "I cannot believe that God plays dice," to which Bohr replied, "Einstein, stop telling God what to do."

There is as yet no experiment that contradicts the statement that all we can do is measure or calculate probabilities.

§7. *Why Did We Think that the World is Deterministic?* If this sort of indeterminism is so fundamental and conspicuous in the atomic world, why did no one notice it before quantum mechanics was discovered? The answer is simple: no one was able to perform experiments with a single atom and a single projectile. All the experiments were done with a very large number of molecules and a large flux of incident projectiles. This masked the probabilistic behavior. To see what I mean, let us examine our electron-scattering experiment as is done in the laboratory, where we send a stream of electrons, all having the same initial energy, to pass through a gas containing a huge number of molecules all in the ground state (I assume this, for simplicity).

The collision of any one electron with a molecule is independent of the collision of any other electron with another molecule. In other words, the energy given by electron A to the molecule α is not affected by the energy given by electron B to the molecule β. This means that we are actually performing, in parallel, a huge number of collisions of one electron with one molecule. The outcome of every one of these individual collisions is uncertain, as described earlier. However, because the number of collisions taking place in the gas is large, the frequencies of various excitations are equal to their probabilities. Quantum mechanics says that these probabilities are predictable and predetermined by the way we do the experiment. If we send in N electrons, the number N_0 of electrons that do not lose energy is equal to $p_0 N$, the number of electrons N_1 that lose the energy $E_1 - E_0$ is equal to $p_1 N$, etc. Here $p_0, p_1, \ldots$ are the probabilities that in one collision the electron loses no energy, or the energy $E_1 - E_0$, etc. In the experiment we measure N_0, N_1, etc. Since we control N, and p_α is predetermined by the laws of quantum mechanics, we get the same result N_0, N_1, … whenever we repeat the experiment, as long as N is large. In such experiments, which produce a large number of collisions simultaneously, Nature appears to be deterministic.

If N is small, the frequencies ν_v differ from experiment to experiment, and so do N_0, N_1, …. The fact that Nature is probabilistic, not deterministic, becomes observable.

Exercise 5.1

Flip a coin N times. Denote by n_0 the number of times you get heads and by n_1 the number of times you get tails. Plot n_0 and n_1 versus N. Determine when n_0/N and n_1/N approach $p_0 = p_1 = 1/2$. If you use **Mathematica**, use the function **Random[Integer,0,1]** to randomly generate the numbers 0 (for heads) and 1 (for tails).

A Discussion of Photon Absorption Measurements

§8. The results explained above are not confined to electron-energy-loss experiments with vibrating diatomics. They are general and are observed in all kinds of collision processes. Whenever several outcomes are possible, we can only predict the probability of each outcome.

To be sure that you understand what this means, I give more examples. Suppose that a molecule has the energy eigenvalues $E_0, E_1, E_2, E_3, \ldots$. I perform an experiment in which the molecule collides with a photon having energy $\hbar\Omega$ (where Ω is the frequency of the photon). Quantum mechanics can be used to calculate the outcome of such an experiment and below I give you the result of such a calculation. Then I will discuss a real experiment where a stream of photons is sent to interact with an ensemble of molecules (gas, liquid, or solid).

A photon coming in contact with a molecule may be absorbed (the photon disappears and the molecule is excited) or it can pass through the sample without affecting it. Energy conservation for the *absorption* process requires that

$$\hbar\Omega + E_i = E_f \tag{5.5}$$

Here $\hbar\Omega + E_i$ is the total energy before the collision and E_f is the energy of the molecule after the collision.

I pick Ω so that energy conservation is satisfied for the excitation of the molecule from a state with energy E_0 to one with energy E_2. This means that

$$\hbar\Omega = E_2 - E_0$$

Therefore, this experiment has only two possible outcomes: either the photon is absorbed (and disappears) and the molecule ends up with energy E_2; or the photon is not absorbed, the molecule remains in E_0, and the photon passes through the sample.

Quantum mechanics says that it is impossible to tell in advance which of these possibilities will occur in an experiment in which one photon interacts with one molecule. One can only calculate the probability π_0 that the photon has no effect, and the probability π_1 that the photon is absorbed.

If I repeat the experiment identically a few times, the results are unpredictable and irreproducible. If I repeat the experiment with N molecules, and N is large, I can predict with a high degree of certainty that $\pi_1 N$ molecules are excited and $\pi_0 N$ are left unchanged. The certainty of this result increases as N is increased.

§9. *Why is the Outcome of Most Absorption Experiments Certain?* To perform an absorption experiment, I shine light of known intensity on a sample containing absorbing molecules dissolved in a solvent. The initial light intensity I_0, the concentration c of the molecule, and the length L of the sample are fixed, and the intensity of the light I_f coming out of the sample is measured. Whenever I do this experiment under identical conditions, I get the same value for I_f. The world is deterministic (i.e. the values of I_0, c, and L determine that of I_f) and predictable (i.e. I can use the Lambert–Beer law to calculate I_f). This is fine with quantum mechanics. We are performing, in parallel, a huge number of one-photon/one-molecule experiments, and the total number of photons passing through is predictable (it is $\pi_0 N$, where N is the number of photons sent on the sample).

A Discussion of Photon Emission Measurements

§10. Now let us follow what happens after we have performed an experiment that leaves the molecule in an excited state. It does not matter whether the excitation took place because of photon absorption or a collision with an electron.

Molecules "dislike" having high energy and will use any available route that will take them to the ground state. One of these routes is to emit a photon. If the energies of the molecule are

$$E_0 < E_1 < E_2$$

and the initial energy of the molecule is E_2, then energy conservation allows two transitions. In one, the molecule emits a photon of frequency

$$\hbar\Omega_{20} = E_2 - E_0 \tag{5.6}$$

and the molecule ends up in state E_0. In the other, the molecule emits a photon of frequency

$$\hbar\Omega_{21} = E_2 - E_1,$$

and the molecule ends up in state E_1.

These two possibilities are shown in the energy level diagram in Fig. 5.1. The arrows indicate transitions between the states of the molecule. An arrow from a low energy to a higher one indicates the excitation of the molecule by photon absorption or a collision with an electron. An arrow from a higher energy to a lower energy indicates photon emission. The notation for the frequency of the photon involved in a transition is written next to the arrow.

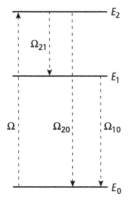

Figure 5.1 Energy level diagram showing the absorption and emission processes.

By now we know that nothing in quantum mechanics is certain. If we experiment with one molecule and one photon, the best we can do is to calculate (or measure) the probability of each possible event. Let us denote by p_{21} the probability of the transition from E_2 to E_1, by p_{20} the probability of the transition from E_2 to E_0, etc. We can determine which event takes place in the experiment if we measure the frequency of the emitted photon.

§11. *A One-Photon, One-Molecule Experiment.* Let us now examine in detail what happens if we send a photon of frequency Ω to interact with the molecule. The frequency is chosen to satisfy Eq. 5.6 and this means that either the photon passes by, with no effect on the molecule, or that it is absorbed and excites the molecule to a state with energy E_2. We measure the frequency of the photons coming out of the gas.

A photon of frequency Ω is detected when the molecule does not absorb the incident photon *or* when absorption to E_2 is followed by emission to E_0. The probability of observing a photon of frequency Ω is then

$$p(\Omega) = \pi_0 + \pi_1 p_{20}$$

The first term is the probability π_0 that the photon was unaffected by the molecule; the second is the probability that photon absorption (having the probability π_1) is followed by the emission of a photon, with the probability p_{20}, which puts the molecule in the state with energy E_0.

The probability of observing a photon of frequency Ω_{21} is

$$p(\Omega_{21}) = \pi_1 p_{21}$$

This is the probability π_1 of the transition $E_0 \rightarrow E_2$ times the probability p_{21} of the transition $E_2 \rightarrow E_1$.

If you observe a photon of frequency Ω_{21}, you are guaranteed to observe a photon of frequency Ω_{10} after it! Once in the state E_1, the molecule will emit a photon to go to the lowest state E_0. In this simple example there is no other choice (see the energy level diagram in Fig. 5.1). (Here I ignore what are called radiationless transitions.)

§12. *We Can Do This Experiment.* This experiment is very similar to the others we discussed here: the same probabilistic game, the same "throw of dice" that Einstein disliked until the end of his life. This probabilistic behavior is a postulate of quantum mechanics. It was not observed in older experiments because we could not work with only one molecule. The experiments in the laboratory appeared to be causal because they were performed with a very large number of molecules.

Nowadays, due to improvements in technology and in our cleverness, we can perform an absorption–emission experiment with a single molecule! Several tricks allow us to do this. First, we make a very dilute solution of molecules and spread it on a piece of glass. We send light through the objective of a microscope so we can focus it very tightly. The spot of light in the focus of the microscope is about 500 Å in diameter. Because the solution is dilute and the spot of light is small it is very likely that there is only one molecule in the illuminated area. The time it takes an excited molecule to emit a photon is about 10^{-9} s. If we can send a stream of photons onto the molecule (through the microscope's objective), some of them will be absorbed and excite the molecule. We can use a filter and allow only photons of frequency Ω_{21} (or Ω_{10}) to reach the detector. Today's photon detectors are so sensitive that they can detect one photon. They are also very fast and can distinguish between two photons reaching the detector at almost the same time. We can determine the times when photons of frequency Ω_{21} reach the detector.

Quantum mechanics also tells us that the time when an excited molecule emits a photon cannot be predicted accurately. We can only calculate or measure the probability that a photon is absorbed at a given time and the probability that it is emitted at a time t after the molecule reached the excited state. It can also calculate the probability that the time between two successive photon emissions is t. This time can be measured.

In our experiments the probability that a photon is emitted at a time between t and $t + \Delta t$ after the photon emitted previously is the number of times $N(t)$ a photon of frequency Ω_{21} was detected in the time interval between t and $t + \Delta t$, divided by the total number of photons of frequency Ω_{21} detected. The probability measured in this way agrees with that calculated by theory. This confirms again that the world is not deterministic and indicates that the quantum mechanical prescription for calculating probabilities is correct.

§13. *How About the Probabilities of Different Events?* The experiment discussed above refers to the times between successive emission of photons of a specified frequency. What is the probability that the molecule emits a photon of frequency Ω_{21}, or one of frequency Ω_{10}, or one of frequency Ω? Can we measure this quantity in a one-molecule experiment? Yes. We can measure how many photons of each kind have been emitted in a given time. Dividing these numbers to the total number of photons detected gives the probability that photons of a specific frequency are emitted (if the total number of photons detected is large). The measured probabilities are in agreement with those calculated by quantum mechanics.

§14. *Summary.* Quantum mechanics states that most of the questions that have a precise answer if classical physics were correct have an evasive answer in reality. Instead of making firm predictions quantum mechanics can only predict the probability that a certain event will take place. We have now to redefine causality: performing an experiment over and over, under identical conditions, does not guarantee that we get, over and over, the same result. Only the probability of getting a given result is the same in two experiments performed under identical conditions. The causality principle works for probabilities, but not for the results of an individual experiment.

In an experiment in which several transitions could take place, we cannot say with certainty which will be observed, nor can we tell the time when the transition takes place. But we can calculate the probability that a certain transition can take place at a certain time.

The world appeared deterministic and causal for two reasons: either the experiments are carried out on large ensembles of molecules and the probabilities lead to precise results; or the difference between various alternatives are so small that we perceive all of them as the same event. In this case the experiment does not offer a choice between various outcomes. If there is one outcome only, the probability of observing it is 1, and the world appears to be filled with certainty.

6

THE PHYSICAL INTERPRETATION OF THE WAVE FUNCTION

§1. So far we have focused on the physical interpretation of eigenvalues and the transitions between them. But what about eigenfunctions? What kind of physical information do they carry?

By far the most important eigenvalue equation in quantum mechanics is the Schrödinger equation

$$\hat{H}\psi_n(\mathbf{r}(1),\dots,\mathbf{r}(N)) = E_n\psi_n(\mathbf{r}(1),\dots,\mathbf{r}(N)), \quad n = 0, 1, 2, \dots \qquad (6.1)$$

Here $\hat{H}$ is the Hamiltonian operator, and ψ_n and E_n are its eigenfunctions and eigenvalues. Several names are frequently used for the eigenfunctions $\psi_n(\mathbf{r}(1),\dots,\mathbf{r}(N))$ of $\hat{H}$: ψ_n is the wave function of the N-particle system having the energy E_n; or the energy eigenfunction corresponding to the energy E_n; or we say that the system has a state ψ_n with the energy E_n.

It is a postulate of quantum mechanics that, if the system is in the state $\psi_n(\mathbf{r}(1),\dots,\mathbf{r}(N))$, then the quantity

$$p_n(\mathbf{r}(1),\dots,\mathbf{r}(N))d\mathbf{r}(1)\cdots d\mathbf{r}(N) = |\psi_n(\mathbf{r}(1),\dots,\mathbf{r}(N))|^2 d\mathbf{r}(1)\cdots d\mathbf{r}(N) \qquad (6.2)$$

is the probability that particle 1 is located in a cube of volume $d\mathbf{r}(1) = dx(1)dy(1)dz(1)$ centered around $\mathbf{r}(1)$, and particle 2 is located in a cube of volume $d\mathbf{r}(2) = dx(2)dy(2)dz = (2)$ centered around $\mathbf{r}(2)$, and .... In what follows, $|\psi|^2 \equiv \psi^*\psi$ where ψ^* is the complex conjugate of ψ.

This information is valid only if we are sure that the system is in the state $\psi_n(\mathbf{r}(1), \ldots, \mathbf{r}(N))$. To be sure of that, we must perform an experiment that places the system in that state. For example, we can force a molecule that is initially in the ground state E_0 to absorb a photon and be excited to the state $\psi_2(\mathbf{r}(1), \ldots, \mathbf{r}(N))$. The probability that the atoms in the excited molecule have certain positions $\mathbf{r}(1), \ldots, \mathbf{r}(N)$ is given by Eq. 6.2 with $n = 2$.

I will confine the discussion that follows to the case of a particle moving along a straight line. I do this to simplify some of the equations. What you learn from these examples can be easily extended to many particles moving in three-dimensional space.

If the particle moves along a straight line (one-dimensional motion), its position is specified by one coordinate x. If the particle is in a state whose wave function is $\psi(x)$, then the probability that the particle is located between x and $x + dx$ is

$$p(x)dx = |\psi(x)|^2 dx \tag{6.3}$$

The probability that the particle is located in the region between a and b is then

$$\int_a^b p(x)dx = \int_a^b |\psi(x)|^2 dx; \tag{6.4}$$

we add up the probabilities that the particle is at points between a and b.

The probability that the particle is somewhere along the line is

$$\int_{-\infty}^{+\infty} |\psi(x)|^2 dx = 1 \tag{6.5}$$

This integral is equal to 1 because we are certain that the particle is somewhere on the line; the probability of an event that is certain is equal to 1, by definition.

When a wave function satisfies Eq.6.5, we say that it is normalized. Eq. 6.5 is called the normalization condition. We can generalize this condition to the case of N

particles moving in three-dimensional space:

$$\int_{-\infty}^{+\infty} |\psi(\mathbf{r}(1), \mathbf{r}(2), \ldots, \mathbf{r}(N))|^2 d\mathbf{r}(1) d\mathbf{r}(2) \cdots d\mathbf{r}(N) = 1;$$

the probability that the N particles are somewhere in the space must be equal to 1.

It is this interpretation of the wave function that prompted us to require that all eigenstates that have a physical meaning must be normalized.

§2. *Application to a Vibrating Diatomic Molecule.* To better understand the probabilistic interpretation of the wave function and the manner in which it is used, I will examine the example of a vibrating molecule. The simplest model for the vibrational motion of a diatomic molecule assumes that the interaction between the two atoms is accurately simulated by a spring (see Fig. 6.1). The potential energy of such a spring is

$$V(x) = \frac{1}{2}k\,(r - r_0)^2 \equiv \frac{kx^2}{2} \tag{6.6}$$

where r is the distance between the centers of the atoms and k is called the force constant of the bond (the spring). The distance r_0 corresponds to the lowest potential energy, and it is the bond length. It is not difficult to show that the potential energy $V(x)$ in Eq. 6.6 leads, in classical mechanics, to an oscillatory motion of the distance x between the atoms, with a frequency

$$\omega = \sqrt{\frac{k}{\mu}}$$

where

$$\mu = \frac{m_1 m_2}{m_1 + m_2}$$

is the effective mass (m_1 and m_2 are the masses of the two atoms).

Figure 6.1 A model for a vibrating diatomic molecule: two atoms connected by a spring.

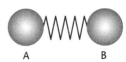

A   B

Since we know the potential energy, we can write the Schrödinger equation for this system and solve it. For the ground state (the lowest energy eigenstate), the wave function is

$$\psi_0(x) = \left(\frac{\mu\omega}{\pi\hbar}\right)^{1/4} \exp\left[-\frac{\mu\omega x^2}{2\hbar}\right] \tag{6.7}$$

and the ground state energy is

$$E_0 = \frac{\hbar\omega}{2} \tag{6.8}$$

The wave function of the first excited state is

$$\psi_1(x) = \left(\frac{4}{\pi}\left(\frac{\mu\omega}{\hbar}\right)^3\right)^{1/4} x \exp\left[-\frac{\mu\omega}{2\hbar}x^2\right] \tag{6.9}$$

with the energy

$$E_1 = \frac{3\hbar\omega}{2} \tag{6.10}$$

We will study harmonic oscillators in more detail in Chapter 16. The information given above is all we need here.

§3. *Data for HCl.* To perform calculations with the wave functions given above, we need to work with a specific molecule. I choose here HCl, for which I have the following data.

The reduced mass is $\mu = m_1 m_2/(m_1 + m_2)$. For $m_1 = 1.0078$ amu $= 1.0078 \times 1.6605 \times 10^{-24}$ g and $m_2 = 34.968$ amu $= 34.968 \times 1.6605 \times 10^{-24}$ g (I use the conversion factor 1 amu $= 1.6605 \times 10^{-24}$ g; see Appendix 1), I have $\mu = 0.979568 \times 1.6605 \times 10^{-24}$ g (see Workbook QM6.1).

Since the force constant[a] is $k = 4.9 \times 10^5$ dyne/cm, the frequency ω has the value

$$\omega = \sqrt{\frac{k}{\mu}} = \sqrt{\frac{4.9 \times 10^5 \text{ erg}}{0.979568 \times 1.6605 \times 10^{-24} \text{ g}}} = 5.4886 \times 10^{14} \text{ s}^{-1}$$

[a] D. A. McQuarrie, *Statistical Mechanics*, Harper Collins, New York, 1976, p. 95.

Using $\hbar = 1.0546 \times 10^{-27}$ erg s (see Appendix 1) and the values of μ and ω just calculated, we can write

$$\psi_0(x) = \left(\frac{0.979 \times 1.66 \times 10^{-24} \times 5.489 \times 10^{14}}{\pi \times 1.055 \times 10^{-27}} \frac{g}{s \text{ erg s}} \right)^{1/4}$$

$$\times \exp\left[-\frac{0.979 \times 1.66 \times 10^{-24} \times 5.489 \times 10^{14}}{2 \times 1.055 \times 10^{-27}} \frac{g \text{ cm}^2}{s \text{ erg s}} \left(\frac{x}{\text{cm}}\right)^2 \right]$$

$$(6.11)$$

Let us look at units. The unit of the wave function is $(g/\text{erg s}^2)^{1/4}$ (see Eq. 6.7) where g stands for grams. Since erg $= g \text{ cm}^2 \text{ s}^{-2}$, we have

$$\frac{g}{s^2 \text{ erg}} = \frac{g}{s^2 g \frac{\text{cm}^2}{s^2}} = \frac{1}{\text{cm}^2},$$

and the wave function has units of

$$\left[\frac{g}{s^2 \text{ erg}} \right]^{1/4} = \frac{1}{\text{cm}^{1/2}}$$

But cm is a cumbersome unit for length in molecular physics, because it is much too large. I will use Å instead. This means that

$$\left[\frac{g}{s^2 \text{ erg}} \right]^{1/4} = \frac{1}{\text{cm}^{1/2}} = \frac{1}{10^4 \text{Å}^{1/2}}$$

and the factor in front of the exponential in Eq. 6.11 is

$$\left(\frac{0.979 \times 1.66 \times 10^{-24} \times 5.489 \times 10^{14}}{\pi \times 1.055 \times 10^{-27}} \right)^{1/4} \frac{1}{10^4 \text{Å}^{1/2}} = \frac{2.27837}{\text{Å}^{1/2}} \qquad (6.12)$$

The exponent in Eq. 6.11 is dimensionless (as it should be) since

$$\frac{g \text{ cm}^2}{s^2 \text{ erg}} = 1$$

But I want to use Å as a unit of length. Therefore the exponent is

$$-\frac{0.979 \times 1.66 \times 10^{-24} \times 5.489 \times 10^{14}}{2 \times 1.055 \times 10^{-27}} \left(\frac{x}{10^8 \text{Å}}\right)^2 = -42.327 \left(\frac{x}{\text{Å}}\right)^2 \quad (6.13)$$

In this expression I must use x in Å.

Putting it all together gives

$$\psi_0(x) = \frac{2.27837}{\text{Å}^{1/2}} \exp\left[-42.327 \left(\frac{x}{\text{Å}}\right)^2\right] \quad (6.14)$$

The probability that the difference $x = r - r_0$, between the interatomic distance r and the equilibrium bond length r_0, takes a value between x and $x + dx$, when the vibrating molecule is in the ground state (see §1), is

$$p_0(x)dx = \left(\frac{2.27837}{\text{Å}^{1/2}} \exp\left[-42.327 \left(\frac{x}{\text{Å}}\right)^2\right]\right)^2 dx \quad (6.15)$$

Because the wave function has units of $\text{Å}^{-1/2}$, dx in this formula must have units of Å and $p(x)dx$ is dimensionless (as it should be).

A plot of the probability distribution function $p_0(x)$ versus x is shown in Fig. 6.2 as a solid line. As you can see, when the molecule is in the ground state, it is more

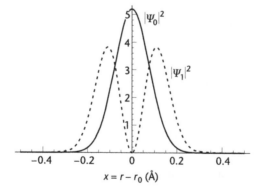

Figure 6.2 Probability distributions for the bond length x when the oscillator is in the ground state (solid line) and the first excited state (dashed line).

likely that the distance between the atoms is equal to the equilibrium distance r_0. However, the probability that the bond length deviates from this value (i.e. that $x = r - r_0 \neq 0$) is not negligible.

The figure also shows (dotted line) the probability distribution function

$$p_1(x) = \psi_1(x)^*\psi_1(x) \equiv |\psi_1(x)|^2$$

when the molecule is excited in the first vibrational state. Here ψ_1 is given by Eq. 6.9. $p_1(x)dx$ is the probability that $r - r_0$ takes values between x and $x + dx$, when we know that the oscillator is excited to the state ψ_1.

§4. *Interpretation.* Here we go again: probabilities instead of certainty. The laws of classical mechanics place no restrictions, in principle, on position measurements. Only technical difficulties prevent us from measuring the position with arbitrary accuracy. Quantum mechanics tells us that if we were to measure the interatomic distance r, for a single oscillator, we would obtain different results in different measurements, performed under identical conditions.

As in the case of the measurements discussed in the previous chapter, knowing the probability with which something takes place is useful when we make a large number N of identical measurements. If N is large enough, we can be sure that for

$$Np_0(x)dx$$

oscillators in the ground state $\psi_0(x)$, $x \equiv r - r_0$ takes values between x and $x + dx$.

§5. *Average Values.* The probability is also useful for calculating various averages. For example, the average distance between the atoms, when the diatomic molecule is in the ground state, is

$$\langle r \rangle = \int_{-\infty}^{\infty} rp_0(r)dr = \int_{-\infty}^{\infty} (r_0 + x)\,p_0(x)dx$$

$$= r_0 \int_{-\infty}^{\infty} p_0(x)dx + \int_{-\infty}^{\infty} xp_0(x)dx$$

$$= r_0$$

In this calculation we have used the fact that $x = r - r_0$ and

$$p_0(r)dr = p_0(x)dx$$

Here $p_0(r)dr$ is the probability that the distance between the atoms takes values between r and $r + dr$ when the molecule is in the ground state. We have also used the facts that

$$\int_{-\infty}^{\infty} p_0(r)dr = 1$$

and

$$\int_{-\infty}^{\infty} xp_0(x)dx = 0$$

You can calculate these integrals fairly easily by hand, or by **Mathematica** or **Mathcad**.

Exercise 6.1

You have a sample of 10^{12} HCl molecules. The force constant is $k = 4.9 \times 10^5$ dyne/cm and the equilibrium bond length is $r_0 = 1.27460$ Å. Calculate how many molecules that are in the ground state have a bond length between 1.24 Å and 1.29 Å.

Exercise 6.2

Do the same type of calculation for I_2. For this molecule, $r_0 = 2.666$ Å and $\omega = 214.57$ cm^{-1}. (*Data are from Herzberg.*[b]) Calculate how many molecules, out of 10^{12}, have an interatomic distance between 2.53 Å and 2.66 Å. (*Note*: $\omega = 214.57$ cm^{-1} is a strange unit. In Appendix 2, there is a table that tells you that 1 cm^{-1} = 1.9862×10^{-16} erg. This means that $\hbar\omega = 1.9862 \times 10^{-16} \times 214.57$ erg. This information can be used to calculate the frequency: since $\hbar = 1.0546 \times 10^{-27}$ erg s, I have $\omega = \hbar\omega/\hbar = \left(1.9862 \times 10^{-16} \times 214.57\right) / \left(1.0546 \times 10^{-27}\right) = 4.041 \times 10^{13}$ s^{-1}.)

§6. *The Effect of Position Uncertainty on a Diffraction Experiment.* Let us look at a system consisting of diatomic molecules adsorbed on a surface. Under certain conditions, these molecules position themselves to form a periodic array

---

[b] G. Herzberg, *Molecular Spectra and Molecular Structure. I. Spectra of Diatomic Molecules*, Van Nostrand, New York, 1950, p. 540.

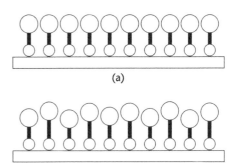

Figure 6.3 Upper figure shows the surface atoms undisturbed by uncertainty in positions. Lower figure shows the real instantaneous positions.

(see Fig. 6.3). If all the molecules have the same bond length, they form a perfect "grating" on the surface (Fig.6.3a). If we scatter from the surface a beam of X-rays (or electrons) having a certain wavelength, the grating will diffract the beam and produce a diffraction pattern. From this pattern, we can, by using diffraction theory, determine the positions of atoms.

However, we know that the molecules will not have the same bond length. According to quantum mechanics, the bond length in each molecule has a given value with a given probability. The instantaneous positions of the atoms are shown, with some exaggeration, in Fig. 6.3b. The grating formed by the molecules is not perfect. We can calculate the diffraction pattern produced by the molecules in Fig. 6.3a and by those in Fig. 6.3b. When calculating the diffraction pattern produced by Fig. 6.3b, we use the probability $p_0(x)dx$. The experiments find that the observed pattern is that calculated for the imperfect grating. This is an indirect confirmation of the quantum mechanical postulate that different distances between atoms are observed with different probabilities; moreover, the agreement with experiment is quantitative, confirming the formula used for computing the probability that the molecule has a given bond length.

§7. *The Effect of Position Uncertainty in an ESDAID Experiment.* Consider another experiment, performed by John Yates and Ted Madey, when they were working at NIST. They started with a perfect metal surface whose atoms are arranged in a nearly perfect periodic array (see Fig. 6.4). Then they adsorbed atoms or molecules on the surface. Fig. 6.4, shows one such atom, adsorbed on top of a surface atom. The most likely position of the adsorbed atom is shown by heavy lines in Fig. 6.4. However, the positions shown with dotted lines, and those between them, also occur with some probability.

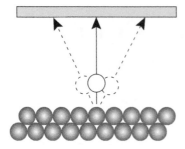

Figure 6.4 The dotted lines show the locations the atom can reach due to uncertainty in position.

This surface is bombarded with high-energy electrons, which break the metal–atom chemical bond. When the bond is broken the adsorbed atoms come off the surface. The atoms whose bond is perpendicular to the surface leave it, when the bond is broken, in a direction perpendicular to the surface (see the solid line and arrow in the figure). The ones that are slanted come off, when the bond is broken, at an angle with the surface. The experiment detects the point where the atoms leaving the surface hit a position-sensitive detector (represented by a slab at the top of the figure). An atom that leaves the surface on a trajectory perpendicular to the surface hits the center of the detector. One that leaves the surface on a slanted trajectory hits the detector off-center. From the distribution of the points of impact on the detector, it is possible to calculate the probability of the initial position of the atom. The result agrees with quantum mechanical calculations. I should caution you that the real experiment is close in spirit to the description just given, but is more complicated.

§8. *Why Did We Think that We Could Measure Position Accurately?* You have just learned that we cannot know with certainty the positions of the particles in a system. Why is this not true for the large objects we encounter in everyday life? Quantum mechanics has a simple answer: the larger a particle, the less uncertain its position.

Let us see how we reach this conclusion, in the case of the bond length of a vibrating diatomic molecule. If the oscillator is in the ground state, the probability that $x = r - r_0$ has a value between x and $x + dx$ is

$$|\psi_0(x)|^2 dx = \left(\frac{\mu\omega}{\pi\hbar}\right)^{1/2} \exp\left[-\frac{\mu\omega x^2}{\hbar}\right] dx \qquad (6.16)$$

I can write Eq. 6.16 as

$$|\psi_0(x)|^2 dx = \left(\frac{1}{\pi\sigma^2}\right)^{1/2} \exp\left[-\left(\frac{x}{\sigma}\right)^2\right] dx \qquad (6.17)$$

where

$$\sigma^2 \equiv \frac{\hbar}{\mu\omega} \qquad (6.18)$$

is a quantity with units of length squared.

From Eq. 6.18 you can see that if $x \gg \sigma$, then $|\psi_0(x)|^2$ is practically zero. Therefore I am sure that the bond length r is never much longer than $r_0 + \sigma$, nor is it much shorter than $r_0 - \sigma$. The smaller σ is, the more accurate is my knowledge of the bond length. Thus σ is a good descriptor of the accuracy with which I can know the bond length.

Let us calculate σ for a diatomic whose reduced mass is a grams and whose frequency is $\omega = 10^{14}$ s^{-1}. In this case

$$\sigma^2 = \frac{\hbar}{\mu\omega} = \frac{1.0546 \times 10^{-27} \text{ erg s}}{1 \text{ gr} \times 10^{14} \text{ g/s}}$$

I find that $\sigma^2 \cong 10^{-41}$ cm^2 and $\sigma \cong \sqrt{10} \times 10^{-20}$ cm.

This uncertainty in position is beyond the accuracy of any conceivable instrument. Because of this, in the classical world (large mass), we are entitled to claim that it is possible to know the position of a particle with "unlimited" accuracy. We know that this is not quite right, but no one will be able to prove us wrong in our lifetime.

The statement that position is precisely known when the mass is large is a dumb statement, even though it is made frequently. The word "large" makes sense only if we compare one mass to another. When I say that a friend is heavy, I am implicitly comparing him to the average person, or comparing his weight to what it was twenty years ago. So what do I mean when I say that for a heavy oscillator, the position is known accurately? Heavy compared to what?

This is easy to answer for the case of the oscillator. Let us say that the highest accuracy of the best position measurement is σ_0. If

$$\sqrt{\frac{\hbar}{\mu\omega}} \ll \sigma_0 \qquad (6.19)$$

then the quantum uncertainty in the position of the oscillator is not detectable. The inequality (Eq. 6.19) is satisfied if the mass satisfies

$$\mu \gg \sqrt{\frac{\hbar}{\omega \sigma_0^2}}$$

If $\sigma_0 = 10^{-8}$ cm, then

$$\mu > \sqrt{\frac{1.0546 \times 10^{-27}}{10^{14} \times 10^{-16}}} \, g = \sqrt{10^{-25}} \, g = \sqrt{10} \times 10^{-12} \, g$$

For a diatomic with reduced mass exceeding 3×10^{-12} g, the difference in the bond length between different copies of the same molecule is not detectable. Because classical physics deals with objects heavier than 10^{-12} g, we can claim that position is accurately determined by the laws of mechanics.

Exercise 6.3

One way of describing the uncertainty in the position is to use the mean-square error

$$\ell^2 \equiv \int_{-\infty}^{+\infty} x^2 |\psi_0(x)|^2 dx$$

Calculate this quantity and show that ℓ is of order $\sigma = \sqrt{\hbar/\mu\omega}$.

7

TUNNELING

§1. *The Classically Forbidden Region.* The example of the vibrating molecule allows me to illustrate a phenomenon, called tunneling, that has no analog in classical mechanics. I proceed in two steps. First, I show that in classical mechanics the coordinate x of an oscillator, having a fixed energy E_0, can only reach a certain region.

This is a trivial statement. An oscillating diatomic molecule behaves like a barbell whose balls are connected by a spring. If I use a certain amount of energy to stretch the barbell and then let it go, the barbell will oscillate (this barbell is free-floating in space). Now, it would not shock you to hear that during this oscillation, the distance r between the balls will not grow larger than that caused by my initial pull. This happens because the energy of the barbell was given to it when I first stretched it. The spring can stretch farther than I initially stretched it only if it gets additional energy from some other source. Since there is no such source, the change in r is limited by the amount of energy given initially.

§2. *How Large is the Accessible Region?* How do I make such a statement quantitative? Consider a diatomic molecule (or a barbell) whose total energy is E. In classical mechanics I have

$$E = \frac{\mu v^2}{2} + V(r) \tag{7.1}$$

Here v is the relative velocity of one atom with respect to the other ($v = dr/dt$), r is the distance between atoms, and $V(r)$ is the potential energy. Solving for v gives

$$v = \pm\sqrt{\frac{2(E - V(r))}{\mu}} \tag{7.2}$$

I have two signs here because knowing the energy of the oscillator determines the *magnitude* of the velocity but not its *direction.*

If $V(r) > E$, the sign of the expression under the square root becomes negative and the velocity is imaginary. This is impossible in mechanics: velocity must be a real number. This means that I must have

$$E - V(r) \geq 0 \tag{7.3}$$

The distance r between the atoms can only reach values for which $E - V(r) \geq 0$. These values are said to be *classically allowed* when the particle has total energy E.

The points r for which $E - V(r) < 0$ are said to be *classically forbidden.*

The points r for which

$$E - V(r) = 0 \tag{7.4}$$

are called the *turning points.* They are at the border between the classically allowed and classically forbidden regions.

§3. *The Classically Allowed Region for an Oscillator.* If I apply these general considerations to an oscillator having the potential energy $V(r) = k(r - r_0)^2/2$, and set $x = r - r_0$, the turning points are the solutions of the equation (derived from Eq. 7.4)

$$E - \frac{kx^2}{2} = 0 \tag{7.5}$$

They are given by

$$x = \pm\sqrt{\frac{2E}{k}} \tag{7.6}$$

The coordinate x of the *classical oscillator* with total energy E is allowed to take values only in the range

$$-\sqrt{\frac{2E}{k}} \le x \le +\sqrt{\frac{2E}{k}} \qquad (7.7)$$

§4. *Tunneling.* If this holds true for quantum mechanics, it would mean that the probability to find the particle in the classically forbidden region is zero:

$$p(x) = 0 \text{ for } x < -\sqrt{\frac{2E}{k}} \text{ or } x > \sqrt{\frac{2E}{k}} \qquad (7.8)$$

Here $p(x)dx$ is the probability that $x = r - r_0$ takes values between x and $x + dx$.

Let us see whether this is true for an HCl molecule in the ground state.

I start by calculating the turning points predicted by classical mechanics. For the HCl molecule, $\omega = \sqrt{k/\mu} = 5.4886 \times 10^{14}$ s^{-1} (see Chapter 6, §3). The ground state energy is

$$E_0 = \frac{\hbar\omega}{2} = 1.0546 \times 10^{-27} \text{ erg s} \times 5.4886 \times 10^{14} \text{ s}^{-1}$$

I convert this energy to eV by using 1 erg = 6.2420×10^{11} eV (see Appendix 2). The result is

$$E_0 = 0.181 \text{ eV}$$

The potential energy is

$$V(x) = \frac{kx^2}{2} = 4.9 \times 10^5 \frac{\text{dyne}}{\text{cm}} \text{ cm}^2 \frac{x^2}{\text{cm}^2} = 4.9 \times 10^5 \text{ erg} \left(\frac{x^2}{\text{cm}^2}\right)$$

If I use x in cm, I get $V(x)$ in erg. To get it in eV, I use

$$V(x) = 6.2420 \times 10^{11} \times 4.9 \times 10^5 \text{ eV} \left(\frac{x}{\text{cm}}\right)^2$$

In this equation I must use x in cm. However, the Ångstrom (1 Å = 10^{-8} cm) is a more convenient unit for x. Therefore I use

$$V(x) = 6.2420 \times 10^{11} \times 4.9 \times 10^5 \text{ eV} \left(\frac{x}{10^8 \text{ Å}}\right)^2 = 15.2929 \left(\frac{x}{\text{Å}}\right)^2 \text{ eV}$$

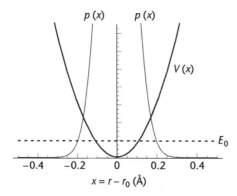

Figure 7.1 The thick line shows the potential energy, the thin line shows the probability distribution $p(x)$, and the dashed line shows the ground-state energy E_0. The turning points are at the intersection of the horizontal dashed line and the parabola representing $V(x)$.

In this equation, x must be in Å.

Figure 7.1, shows a plot of $V(x)$ and E_0. The two curves intersect when $\mu v^2/2 = E_0 - V(x) = 0$; therefore, this intersection gives the turning points. I can calculate them from

$$15.2929 \left(\frac{x}{\text{Å}} \right)^2 \text{eV} = E_0 = 0.181 \text{ eV},$$

which gives the turning points at

$$x_1 = -0.1089 \text{ Å}$$

$$x_2 = 0.1089 \text{ Å}$$

Figure 7.1 also shows a plot of $p(x)$. Obviously $p(x)$ *is not zero* for $x < x_1$ or $x > x_2$, as classical mechanics would have it. You can see clearly that values of x larger than 0.1089 Å and smaller than −0.1089 Å have a fair probability. The *particle penetrates into the classically forbidden region*. This phenomenon is called tunneling.

I can calculate the probability that $x < -0.1089$ Å from

$$P \equiv \int_{-\infty}^{-0.1089} |\psi_0(x)|^2 \, dx$$

I can perform the integral by using the ground-state wave function $\psi_0(x)$ given in Chapter 6, §3 (see Workbook QM.7). The result is

$$P = 0.07826$$

By symmetry, I find that the probability that $x > 0.1089$ Å is

$$\int_{0.1089}^{\infty} |\psi_0(x)|^2 \, dx = 0.07826$$

The probability of finding a particle in the classically forbidden region is 2×0.07826.

§5. *Tunneling Depends on Mass and Energy.* The distance that a particle tunnels into the forbidden region depends on its mass and on the relationship between its total energy and $V(x)$. The lighter the particle, or the higher its total energy, the further it tunnels. A particle penetrates deeper into the forbidden region when the difference $|V(x) - E|$ is smaller. Under favorable conditions, electrons can tunnel as far as 50 Å.

Exercise 7.1

Calculate the probability to tunnel into the forbidden region for an oscillator with the mass of an electron if the force constant is the same as that for HCl. Note that ω changes. Calculate the same probability for an object with $\mu = 500$ amu.

Exercise 7.2

Calculate the probability to penetrate in the classically forbidden region for HCl in the excited states $\psi_1(x)$ and $\psi_2(x)$.

Exercise 7.3

To describe how far an oscillator in the state $\psi_n(x)$ penetrates into the forbidden region, we can use

$$\bar{x} \equiv \int_{x_t}^{+\infty} |\psi_n(x)|^2 (x - x_t) \, dx$$

where x_t is the turning point having $x_t > 0$. Calculate $\bar{x}$ for an HCl oscillator for $n = 0, 1, 2$.

§6. *Tunneling Junctions.* Tunneling is not an accident confined to oscillators. It is a general quantum mechanical phenomenon.

An important example of tunneling is the gadget shown in Fig. 7.2. At the top of Fig. 7.2 I show a very thin oxide layer sandwiched between two metallic films. The oxide layer is 15 to 30 Å thick. It so happens that the electron has a lower potential energy $V(x)$ in the metal than inside the oxide. This is what the graph in the lower part of the figure is showing (the heavy line is the potential energy $V(x)$). x is the electron coordinate in a direction perpendicular to the surface of the film.

Consider now an electron located in the metal at the left, having a total energy E (dotted line in the figure). Let us see what classical mechanics has to say about this system. Inside the metal, $E > V(x)$ and the electron moves with the kinetic energy $mv^2/2 = E - V(x)$. Inside the oxide, $E < V(x)$ and the kinetic energy is negative. The electron is *classically forbidden from penetrating into the oxide.* The potential energy acts as a barrier preventing electron penetration.

Quantum mechanics says that this "ain't necessarily so": the electron can tunnel into the forbidden region. This means that there is some probability that the electron "wanders" inside the oxide. If the oxide layer is thin, the electron can go through it, to emerge in the metal film at the right, where it is classically allowed. This means that if we put an infinitesimal voltage across the junction, we can pass current through it, even though the oxide is an insulator. The existence of this current means that electrons pass through the oxide layer (the classically forbidden region), from the left metal to the right metal. To do so, they must tunnel through the oxide film.

Tunneling junctions are used in many solid state electronic devices.

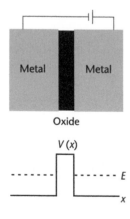

Figure 7.2 Upper: a tunneling junction. Lower: the solid line shows the potential energy as a function of electron position. The dotted line shows the total energy.

§7. *Scanning Tunneling Microscopy.* Tunneling makes possible an instrument that has revolutionized the way we study solid surfaces. It is called a scanning tunneling microscope (STM) and was discovered by Binning and Rohrer in the 1980s. I do not say "invented," because they discovered it while trying to do something else. If you want to be successful in research, you need luck but this is useless without the ability of recognizing that you have been lucky.

Let us ask ourselves how we can determine, with high resolution, the shape of a metal surface. I make a very sharp metal tip, put it in contact with the surface, and move the tip along the surface (see Fig. 7.3). The tip moves up when it encounters a bump and down when it goes over a pit. If I measure how the height of the upper end of the tip changes, as the tip scans along the surface, I get a picture of the surface shape along the line that was scanned.

Before Binning and Rohrer, this device worked poorly because of several difficulties. Firstly, to find very fine details on the surface, a very precise method is required for measuring the tip height. If the tip goes over a bump that is 5 Å high, I will "see" the bump only if I can detect a 5-Å upward movement of the tip. Secondly, to see a pit of 5-Å diameter, the tip must be fairly sharp; a broad tip cannot lower itself into the pit. Finally, if the tip is dragged over the surface, it is blunted in the process and it may also modify the surface. For example, it may sweep, like a broom, the molecules adsorbed on the surface.

In principle, these difficulties are easy to solve. To avoid scratching the surface and blunting the sharp tip, I will have to keep the tip above the surface. This is a fine idea, but if the tip does not touch the surface, how can I tell what the surface shape is? In principle, that is simple, too: keep the distance between tip and surface constant. Then, if the surface has a bump, the tip will go up. This would work well, if I knew how to do it. But even so, I am left with the other problem: to detect

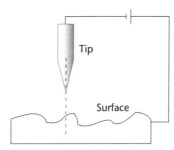

Figure 7.3 A simple surface scanner.

very small bumps, it is necessary to measure the position of the tip with very high accuracy.

Binning and Rohrer used tunneling and piezoelectricity to solve all these problems. They put a voltage between the tip and the metal surface and passed an electric current through the circuit (see Fig. 7.3). This device is a sort of tunneling junction. One metal film in the junction is replaced by the tip, and the oxide is replaced by air (or vacuum, if the instrument is in a vacuum chamber). Let us pursue this obvious analogy. What is the potential energy of an electron moving along the dotted line in Fig. 7.3? It is similar to that shown at the bottom of Fig. 7.2. It is lower when the electron is in the metal (either the surface or the tip) and higher when it is in the air. The space between the two metals is classically forbidden. However, the electrons can tunnel through the vacuum, and if I put a voltage across the junction I will get a tunneling current.

Why would this solve our problems? The magnitude of the tunneling current is very sensitive to the tip-to-surface distance d. When the tip is close to the surface, the current increases considerably, and vice versa. Changing the tip-to-surface distance by 0.1 Å will cause a detectable change in the current!

Binning and Rohrer connected the circuit to a computer which is "instructed" to keep the voltage and the current constant, as the tip moved along the surface. As the tip approaches a bump, the distance to the surface becomes smaller and the current increases. But the computer, instructed to keep the current constant, will then increase the tip-to-surface distance to bring the current to the desired constant value. The opposite happens when the tip goes over a pit.

By keeping the current and the voltage constant during the scan, the computer is keeping the tip-to-surface distance constant. The tip moves along the surface but it does not touch it. Since the computer records how much it pulled the tip up or down, it obtains a "picture" of the surface shape.

This sounds great, but I can see a further complication. How am I going to move the tip a distance of 0.1 Å? How am I going to keep the jiggling and trembling of the tip (due to building vibrations, trucks on the road, etc.) from messing up the measurement?

Take one problem at a time. There is a wonderful phenomenon called the piezo-electric effect. Some crystals will change dimensions if an electric field is applied to them. If the field is small, the size of the crystal changes a little. By gluing the solid whose surface I wish to study to such a crystal, I can move it up, down or sideways, by applying an electric field to the crystal. Since the displacement of the crystal face is proportional to the electric field, I can determine how far the computer pulled

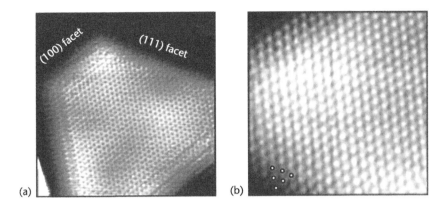

Figure 7.4 Left figure: a facet of a very small cluster of Pd atoms supported on an Al_2O_3 surface. Right: the atoms seen with higher resolution.

the sample down or moved it up (to keep the distance to the surface constant over a bump) by how large a field was needed to perform that movement.

The vibrations of the system are quenched by suspending the tip–surface assembly in an arrangement of springs. The whole assembly may vibrate but the tip–surface distance does not.

To their surprise, Binning and Rohrer found that with this device they could "see" the contour of the atoms on a metal or a semiconductor surface. In Fig. 7.4 you can see the atoms on the surface of a Pd cluster sitting on an Al_2O_3 surface. The image was taken in a collaboration between the groups of Hajo Freund, at the Fritz Haber Institute in Berlin, and Flemming Besenbacher, at the University of Aarhus in Denmark. The upper picture shows a facet of a cluster, the lower one shows a high-resolution scan of the facet. The bumps in the picture are Pd atoms.

By extraordinarily well-designed experiments, Wilson Ho (University of California, Riverside) has managed to use STM to measure the vibrational energy of single molecules and to "see" how a single molecule is dissociated by the electrons tunneling through it. He has also used the tip to nudge molecules close to each other and make them react or move gold atoms on a surface to make wires that have a width of one atom and to measure their properties.

8

PARTICLE IN A BOX

Define the System

§1. *Introduction.* In this chapter I solve the Schrödinger equation for a very simple system, called a particle in a box. This was invented to illustrate how quantum mechanics works, without having to wade through complicated or subtle mathematics. In the past 20 years, scientists have learned how to make systems in which electrons behave as if they are trapped in a box. Unexpectedly, this toy model, meant to be a teaching tool, is being used in important applications: the lasers used at supermarket counters and in your CD player are based on it.

The system we are going to study consists of a particle (e.g. an electron) inside a chamber shaped like a parallelepiped. When the particle hits the walls of the chamber, it is reflected without losing or gaining energy. This is an idealization; the wall is made of atoms and when the particle collides with them, its energy may change. As it happens, this energy exchange has a small effect on many of the phenomena of interest to us, and it can be ignored.

Most phenomena in which our particle may be involved can be understood once we know its energy eigenvalues and eigenfunctions. To obtain these I follow a general recipe. First, I determine the *classical Hamiltonian* (i.e. the total energy) of the system. Then, I *quantize* it, by replacing the classical variables with the appropriate operators; this gives me the energy operator (or the Hamiltonian) of the particle. The Schrödinger equation is the eigenvalue problem for this operator; solving it provides the energy eigenvalues and eigenstates. These can be used, as described in the previous chapters, to find how the particle emits or absorbs light or how it is excited by electron scattering.

To solve the Schrödinger equation I use a method, called *the separation of variables,* that has been applied to solve many problems in quantum mechanics, electrodynamics, fluid mechanics, and the theory of elasticity.

§2. *The Classical Hamiltonian.* The classical Hamiltonian is the total energy of the particle (kinetic plus potential). It is not difficult to find the potential energy. The model assumes that, as long as the particle is inside the box, no force acts on it (i.e. the force is zero). The potential energy $V(r)$ is connected to the force $F(r)$ acting on the particle through (see your physics book)

$$\mathbf{F}(\mathbf{r}) = -\frac{\partial V}{\partial \mathbf{r}} \equiv -\left[\mathbf{i}\,\frac{\partial V}{\partial x} + \mathbf{j}\,\frac{\partial V}{\partial y} + \mathbf{k}\,\frac{\partial V}{\partial z}\right] \tag{8.1}$$

The middle expression is a shorthand notation for the right-hand side. I use here a Cartesian system of coordinates with the unit vectors $\mathbf{i}$ (in the x-direction), $\mathbf{j}$ (in the y-direction) and $\mathbf{k}$ (in the z-direction). According to Eq. 8.1, zero force means a constant potential energy (because $\partial \text{constant}/\partial x = 0$, $\partial \text{constant}/\partial y = 0$, $\partial \text{constant}/\partial z = 0$).

Newton's equation of motion

$$m\frac{d^2\mathbf{r}}{dt} = -\frac{\partial V}{\partial \mathbf{r}} \tag{8.2}$$

tells me that adding a constant C to $V(r)$ causes no change in the way the particle moves, because $\partial(C+V)/\partial \mathbf{r} = \partial V/\partial \mathbf{r}$. Two particles, one with the potential energy $V(\mathbf{r})$ and the other with the potential energy $V(\mathbf{r}) + C$, have the same Newton equation and therefore move in exactly the same way. Because of this, I can add a constant to the potential energy without changing the way the particle moves. In what follows I assume that this constant was chosen to make the potential energy of the particle in the box equal to zero.

The kinetic energy of the particle is (in Cartesian coordinates)

$$\frac{\mathbf{p}\cdot\mathbf{p}}{2m}$$

where m is the mass of the particle, $\mathbf{p}$ is the momentum vector and p_x, p_y and p_z are its Cartesian components. The total energy is therefore

$$H = \frac{\mathbf{p} \cdot \mathbf{p}}{2m} + V = \frac{\mathbf{p} \cdot \mathbf{p}}{2m} = \frac{p_x^2 + p_y^2 + p_z^2}{2m} \tag{8.3}$$

Exercise 8.1

Pick $V(\mathbf{r}) = V_0$ where V_0 is a constant. Study the consequences of this choice by performing all the calculations done in this chapter, for your choice of V_0. Will any measurable property of the system change when you replace $V(\mathbf{r}) = 0$ with $V(\mathbf{r}) = V_0 \neq 0$?

§3. *Quantize the System.* Now I follow the rules of Chapters 2 and 3 to obtain the Hamiltonian operator from the expression (Eq. 8.3) for the classical energy: the components p_x, p_y and p_z of classical momentum, are replaced with the operators

$$\hat{p}_x = \frac{\hbar}{i} \frac{\partial}{\partial x} \tag{8.4}$$

$$\hat{p}_y = \frac{\hbar}{i} \frac{\partial}{\partial y} \tag{8.5}$$

$$\hat{p}_z = \frac{\hbar}{i} \frac{\partial}{\partial z} \tag{8.6}$$

Replacing the momenta, in the classical Hamiltonian (Eq. 8.3), with the corresponding operators gives:

$$\hat{H} = -\frac{\hbar^2}{2m} \left(\frac{\partial^2}{\partial x^2} + \frac{\partial^2}{\partial y^2} + \frac{\partial^2}{\partial z^2} \right) \tag{8.7}$$

In obtaining this equation I use the rules for operator multiplication (see Chapter 2) and the fact that $i^2 = -1$.

The eigenvalue equation for this Hamiltonian operator is:

$$-\frac{\hbar^2}{2m} \left(\frac{\partial^2 \psi}{\partial x^2} + \frac{\partial^2 \psi}{\partial y^2} + \frac{\partial^2 \psi}{\partial z^2} \right) = E\,\psi(x, y, z) \tag{8.8}$$

This is the Schrödinger equation for the particle in a box, which is satisfied by the energy eigenfunctions (or wave functions) and eigenvalues (see Chapter 3). As you learned in Chapter 4, these eigenvalues are the only energies the particle can have.

§4. *The Boundary Conditions.* When the particle is inside the box, its coordinates satisfy

$$0 \leq x \leq L_x \ \text{ and } \ 0 \leq y \leq L_y \ \text{ and } \ 0 \leq z \leq L_z \tag{8.9}$$

where L_x, L_y, and L_z are the lengths of the box edges (the box is a parallelepiped). The origin of the coordinate system is in a corner of the box (see Fig. 8.1). The fact that the particle cannot escape from the box, no matter how high its energy, means that the probability

$$|\psi(x, y, z)|^2$$

of finding the particle at the point $\mathbf{r} = \{x, y, z\}$ is zero if the point is *outside the box*.

Because the eigenfunctions must be continuous, the wave function must be zero when the particle is at the border. This means that we must have

$$\psi(x = 0, y, z) = \psi(x = L_x, y, z) = 0 \tag{8.10}$$

$$\psi(x, y = 0, z) = \psi(x, y = L_y, z) = 0 \tag{8.11}$$

$$\psi(x, y, z = 0) = \psi(x, y, z = L_z) = 0 \tag{8.12}$$

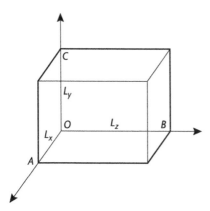

Figure 8.1 The coordinate system and the box in which the particle is confined.

These are the *boundary conditions*. As you have seen in Chapter 3, the wave function must satisfy both the Schrödinger equation and the boundary conditions.

§5. *Can we Really have such a Box?* Before going on with the mathematics, I pause for a bit of physics. Can such a box be made in the laboratory, or is it pure science fiction?

It is likely that you have learned in one of your chemistry classes that if you reduce a Au salt in solution, under the right conditions, you can form very small, nearly spherical Au particles. Their size depends on the method of preparation and could be of the order of 10^{-6} cm. You can tell that these particles are present, because upon reduction the color of the solution changes to a beautiful red. The color depends on the size of the particles and this property has been used for centuries to make colored glass. One can make such *colloidal particles* from many metals or semiconductors.

You have also learned in your physics class that some of the electrons in a metal are not tied to a specific atom, but move freely through the sample. It is because of these "free electrons" that metals have high electrical conductivity. The electrons are fairly happy inside the metal particle; they do not spill out into the solution because they are attracted by the positive ions in the metal. They are confined to the "box" made by the positive ions.

This colloidal particle differs from our box. First, if we give the electrons sufficient energy, by illuminating the colloidal particle with UV light, they will leave the particle. This is the *photoelectric effect*, whose discovery and explanation earned Lenard and Einstein separate Nobel Prizes. The box that we consider does not have this property: it is "constructed" so that the particle can never leave it (the particle trapped in it has a life sentence with no possibility of parole). However, the electrons in the colloidal particle behave like the electrons in our box, as long as they are not exposed to a source of energy. They are confined for life in the colloidal particle (just as they are in our box) not because the jail is perfect (as in our model) but because the prisoner doesn't have enough energy to jump over the fence.

Second, in our model the electron is *alone* in the box, while a colloidal particle has many "free" electrons. If these electrons do not interact with each other, they behave as if they are alone in the colloid. This is less likely to happen if the particle is a metal, because the density of the "free" electrons is high. It is possible, however, to make semiconductor particles with very few free electrons and the interaction between them can be neglected. They behave as if they are alone in the box, just like in our model.

Finally, we have assumed that the potential energy of the particle in our box is the same in every place in the box. The electron in a colloidal particle must interact with the ions (Au ions in a Au particle), so we expect the potential energy of an electron near an ion to be different from the potential energy of an electron located between the ions. It turns out that, for very subtle reasons that we cannot go into here, to a reasonable approximation we can ignore the dependence of the potential energy of the electron on its location.

So, to some extent, under the right conditions, the electrons in a colloidal particle (especially one made of a semiconducting material such as CdSe, GaAs or HgTe) behave as if they are enclosed in a box of the kind studied here. It is important to realize that the properties of these electrons *depend on the shape of the box*: the colloidal particles are nearly spherical, while ours is a parallelepiped. If you are brave and know your mathematics you may try to solve the Schrödinger equation for a particle in a spherical box, to see how much of a difference the shape of the box makes. (*Hint.* Use spherical coordinates and compare the equation for the radial wave function to the Bessel equation.)

In Chapter 9 I will describe a method for making parallelepipeds in which the electrons behave as they do in our model.

Solving the Eigenvalue Problem (the Schrödinger Equation) for the Particle in a Box

§6. *Separation of Variables.* I will solve the eigenvalue problem Eq. 8.8, with the conditions given by Eqs. 8.10–8.12, by a method called the *separation of variables*.

This takes advantage of the fact that in the operator

$$-\frac{\hbar^2}{2m}\left(\frac{\partial^2}{\partial x^2} + \frac{\partial^2}{\partial y^2} + \frac{\partial^2}{\partial z^2}\right)$$

$-\frac{\hbar^2}{2m}\frac{\partial^2}{\partial x^2}$, depends *only on x*, $-\frac{\hbar^2}{2m}\frac{\partial^2}{\partial y^2}$, depends *only on y*, and $-\frac{\hbar^2}{2m}\frac{\partial^2}{\partial z^2}$, depends *only on z*. There are no "cross" terms such as $\frac{\partial^2}{\partial x\partial y}$ or $V(x,y,z)$.

For any equation having this property I can seek a solution of the form

$$\psi(x,y,z) = \phi(x)\,\eta(y)\,\lambda(z) \tag{8.13}$$

where ϕ, η, and λ are *unknown functions*. We will determine them by forcing the function ψ, given by Eq. 8.13, to satisfy the eigenvalue equation Eq. 8.8 and the boundary conditions Eqs. 8.10–8.12.

To force $\psi(x, y, z)$ to satisfy Eq. 8.8, I insert it into that equation and divide the result by $\phi(x)\eta(y)\lambda(z)$. I obtain:

$$-\frac{\hbar^2}{2m\phi(x)}\frac{\partial^2\phi(x)}{\partial x^2} - \frac{\hbar^2}{2m\eta(y)}\frac{\partial^2\eta(y)}{\partial y^2} - \frac{\hbar^2}{2m\lambda(z)}\frac{\partial^2\lambda(z)}{\partial z^2} = E \qquad (8.14)$$

Note an interesting thing: the first term on the left-hand side depends only on x, the second depends only on y, the third depends only on z, and the right-hand side is a constant. The equality in Eq. 8.14 is possible only if each term in the left-hand side is a constant.

In other words, I must have

$$-\frac{\hbar^2}{2m\phi(x)}\frac{\partial^2\phi(x)}{\partial x^2} = E_x \qquad (8.15)$$

$$-\frac{\hbar^2}{2m\eta(y)}\frac{\partial^2\eta(y)}{\partial y^2} = E_y \qquad (8.16)$$

$$-\frac{\hbar^2}{2m\lambda(z)}\frac{\partial^2\lambda(z)}{\partial z^2} = E_z \qquad (8.17)$$

where E_x, E_y, and E_z are unknown constants. Because of Eq. 8.14, I must also have (add Eqs. 8.15–8.17 and compare the result to Eq. 8.14)

$$E_x + E_y + E_z = E \qquad (8.18)$$

This procedure reduces the three-dimensional problem (Eq. 8.8) to three one-dimensional problems (Eqs. 8.15–8.17). If I solve all three of them, I can insert the results in Eq. 8.13 and obtain a solution to Eq. 8.8.

§7. *Boundary Conditions.* This is, not enough, however: ψ must also satisfy the boundary conditions (Eqs. 8.10–8.12).

Condition (8.10) and the fact that $\psi(x, y, z) = \phi(x)\eta(y)\lambda(z)$ (see Eq. 8.13) lead me to

$$\psi(x = 0, y, z) = \phi(x = 0)\eta(y)\lambda(z) = 0 \qquad (8.19)$$

and

$$\psi(x = L_x, y, z) = \phi(x = L_x)\eta(y)\lambda(z) = 0 \qquad (8.20)$$

These two equations must be true for any values of y and z satisfying $0 \leq y \leq L_y$ and $0 \leq z \leq L_z$. Since we do not allow the possibility that either η or λ is identically zero when the particle is in the box, this means that I must have

$$\phi(x = 0) = 0 \quad \text{and} \quad \phi(x = L_x) = 0 \tag{8.21}$$

A similar analysis leads to

$$\eta(y = 0) = 0 \quad \text{and} \quad \eta(y = L_y) = 0 \tag{8.22}$$

as well as

$$\lambda(z = 0) = 0 \quad \text{and} \quad \lambda(z = L_z) = 0 \tag{8.23}$$

By this procedure, I have turned the boundary conditions Eqs. 8.10–8.12 for $\psi(x, y, z)$ into the boundary conditions Eqs. 8.21, 8.22, and 8.23 for $\phi(x)$, $\eta(y)$, and $\lambda(z)$, respectively.

§8. *Summary.* The function

$$\psi(x, y, z) = \phi(x)\eta(y)\lambda(z)$$

is an eigenfunction of the Hamiltonian (i.e. satisfies Eq. 8.8) if

(a) $\phi(x)$ satisfies Eq. 8.15 with the boundary conditions Eq. 8.21, and

(b) $\eta(y)$ satisfies Eq. 8.16 with the boundary conditions Eq. 8.22, and

(c) $\lambda(z)$ satisfies Eq. 8.17 with the boundary conditions Eq. 8.23.

If I solve these equations for ϕ, η, and λ, I determine the eigenfunctions and the eigenvalues of the particle in the three-dimensional box.

§9. *Solve Eq. 8.15 for $\phi(x)$.* I can write Eq. 8.15 as

$$-\frac{\hbar^2}{2m} \frac{\partial^2 \phi(x)}{\partial x^2} = E_x \phi(x) \tag{8.24}$$

This is an eigenvalue problem for the kinetic energy operator $(-\hbar^2/2m)(\partial^2/\partial x^2)$ of a particle moving in one dimension, along the x axis. This problem, with the boundary conditions in Eq. 8.21, was solved in Chapter 3, §20. There I found that the eigenfunctions of Eq. 8.24 are (see Eqs. 3.40 and 3.41)

$$\phi_n(x) = \sqrt{\frac{2}{L_x}} \sin\left(\frac{\pi n x}{L_x}\right), \quad n = 1, 2, 3, \ldots \tag{8.25}$$

and the corresponding eigenvalues are

$$E_x(n) = \frac{\hbar^2}{2m} \left(\frac{\pi n}{L_x}\right)^2, \quad n = 1, 2, 3, \ldots \tag{8.26}$$

As noted in Chapter 3, negative values for the integer n can be discarded: the solutions $\phi_n(x)$ and $\phi_{-n}(x)$ are not physically distinct. The argument is as follows. Because $\sin(-n\pi x/L_x) = -\sin(n\pi x/L_x)$, I have $\phi_n(x) = -\phi_{-n}(x)$. It so happens that whenever I calculate observable quantities in quantum mechanics, the wave function appears *squared*. Therefore the wave functions $\psi(x)$ and $-\psi(x)$ lead to identical physical properties. Because of this, I ignore the solutions $\phi_n(x)$ with $n < 0$; they do not represent states that are physically distinct from $\phi_n(x)$ with $n > 0$.

I can solve Eq. 8.16 for $\eta(y)$ with the boundary conditions (Eq. 8.22) in the same way that I solve for $\phi(x)$. The result is (compare to Eqs. 8.25 and 8.26)

$$\eta_j(y) = \sqrt{\frac{2}{L_y}} \sin\left(\frac{\pi j y}{L_y}\right), \quad j = 1, 2, 3, \ldots \tag{8.27}$$

and

$$E_y(j) = \frac{\hbar^2}{2m} \left(\frac{\pi j}{L_y}\right)^2, \quad j = 1, 2, 3, \ldots \tag{8.28}$$

Finally, Eq. 8.17 for $\lambda(z)$, with the boundary conditions Eq. 8.23 gives

$$\lambda_k(z) = \sqrt{\frac{2}{L_z}} \sin\left(\frac{\pi k z}{L_z}\right), \quad k = 1, 2, 3, \ldots \tag{8.29}$$

and

$$E_z(k) = \frac{\hbar^2}{2m} \left(\frac{\pi k}{L_z}\right)^2, \quad k = 1, 2, 3, \ldots \tag{8.30}$$

§10. *Putting Humpty-Dumpty Together Again.* We have solved the three eigenvalue problems, for $\phi(x)$, $\eta(y)$, and $\lambda(z)$. According to Eq. 8.13, the functions

$$\psi_{n,j,k}(x,y,z) = \left(\frac{8}{L_x L_y L_z}\right)^{1/2} \sin\left(\frac{\pi n x}{L_x}\right) \sin\left(\frac{\pi j y}{L_y}\right) \sin\left(\frac{\pi k z}{L_z}\right) \tag{8.31}$$

with

$$\left. \begin{array}{l} n \ = 1,2,3,\ldots \\[4pt] j \ = 1,2,3,\ldots \\[4pt] k \ = 1,2,3,\ldots \end{array} \right\} \tag{8.32}$$

are eigenfunctions of the Hamiltonian of a particle in a box.

The corresponding eigenvalues are (see Eq. 8.18)

$$E(n,j,k) = E_x(n) + E_y(j) + E_z(k)$$

$$= \frac{\hbar^2 \pi^2}{2m} \left[\left(\frac{n}{L_x}\right)^2 + \left(\frac{j}{L_y}\right)^2 + \left(\frac{k}{L_z}\right)^2 \right] \tag{8.33}$$

where n, j, and k take the values specified by (8.32).

The integers n, j, and k are called *the quantum numbers.* They specify the state of the system. The state $\{n,j,k\} = \{1,1,1\}$ is the ground state (the lowest energy). Its energy is obtained from Eq. 8.33 by making $n = 1, j = 1$, and $k = 1$. Its wave function is obtained from Eq. 8.31 by making $n = 1, j = 1$, and $k = 1$.

Exercise 8.2

(a) Show that Eq. 8.31 and Eq. 8.33 satisfy Eq. 8.8 for any numerical values of n, j, and k, not just integers. (b) Show that $\psi_{n,j,k}(x,y,z)$ satisfies the boundary conditions only if n, j, k are integers.

The Behavior of a Particle in a Box

§11. Let us see how we can use the eigenfunctions and the eigenvalues of the Hamiltonian to study the behavior of the particle in a box. I will examine the values that the energy can take and the probability of finding the particle at a specified place in the box. In Chapter 12 we examine how the particle adsorbs and emit light.

In Chapter 4, I showed that the energies observed in experiments must be eigenvalues of the Hamiltonian. Let us calculate the values of these energies, for a particle in a box, to see how large they are and how they depend on the parameters of the system. In a later chapter we will use these values to study light absorption and emission by a particle in a box.

§12. *The Ground State Energy is not Zero.* The state $\{n, j, k\} = \{1, 1, 1\}$ has the lowest energy,

$$E(1, 1, 1) = \frac{\hbar^2 \pi^2}{2m} \left[\frac{1}{L_x^2} + \frac{1}{L_y^2} + \frac{1}{L_z^2} \right] \tag{8.34}$$

A "classical mechanic" would find this result puzzling. The reasoning would be as follows. By the definition of the problem, the potential energy of the particle in the box is zero. The total energy E is therefore equal to the kinetic energy: $E = mv^2/2$. The particle has the lowest energy, $E = 0$, when it does not move (i.e. when $v = 0$).

Quantum mechanics says that this is not true: the lowest energy is given by Eq. 8.34 and it is *always larger than zero*. It goes towards zero only if *all sides of the box become infinite or if the mass of the particle becomes infinite*.

Since we never work in an infinite laboratory, nor do we handle objects with infinite mass, the lowest energy is never zero. If this is true (and it is), why was this not noticed before quantum mechanics was discovered?

Let us think about this. Suppose that ϵ is the smallest energy that the best available instrument can detect. If $E(1, 1, 1) < \epsilon$, it will appear that $E(1, 1, 1) = 0$. The relation

$$\frac{\hbar^2 \pi^2}{2m} \left[\frac{1}{L_x^2} + \frac{1}{L_y^2} + \frac{1}{L_z^2} \right] < \epsilon$$

gives the conditions under which the measurements will tell me that the lowest energy of the particle is equal to zero. Obviously, it will appear that classical mechanics is correct when either m, or all the dimensions L_x, L_y, and L_z are large. How large? It depends on $\hbar$ and ϵ.

Workbook

Let us look at an example. I take $m = 10^{-6}$ g and $L_x = L_y = L_z = 10{,}000$ Å $= 10^4 \times 10^{-8}$ cm $= 10^{-4}$ cm. Using $\hbar = 1.0546 \times 10^{-27}$ erg s gives (see Workbook QM8.1)

$$\frac{\hbar^2\pi^2}{2m}\left[\frac{1}{L_x^2} + \frac{1}{L_y^2} + \frac{1}{L_z^2}\right] = \frac{(1.0546 \times 10^{-27})^2\pi^2}{2 \times 10^{-6}}\left[\frac{3}{(10^{-4})^2}\right]$$

$$= 1.65 \times 10^{-39} \text{ erg} = 6.242 \times 10^{11} \times 1.65 \times 10^{-39} \text{ eV}$$

$$= 10.3 \times 10^{-28} \text{ eV}$$

(To convert erg to eV, use 1 erg $= 6.242 \times 10^{11}$ eV.) This energy is much too small to be detected by any available instrument. If I measured the energy of this system, I would conclude that it is zero, and think that classical mechanics is correct.

The dimensionless quantity that controls whether we see quantum or classical behavior is

$$\frac{\hbar^2\pi^2}{2mL^2\epsilon}$$

where L is the smallest dimension of the parallelepiped and ϵ is the accuracy with which we can measure the energy. If this quantity is greater than 1, the measurements will show that classical mechanics is invalid.

The fact that the lowest energy is not zero is not peculiar to a particle in a box. It is common to all systems consisting of particles confined to move within a finite volume. In all cases, this energy is higher if the confining volume or the mass of the particle is smaller. One of the strategies of nano-technology (a branch of chemical physics which is currently fashionable) is to make objects whose size is so small that the electrons in them have a pronounced quantum behavior. By manipulating the size of the object we can manipulate the energies of the electrons in it, the frequency of the light that the object can absorb and emit and many other properties.

§13. *Other Energies.* Let us calculate the energies $E(n,j,k)$ for other states, for an electron in a cubic box with $L_x = L_y = L_z = 100$ Å. I use $\hbar = 1.05457 \times 10^{-34}$ J s, m = 9.109×10^{-31} kg for the mass of the electron, and 1 Å = 10^{-10} m. The energy is (see Eq. 8.34)

$$E(n,j,k) = \frac{(1.05457 \times 10^{-34} \text{ J s})^2\pi^2}{2 \times 9.109 \times 10^{-31} \text{ kg } (100 \times 10^{-10})^2 \text{ m}^2} \times \left[n^2 + j^2 + k^2\right] \quad (8.35)$$

Table 8.1 The energies $E(n,j,k)$ in eV, from Eq. 8.35, for an electron in a cubic box of side $L = 100$ Å.

n	j	k	energy (eV)	n	j	k	energy (eV)	n	j	k	energy (eV)
1	1	1	0.01128	2	1	1	0.02256	3	1	1	0.04137
1	1	2	0.02256	2	1	2	0.03385	3	1	2	0.05265
1	1	3	0.04137	2	1	3	0.05265	3	1	3	0.07145
1	2	1	0.02256	2	2	1	0.03385	3	2	1	0.05265
1	2	2	0.03385	2	2	2	0.04513	3	2	2	0.06393
1	2	3	0.05265	2	2	3	0.06393	3	2	3	0.08274
1	3	1	0.04137	2	3	1	0.05265	3	3	1	0.07145
1	3	2	0.05265	2	3	2	0.06393	3	3	2	0.08274
1	3	3	0.07145	2	3	3	0.08274	3	3	3	0.10154

The units are

$$\frac{J^2 \, s^2}{kg \, m^2} = \frac{J^2}{\frac{kg \, m^2}{s^2}} = \frac{J^2}{J} = J$$

To obtain the energy in eV, which is a convenient unit, use the conversion factor $1\,J = 6.242 \times 10^{19}$ eV. The energies calculated from Eq. 8.35 (see Workbook QM8.2) are shown in Table 8.1.

Workbook

Exercise 8.3

Calculate the energies of an electron in a box with sides $L_x = 10$ Å, $L_y = 10$ Å, $L_z = 1000$ Å. Does this system behave as if it is one-dimensional? What exactly do I mean when I answer yes?

§14. *Degeneracy.* It is interesting to note that some of the states in Table 8.1 have the same energy. For example, the energy of the state $\{n,j,k\} = \{1,1,2\}$ is 0.02256 eV and this is equal to the energy of the state $\{n,j,k\} = \{1,2,1\}$, and to that of the state $\{n,j,k\} = \{2,1,1\}$. Since the energy of a particle in a cubic box is proportional to $n^2+j^2+k^2$, all states $\{n,j,k\}$ that give the same value for this expression have the same energy. According to the terminology introduced in Chapter 3, we say that these states (that have the same energy) are *degenerate*. Note, however, that the particular identity of degenerate states depends on the shape of the box.

Let us look more closely at the degenerate states $\{3,1,1\}$ and $\{1,3,1\}$ (you can see in Table 8.1 that these two states have the same energy) to find out how they differ from each other. They have the same energy, but their wave functions *are different*. The wave function of the state $\{3, 1, 1\}$ is (use Eq. 8.31 with $n = 3, j = 1$, and $k = 1$, and $L_x = L_y = L_z = L$):

$$\psi_{3,1,1}(x, y, z) = \left(\frac{8}{L^3}\right)^{1/2} \sin\left(\frac{3\pi x}{L}\right) \sin\left(\frac{\pi y}{L}\right) \sin\left(\frac{\pi z}{L}\right) \tag{8.36}$$

and that of the state $\{1, 2, 1\}$ is

$$\psi_{1,3,1}(x, y, z) = \left(\frac{8}{L^3}\right)^{1/2} \sin\left(\frac{\pi x}{L}\right) \sin\left(\frac{3\pi y}{L}\right) \sin\left(\frac{\pi z}{L}\right) \tag{8.37}$$

What is the physical meaning of the difference between the two eigenstates? Recall from your physics courses that the three-dimensional motion of a particle can be decomposed into three components along the directions x, y, and z. Let us calculate the kinetic energy corresponding to the motion in x-direction, for a particle in the state $\{3, 1, 1\}$, and for a particle in the state $\{1, 3, 1\}$. The kinetic energy operator $\hat{K}_x$ for motion in the x-direction is (see Chapter 3):

$$\hat{K}_x = -\frac{\hbar^2}{2m}\frac{\partial^2}{\partial x^2} \tag{8.38}$$

It is very easy to see (just use Eq. 8.36 for $\psi_{3,1,1}(x, y, z)$ and take the derivatives required by the expression given below) that

$$\hat{K}_x \psi_{3,1,1}(x, y, z) = -\frac{\hbar^2}{2m}\frac{\partial^2 \psi_{3,1,1}(x, y, z)}{\partial x^2} = \frac{3\hbar^2 \pi^2}{2mL^2} \psi_{3,1,1}(x, y, z) \tag{8.39}$$

This equation tells us that the state $\psi_{3,1,1}(x, y, z)$ is an eigenstate of $\hat{K}_x$ with the eigenvalue

$$\frac{3\hbar^2 \pi^2}{2mL^2} \tag{8.40}$$

A similar calculation shows that $\psi_{1,3,1}(x, y, z)$ is an eigenstate of $\hat{K}_x$ with the eigenvalue

$$\frac{\hbar^2 \pi^2}{2mL^2} \tag{8.41}$$

Thus, the two states differ through their kinetic energy in the x-direction.

The same type of calculation (please perform it) shows that the kinetic energy in the y-direction for the state $\psi_{3,1,1}(x,y,z)$ is given by Eq. 8.41, while the kinetic energy in the y-direction of the state $\psi_{1,3,1}(x,y,z)$ is given by Eq. 8.40.

Thus, the particle in state $\psi_{3,1,1}(x,y,z)$ moves fast in the x-direction and slow in the y-direction; in the state $\psi_{1,3,1}(x,y,z)$ the situation is reversed. The two states have *the same total kinetic energy but differ in the way this energy is distributed between the directions x and y.*

Why do we consider these states to be different? Since the box *is cubic*, I can rotate the box so that, after the rotation, the x-direction becomes the y-direction and vice versa. Such a rotation turns the state $\psi_{3,1,1}(x,y,z)$ into $\psi_{1,3,1}(x,y,z)$ and vice versa. Should I consider that states which differ only through the fact that the box is rotated are physically different? The answer is yes, and here is the reason why. In Chapter 5, I told you that a system in an excited state ψ_i, of energy E_i, will spontaneously emit light, to go to a state ψ_f having a lower energy E_f. The frequency of the emitted photon is $\omega = (E_i - E_f)/\hbar$.

Let us apply this knowledge to the states $\psi_{3,1,1}$ and $\psi_{1,3,1}$. Suppose that each will emit a photon and go to the state $\{1,1,1\}$. Since the energies $E_{3,1,1}$ and $E_{1,3,1}$ are equal, the photons emitted by a particle in the state $\{3,1,1\}$ and the one emitted from a particle in the state $\{1,3,1\}$ *will have the same frequency* $\omega = [E(3,1,1) - E(1,1,1)]/\hbar = [E(1,3,1) - E(1,1,1)]/\hbar$. We cannot distinguish these states from each other by measuring the frequency of the emitted photon.

However, quantum mechanics can be used to show that the two states will emit the photon in *different directions*. I can find out whether the initial state was $\{3,1,1\}$ or $\{1,3,1\}$ by measuring the direction of photon emission with respect to the position of the box. Therefore, these two states are physically different (meaning that I can make a measurement to tell which is which).

In almost all practical situations when two states are degenerate, they have the same energy but differ through some other measurable properties.

§15. *Degeneracy is Related to the Symmetry of the System.* In the calculations performed above I have studied a *cubic box* and found that the states $\{2,1,1\}$, $\{1,2,1\}$, and $\{1,1,2\}$ are degenerate. Let us check what happens to the energy of these states if the box is a parallelepiped and $L_x \neq L_y \neq L_z$. In this case the energies are (use Eq. 8.33)

$$E(2,1,1) = \frac{\hbar^2\pi^2}{2m}\left[\left(\frac{2}{L_x}\right)^2 + \left(\frac{1}{L_y}\right)^2 + \left(\frac{1}{L_z}\right)^2\right]$$

$$E(1,2,1) = \frac{\hbar^2 \pi^2}{2m} \left[\left(\frac{1}{L_x} \right)^2 + \left(\frac{2}{L_y} \right)^2 + \left(\frac{1}{L_z} \right)^2 \right]$$

and

$$E(1,1,2) = \frac{\hbar^2 \pi^2}{2m} \left[\left(\frac{1}{L_x} \right)^2 + \left(\frac{1}{L_y} \right)^2 + \left(\frac{2}{L_z} \right)^2 \right]$$

If you inspect these equations you will convince yourself that:

- as long as $L_x \neq L_y \neq L_z$ we have $E_{2,1,1} \neq E_{1,2,1} \neq E_{1,1,2}$, and so the states $\{2,1,1\}$, $\{1,2,1\}$ and $\{1,1,2\}$ *are not degenerate*;
- if $L_x = L_y = L_z$ the states $\{2,1,1\}$, $\{1,2,1\}$ and $\{1,1,2\}$ are degenerate;
- if $L_x = Ly \neq L_z$ the states $\{2,1,1\}$ and $\{1,2,1\}$ are degenerate, but the state $\{1,1,2\}$ has a different energy.

Clearly the degeneracy is related to the symmetry of the box. Breaking the symmetry (e.g. by making $L_x \neq L_y$) removes the degeneracy of the states $\{2,1,1\}$ and $\{1,2,1\}$.

This connection between degeneracy and the symmetry of the physical object whose states are being examined is general: if there is degeneracy there is some symmetry in the system and vice versa. It is very rare, but not impossible, to have degeneracy accidentally, unrelated to symmetry.

Exercise 8.4

Calculate the energies of an electron enclosed in a box with $L_x = 10$ Å, $L_y = 20$ Å, and $L_z = 200$ Å, for $n = 1,2,3$, $j = 1,2,3$, and $k = 1,2,3$. Verify that none of these states is degenerate (see Workbook QM8.3).

Exercise 8.5

Perform the same calculation as in the previous exercise, but use $L_x = 20$ Å, $L_y = 20$ Å, and $L_z = 200$ Å. Is there any degeneracy? Before doing the calculation, try to guess which states will be degenerate.

§16. *The Eigenfunctions are Normalized.* In Chapter 3 I told you that only eigen-functions that are normalized are physically meaningful and that the eigenfunctions of the operators encountered in quantum mechanics are orthogonal to each other. Let us see if the energy eigenstates of a particle in a box (Eq. 8.31) have these properties.

I remind you that a function is normalized if

$$\langle n,j,k|n,j,k\rangle \equiv \int_{-\infty}^{+\infty} dx \int_{-\infty}^{+\infty} dy \times \int_{-\infty}^{+\infty} dz\, \psi_{n,j,k}^*(x,y,z)\psi_{n,j,k}(x,y,z) = 1$$

(8.42)

The first equality defines the symbol $\langle n,j,k|n,j,k\rangle$, called the scalar product of $\psi_{n,j,k}$ with itself.

To check whether the eigenstates Eq. 8.42 are normalized I insert the function $\psi_{n,j,k}(x,y,z)$, given by Eq. 8.31, into Eq. 8.42. This leads to

$$\langle n,j,k|n,j,k\rangle = \left[\frac{2}{L_x}\int_0^{L_x}\sin^2\left(\frac{n\pi x}{L_x}\right)dx\right]$$

$$\times\left[\frac{2}{L_y}\int_0^{L_y}\sin^2\left(\frac{j\pi y}{L_y}\right)dy\right]$$

$$\times\left[\frac{2}{L_z}\int_0^{L_z}\sin^2\left(\frac{k\pi z}{L_z}\right)dz\right] \quad (8.43)$$

An explanation is in order. The integrals in Eq. 8.42 are from $-\infty$ to $+\infty$. However, the function $\psi_{n,j,k}(x,y,z)$ differs from zero only if $0 \leq x \leq L_x$ and $0 \leq y \leq L_y$ and $0 \leq z \leq L_z$, when the particle is inside the box. This is why the integrals in Eq. 8.43 are from 0 to L_x, 0 to L_y, and 0 to L_z.

Because the $\psi_{n,j,k}(x,y,z)$ is a function of x, times a function of y, times a function of z (see Eq. 8.31), the integral in Eq. 8.42 is the product of three single integrals. This is fortunate because one-dimensional integrals are easier to evaluate: they give (see Workbook QM8.4)

$$\langle n|n\rangle \equiv \frac{2}{L_x}\int_0^{L_x}\sin^2\left(\frac{n\pi x}{L_x}\right)dx = 1 - \frac{\sin(2\pi n)}{2\pi n} = 1 \quad (8.44)$$

The last equality follows because n is a positive integer and $\sin(2\pi n) = 0$. The same is true for the other two integrals in Eq. 8.43. This means that

$$\langle n,j,k|n,j,k\rangle = \langle n|n\rangle \langle j|j\rangle \langle k|k\rangle = 1 \qquad (8.45)$$

This is what I expected: the eigenfunctions are normalized.

There is no reason to expect the eigenfunctions of an operator to be normalized. Presently they were normalized because I have normalized them by using the procedure described in Chapter 3. That procedure is general and we can always normalize the eigenstates of an operator representing an observable.

§17. *Orthogonality.* In Chapter 3 I told you that the eigenfunctions of the operators corresponding to measurable quantities (such as energy) are orthogonal. This means that

$$\langle n_1,j_1,k_1|n_2,j_2,k_2\rangle \equiv \int_{-\infty}^{+\infty} dx\,dy\,dz\,\psi^*_{n_1,j_1,k_1}(x,y,z)\psi_{n_2,j_2,k_2}(x,y,z) = 0 \quad (8.46)$$

if $\{n_1,j_1,k_1\}$ and $\{n_2,j_2,k_2\}$ are different states, that is, if at least one of the inequalities $n_1 \neq n_2, j_1 \neq j_2$, or $k_1 \neq k_2$ is correct. Using Eq. 8.31 for the wave functions leads to

$$\langle n_1,j_1,k_1|n_2,j_2,k_2\rangle = \frac{2}{L_x} \int_0^{L_x} dx\,\sin\left(\frac{n_1\pi x}{L_x}\right)\sin\left(\frac{n_2\pi x}{L_x}\right)$$

$$\times \frac{2}{L_y} \int_0^{L_y} dy\,\sin\left(\frac{j_1\pi y}{L_y}\right)\sin\left(\frac{j_2\pi y}{L_y}\right)$$

$$\times \frac{2}{L_z} \int_0^{L_z} dz\,\sin\left(\frac{k_1\pi z}{L_z}\right)\sin\left(\frac{k_2\pi z}{L_z}\right) \qquad (8.47)$$

Each single integral in Eq. 8.47 is easy to evaluate (see Workbook QM8.4):

$$\langle n_1|n_2\rangle \equiv \int_0^{L_x} dx\,\frac{2}{L_x}\sin\left(\frac{n_1\pi x}{L_x}\right)\sin\left(\frac{n_2\pi x}{L_x}\right)$$

$$= \frac{2}{L_x}\left[\frac{L_x\sin((n_1 - n_2)\pi)}{2(n_1 - n_2)\pi} - \frac{L_x\sin((n_1 + n_2)\pi)}{2(n_1 + n_2)\pi}\right] \qquad (8.48)$$

If $n_1 \neq n_2$, both sine functions in the right-hand side of the last equality are zero (since $\sin(\text{integer} \times \pi) = 0$); this means that $\langle n_1 | n_2 \rangle = 0$ if $n_1 \neq n_2$.

Combining Eqs. 8.44 and 8.48 leads to the following result:

$$\langle n_1 | n_2 \rangle \equiv \frac{2}{L_x} \int_0^{L_x} dx \, \sin\left(\frac{n_1 \pi x}{L_x}\right) \sin\left(\frac{n_2 \pi x}{L_x}\right) = \delta_{n_1, n_2} \qquad (8.49)$$

The symbol δ_{n_1, n_2} is called the Kronecker delta and has the definition (for integers ℓ and m)

$$\delta_{\ell, m} = \begin{cases} 0 & \text{if } \ell \neq m \\ 1 & \text{if } \ell = m \end{cases}$$

The other two integrals in Eq. 8.47 can be treated similarly, and as a result Eq. 8.47 becomes

$$\langle n_1, j_1, k_1 | n_2, j_2, k_2 \rangle = \langle n_1 | n_2 \rangle \langle j_1 | j_2 \rangle \langle k_1 | k_2 \rangle = \delta_{n_1, n_2} \delta_{j_1, j_2} \delta_{k_1, k_2} \qquad (8.50)$$

If $n_1 \neq n_2$ or $j_1 \neq j_2$ or $k_1 \neq k_2$, then the integral $\langle n_1, j_1, k_1 | n_2, j_2, k_2 \rangle$ is zero; the eigenfunctions are orthogonal.

§18. *The Position of a Particle in a Given State.* Let us assume that we carried out an experiment that left the particle in the state $\{n, j, k\}$ (this means that its wave function is $\psi_{n,j,k}(x, y, z)$ and its energy is $E(n, j, k)$). What is the position of the particle inside the box?

Quantum theory says that the best we can do is to specify the probability that the particle is located in a volume element $dx \, dy \, dz$, centered on a point (x, y, z). This probability is

$$p_{n,j,k}(x, y, z) \, dx \, dy \, dz = \psi_{n,j,k}^*(x, y, z) \, \psi_{n,j,k}(x, y, z) \, dx \, dy \, dz \qquad (8.51)$$

The function $p_{n,j,k}(x, y, z)$ is called the distribution function of the coordinates.

For the particle in the box, $\psi_{n,j,k}$ is given by Eq. 8.31, so

$$p(x, y, z) \, dx \, dy \, dz = \frac{8}{L_z L_y L_z} \sin^2\left(\frac{n \pi x}{L_x}\right) \sin^2\left(\frac{j \pi y}{L_y}\right)$$

$$\times \sin^2\left(\frac{k \pi z}{L_z}\right) dx \, dy \, dz \qquad (8.52)$$

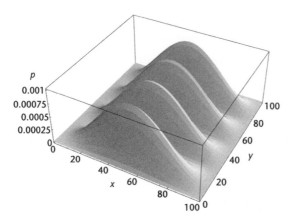

Figure 8.2 The probability $p_{1,4,1}(x, y, z = 50$ Å$)$ plotted as a function of x, y (see Workbook QM8.5) .

Figures 8.2 and 8.3 show plots of the functions $p_{1,4,1}(x,y,z = 50$ Å$)$ and $p_{2,3,1}(x,y,z = 50$ Å$)$ for a box with $L_x = L_y = L_z = 100$ Å. Because we do not know how to plot functions of three variables, I took $z = 50$ Å to turn $\psi_{1,4,1}(x,y,z)$ into a function of two variables. The plot tells me the probability of finding the particle at an arbitrary point of coordinates $\{x,y\}$, in the plane $z = 50$ Å.

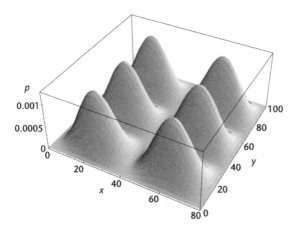

Figure 8.3 The probability $p_{2,3,1}(x, y, z = 50$ Å$, \delta = 5$ Å$)$ plotted as a function of x, y .

For someone who believes in classical mechanics, these graphs are very disturbing. The distribution function has peaks in regions where the particle is most likely to be located, and has valleys at regions that the particle avoids. This makes no sense classically. No force acts on the particle inside the box. Why would a particle prefer a certain region in space if there is no force to push it there?

You may have heard that in quantum mechanics an electron or an atom behaves sometimes like a wave and sometimes like a particle. The behavior of $p_{n,j,k}(x, y, z)$ is typical of a wave. You may have performed the following experiment in the physics laboratory. You had a cylinder on whose walls you put a light powder. Then you produced a sound wave in the cylinder and found that the powder accumulates in some regions on the wall of the tube and is removed from others. The pattern you obtained was similar to the one seen in Figs 8.2 and 8.3. This happens because the equation for the amplitude of the sound wave is very similar to the Schrödinger equation. The energy of the sound is given by its amplitude squared, just like the probability distribution is given by the wave function squared.

9

LIGHT EMISSION AND ABSORPTION: THE PHENOMENA

Introduction

§1. The next four chapters discuss the experiment shown schematically in Fig. 9.1. The source at the left produces light of known frequency Ω. We measure how the intensity of the light passing through the sample (to reach the detector) changes when we change the frequency of the incident light. When the incident light has certain frequencies, the intensity of the light reaching the detector is substantially diminished. We say that at these frequencies the molecules in the sample *absorb* light.

The effect of light absorption on the absorbing molecule depends on the light frequency. Microwave and infrared (IR) radiation heat the sample (think of the microwave oven or a glass of water kept near a fireplace); visible and lower-frequency ultraviolet (UV) light can cause chemical reactions and heating; higher-frequency UV light and X-ray radiation will eject electrons from the molecules, produce heat, and sometimes cause chemical reactions.

Figure 9.1 A schematic representation of an absorption experiment.

But this is not all. After absorbing a photon the molecule can also emit a photon of the same or a different frequency. Moreover, some photons go through the sample without harm to the molecules or to themselves.

We can determine what kind of molecules we have in a sample from the frequencies at which they absorb light. This is why *absorption spectroscopy* is a very common tool in analytical chemistry.

Molecules excited by photon absorption, or by other means (e.g. heat, collision with electrons or ions), emit photons whose frequency depends on the chemical nature of the molecule. *Emission spectroscopy* uses the frequency of the emitted light to identify the nature of the emitting molecules.

The amount of light absorbed or emitted is proportional to the concentration of the molecule in the sample. This is why emission or absorption spectroscopy is often used to determine reaction rates.

This chapter explains the phenomena observed during photon absorption and emission and gives a qualitative description of the molecular processes causing them. Chapter 10 examines a phenomenological theory which uses the methods of chemical kinetics to describe the rates of emission and absorption of light. This theory shows that the ability of a molecule to emit or absorb light is characterized by one quantity: Einstein's B coefficient. Chapter 11 presents the results of quantum theory of light absorption and emission. This connects the rate of absorption and emission to properties of the molecule. This theory is used in Chapter 12 to examine light absorption and emission by an electron in a quantum dot.

Light Absorption and Emission: the Phenomena

§2. *An Absorption Experiment.* There are many ways of performing an absorption experiment and I describe here a generic set-up which displays the essential phenomena, common to all other implementations.

A beam of light of frequency Ω and intensity $I_0(\Omega)$ is sent through a liquid (or it could be a gas or a solid) contained in a vessel with transparent walls (see Fig. 9.1).

I measure the intensity $I(\Omega)$ of the light that passes through the sample. A plot of

$$a(\Omega) = \ln\left(\frac{I(\Omega)}{I_0(\Omega)}\right) \tag{9.1}$$

versus Ω is the *transmission spectrum* of the sample.

The important features of a transmission spectrum are the dips that occur when the incident light has certain frequencies. A schematic spectrum is shown in Fig. 9.2. The number of dips would have been larger had I used a wider frequency range. The dips here are well separated, but this is not always the case: sometimes two or more dips overlap.

Most people prefer to plot the *absorption spectrum*, which is the transmission spectrum multiplied by -1. The absorption spectrum has peaks where the transmission spectrum has dips.

§3. *Characterization of an Absorption Spectrum.* An absorption spectrum is characterized by several quantities. The *peak frequencies*, or the *absorption frequencies*, are the frequencies at which the spectrum has maxima. The *peak intensities* or *absorption intensities* are the maximum heights of the peaks. Finally, each peak is characterized by a *line width* which tells us how broad the peak is.

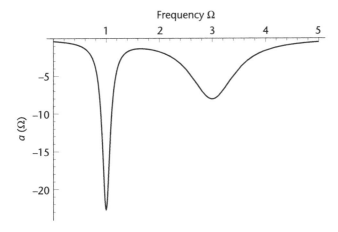

Figure 9.2 Schematic representation of $a(\Omega) = \ln(I(\Omega)/I_0(\Omega))$ as a function of the frequency Ω. The scales on the axes are arbitrary.

Not all peaks have the simple form shown in Fig. 9.2 and sometimes we cannot characterize them by one width. In such cases, we talk about the *line shape*.

The peak frequencies are specific to the molecules in the sample. Analytical chemists routinely use absorption spectra for identifying molecules.

The peak intensities are proportional to the concentration of the molecules in the sample. They are used in chemical kinetics to monitor how the concentration of a specific molecule changes during a reaction. If a compound is consumed in a reaction, the peak intensities in its absorption spectrum decay. By monitoring how fast a peak intensity changes, we can determine the rate of the reaction.

All absorption peaks have a width. There are many processes that "broaden" a spectrum or give it a certain shape. Their theory is rather involved and we do not discuss it here. Most chemists will measure the peak intensity and peak frequency. For them the peak width is a nuisance: it lowers the signal and sometimes makes the peaks overlap, making it hard to determine what the peak frequency is. Several experimental procedures have been designed to "narrow" the peaks in an absorption spectrum, but discussing them is beyond the scope of this book.

Often an absorption spectrum can be fitted by sums of simple functions called Lorentzians:

$$\sum_{i=1}^{n} \ell_i(\Omega) = \sum_{i=1}^{n} \frac{\mu_i}{(\Omega - \omega_i)^2 + \Gamma_i^2} \tag{9.2}$$

Each term in the sum generates a peak when the light frequency Ω matches one of the *peak frequencies* ω_i. The parameter Γ_i controls the width of a peak and μ_i/Γ_i^2 the peak intensity.

Exercise 9.1

Pick up some values for μ, Γ, and ω and plot an absorption spectrum that has three peaks. (a) Choose values of the parameters for which the three peaks are well separated. (b) Find values of these parameters for which the peaks overlap and see what the spectrum looks like.

§4. *Why is the Transmitted Intensity Low at Certain Frequencies?* A dip in the transmission spectrum, at a given frequency, indicates that not all the light entering the sample passes through it. What events take place in the sample to cause this?

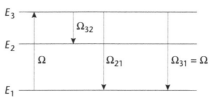

Figure 9.3 Example of an energy-level diagram and the emission and absorption processes.

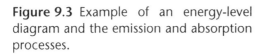

I was telling you in Chapter 4 that when a molecule absorbs light, its energy increases (the molecule is *excited*). The life of a large molecule that has swallowed a photon is rather complicated. What follows is a simplified description of the events triggered by photon absorption.

Let us assume that the molecules in the sample have the energy-level diagram shown in Fig. 9.3. We set the frequency of light Ω to equal $(E_3 - E_1)/\hbar$ and some molecules in the sample will absorb some of the photons streaming through. These photons are removed from the light beam.

A molecule absorbing a photon resembles a dog that has been chasing cars all his life and finally caught one; it will let go as soon as it has a chance. The excited molecules have several ways of getting rid of the energy acquired by absorbing a photon.

One is to undergo a transition from E_3 to E_1 and emit a photon of frequency Ω. The absorbed photon is born again but it is emitted in various directions, with various probabilities: some of them may reach the detector but most of them do not. The intensity of light reaching the detector is *diminished* by this absorption–emission process; the photons of frequency Ω are not destroyed, but are redirected. We call this process *light scattering*. This is not true photon absorption, even though it diminishes the amount of light passing through the sample.

The molecule excited to the third level can also emit photons of frequency $\Omega_{32} = (E_3 - E_2)/\hbar$ (see Fig. 9.3). This process also diminishes the intensity of the light of frequency Ω passing through the sample and reaching the detector; photons of frequency Ω are destroyed and an equal number of photons of frequency Ω_{32} are created.

The molecules that emitted the photon of frequency Ω_{32} "land" in the state of energy E_2, from which they can emit a photon of frequency $\Omega_{21} = (E_2 - E_1)\hbar$. The molecules doing this end with energy E_1 and have no other place to go; they are at the bottom of the barrel, in the lowest energy state. They are now ready to absorb again a photon of frequency Ω.

All processes discussed so far are *radiative*: they involve photon absorption or emission. If you think a little you will realize that no photon energy is lost if only these processes were to take place in the sample. The photons are "split" and/or redirected, but the total energy of the radiation coming out of the sample (in all directions) is the same as that going in. Therefore, some people like to say that these processes are not *true absorption processes*.

True absorption occurs through either *photochemistry* or through a *radiationless transition*. Photochemistry is easy to understand. For example, if we illuminate CH_3I with light of appropriate frequency, the molecule will absorb a photon and then break up into CH_3 and I. The fragments formed in this way have a large kinetic energy (about 1 eV) and CH_3 is vibrationally excited. They collide with other molecules in the system and lose their excess energy, which is spread around as heat. Photochemistry changes the chemical composition of the system, produces heat, and lowers the number of photons coming out of the sample.

Radiationless transitions are processes in which an excited molecule loses energy by giving it to the other molecules in the system. Ultimately, the energy $\hbar\Omega$ of the absorbed photon is turned into heat. There are many kinds of radiationless processes, and we do not have time to examine them here.

§5. *Emission Spectroscopy.* As you have seen above, molecules having the energy level diagram shown in Fig. 9.3 can emit light of frequency Ω_{32} or Ω_{21}. I am sure that you realize that this property is general. If you excite a state of high energy the molecule can emit light by performing transitions to various lower energy levels, until the molecule reaches the ground state.

Exercise 9.2

A molecule that has four energy levels E_1, E_2, E_3 and E_4 is exposed to light of frequency $\Omega = (E_4 - E_1)/\hbar$. Enumerate the frequencies of all photons that can be emitted by this system.

We can disperse the light emitted by the sample, with a prism or a grating. This means that the light coming out of the sample is split into several beams, one for each frequency, which propagate in different directions (see Fig. 9.4). By moving the detector, as shown in the figure, we can measure the intensity of the emitted light having various frequencies.

Let us denote this intensity by $I(\Omega, \bar{\Omega})$. The notation reminds me that the detected intensity depends on both the incident frequency Ω and the frequency $\bar{\Omega}$ of the

Figure 9.4 Light, of three frequencies, that emerges from the sample is split by a prism into three beams moving in different directions. By moving the detector D in the direction shown by the arrow we can measure the intensity of the three beams. The distance the detector moves is related to the frequency of light. The detector will find three peaks, corresponding to the intensity of light of the three frequencies.

emitted radiation. It is customary, in emission spectroscopy, to report the quantity

$$e(\Omega, \bar{\Omega}) = \ln\left(\frac{I(\bar{\Omega})}{I_0(\Omega)}\right) \tag{9.3}$$

Since $e(\Omega, \bar{\Omega})$ is a function of two variables, there are two ways of performing emission measurements. In one we keep the incident frequency Ω constant and measure the dependence of $e(\Omega, \bar{\Omega})$ on $\bar{\Omega}$. This is called an *emission spectrum*. For example, for the three-level system shown in Fig. 9.3, if the incident light has the frequency $\Omega = (E_3 - E_1)/\hbar$, the emission spectrum will have three peaks at the frequencies $\bar{\Omega} = (E_3 - E_2)/\hbar$, $\bar{\Omega} = (E_3 - E_1)/\hbar$, and $\bar{\Omega} = (E_2 - E_1)/\hbar$.

Exercise 9.3

A molecule has six levels of energies $E_1 < E_2 \cdots < E_6$.

1. What is the maximum number of peaks you expect in the emission spectrum, if the incident frequency is $\Omega = (E_6 - E_0)/\hbar$, and what will their frequency be?

2. Answer the same question in the case that the incident frequency is $\Omega = (E_4 - E_1)/\hbar$.

It is also possible to place the detector in a position in which it receives only light of one frequency $\bar{\Omega}$ and measure the emitted intensity as a function of the frequency of the incident laser. Such a spectrum is called an *excitation spectrum*.

For example, if the molecule has seven energy levels, we can set the detector to measure only the intensity of the radiation of frequency $\bar{\Omega} = (E_3 - E_1)/\hbar$. Let us see what the excitation spectrum will look like. As long as the incident frequency is less than $(E_4 - E_1)/\hbar$ our detector will give no signal, except for the case $\Omega = (E_3 - E_1)/\hbar$ when the incident light passing through the sample is detected. When $\Omega = (E_4 - E_1)/\hbar$ the molecule can emit a photon of frequency $\bar{\Omega} = (E_4 - E_3)/\hbar$ (which will not be detected) and then emit a photon of frequency $\bar{\Omega} = (E_3 - E_1)/\hbar$ (which is detected). These two processes are shown schematically, as a, in the energy level diagram of Fig. 9.5. The same figure shows that no signal is detected as the frequency of the incident light is between $\Omega = (E_4 - E_1)/\hbar$ and $\Omega = (E_5 - E_1)/\hbar$; if there is no absorption, there is no emission. When the incident frequency equals $\Omega = (E_5 - E_1)/\hbar$, light of frequency $\bar{\Omega} = (E_3 - E_1)/\hbar$ is emitted (and detected) again, etc.

In general, the emission spectrum, measured at the emission frequency $\bar{\Omega} = (E_3 - E_1)/\hbar$, has peaks every time the incident light resonates with an energy level whose energy exceed E_3. If you think a bit about it, this is the absorption spectrum, "seen through the eyes" of the emission from level E_3. However, the information is different from that provided by the ordinary absorption spectrum. The absorption spectrum depends on the efficiency of the processes indicated by solid arrows in Fig. 9.5; the emission spectrum describes the joint efficiency of the processes shown by solid *and* dotted lines.

We have many means of exciting molecules: we can bombard them with electrons, we can force them to absorb light, we can place them in an electric discharge (like

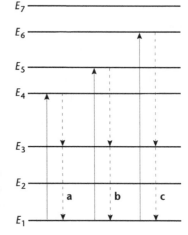

Figure 9.5 The processes leading to the emission of light of frequency $\bar{\Omega} = (E_3 - E_1)/\hbar$, for three excitation frequencies: **a** shows the case when the incident frequency is $\Omega = (E_4 - E_1)/\hbar$, **b** when $\Omega = (E_5 - E_1)\hbar$, and **c** when $\Omega = (E_6 - E_1)\hbar$.

in a neon lamp) or we can heat them to high temperatures. All these processes are used in emission spectroscopy to excite the molecules and analyze the frequency of the emitted light.

The frequencies at which peaks appear in the emission or the excitation spectrum are typical of a given atom or molecule. Because of this, emission spectroscopy is widely used for chemical analysis. For example, if your doctor wants to know if you have mercury poisoning, he might vaporize a few strands of your hair, with a powerful laser. The evaporated material is chemically decomposed. Hell is a cool place compared to a place hit by a powerful laser. Excited Hg atoms are formed and they emit light of characteristic frequencies. The presence of this emission establishes the presence of mercury. The intensity of the emission tells how much mercury is present.

The concepts of peak frequency, line width and emission intensity, in emission spectroscopy, are very similar to those defined for absorption spectroscopy and require no further elaboration.

§6. *How General is This?* The processes described here are very general. One can perform absorption and emission spectroscopy with any kind of radiation, ranging from radio waves to X-rays and even gamma rays. Table 9.1 gives the frequencies

Table 9.1 The names of radiation of different energies. The energy is given in several units used by spectroscopists. The wavelength λ and the wave vector k are given for the photon in vacuum (see §7 for the definition of these quantities). The meanings of the prefixes E, P, T, G, M, n, μ are given in Table 9.2.

	Wavelength, λ	Wave vector, k	Frequency, v	Energy, hv	Energy, hv
γ rays	10 pm	10^9 cm^{-1}	30.0 EHz	19.9×10^{-15} J	1.242×10^5 eV
X-rays	10 nm	10^6 cm^{-1}	30.0 PHz	19.9×10^{-18} J	124.206 eV
Vacuum UV	200 nm	50.0×10^3 cm^{-1}	1.50 PHz	993×10^{-21} J	6.198 eV
Near UV	380 nm	26.3×10^3 cm^{-1}	789 THz	523×10^{-21} J	3.264 eV
Visible	780 nm	12.8×10^3 cm^{-1}	384 THz	255×10^{-21} J	1.591 eV
Near IR	2.5 μm	4.00×10^3 cm^{-1}	120 THz	79.5×10^{-21} J	0.496 eV
Mid IR	50 μm	200 cm^{-1}	6.00 THz	3.98×10^{-21} J	0.0248 eV
Far IR	1 mm	10 cm^{-1}	300 GHz	199×10^{-24} J	1.242×10^{-3} eV
Microwaves	100 mm	0.1 cm^{-1}	3.00 GHz	1.99×10^{-24} J	1.242×10^{-5} eV
Radio waves	1000 mm	0.01cm^{-1}	300 MHz	0.199×10^{-24} J	1.242×10^{-6} eV

range of various kinds of radiation, which is classified according to the energy of the photon in it. The frequencies are given in a variety of energy units used by spectroscopists.

No matter what kind of radiation is used, the outcome is the same: when radiation passes through a sample, some of it is lost. Radiation in different frequency ranges has different effects on the absorbing molecule. Gamma rays excite the internal motion of the nuclei while X-rays will ionize the molecule by knocking an electron from one of the constituent atoms. High-energy UV radiation ionizes the molecule by ejecting either a valence electron or a core electron. Low-energy UV and visible light excite the electrons, and this will often cause the molecule to undergo a chemical reaction. Near-IR photons excite the vibrational motion of the atoms in molecules, far IR and microwaves excite molecular rotations, and radio waves excite the twisting and turning of the molecules in a liquid or solid.

§7. *Units.* In the past, and even today, many people were experts in one kind of spectroscopy and performed experiments in a narrow frequency range. They developed distinct units for the photon energy in the range they were working on. Understanding where these units come from and working with them is a nightmare. Unfortunately, this madness has some rationality at its core: each unit is most convenient in its particular range. So, these units are likely to persist in your lifetime and you will have to learn how to use them.

The practitioners of all branches of science tend to use those units in which the numbers they work with are simple. For example, if you spend your life doing IR (infrared) spectroscopy, you would rather use cm^{-1} as a frequency (or energy) unit. Telling someone to set the frequency to 200 cm^{-1} seems a lot better than telling him to produce photons having an energy of 3.98×10^{-21} J.

To explain how these units were defined I have to review a few equations pertaining to the energy, the *wave vector*, and the *wavelength* of light in *vacuum*.

With modern notation the energy E of a photon of frequency Ω is

$$E = \hbar\Omega \tag{9.4}$$

where $\hbar$ is the "new" Planck constant

$$\hbar = 1.054572 \times 10^{-34} \text{ Js} \tag{9.5}$$

(J = joule).

From Maxwell's theory of electromagnetic radiation we know that

$$\Omega = kc \tag{9.6}$$

where c is the speed of light

$$c = 299,794,458 \text{ m/s} \tag{9.7}$$

and k is its wave vector. In turn,

$$k = 2\pi/\lambda \tag{9.8}$$

where λ is the wavelength of light.

Because of these formulae, if you know one of the quantities E, Ω, k or λ, you can calculate all others. This is what started the unit proliferation. In the SI system, E is given in joule, Ω in hertz (this is s^{-1}), λ in meter, k in m^{-1}. When I tell you that a photon has an energy of 10^4 m^{-1}, I am giving you the magnitude of the wave vector k. From Eqs 9.4–9.8 you can calculate the values of E, Ω, and λ corresponding to this value of k.

Giving the energy by giving Ω or k or λ is bad enough. Things are made worse by the fact that the nomenclature was developed during the days of "old" quantum theory. At that time people used the "old" Planck constant

$$h = 2\pi\hbar \tag{9.9}$$

The photon energy was written as

$$E = h\nu \tag{9.10}$$

where ν might be called the "old" frequency of light. If you compare Eq. 9.4 with Eq. 9.10, you get

$$E = h\nu = \hbar\Omega \tag{9.11}$$

Using $\hbar = h/2\pi$ from Eq. 9.11 gives

$$\nu = \frac{\Omega}{2\pi} \tag{9.12}$$

This connects the "old" frequency ν to the "new" frequency Ω.

From Eqs 9.6 and 9.12 I get

$$\nu = \frac{kc}{2\pi} = \frac{2\pi}{\lambda}\frac{c}{2\pi} = \frac{c}{\lambda} \tag{9.13}$$

To get the second equality I used Eq. 9.8 for k. Eq. 9.13 connects the "new" frequency ν to the wavelength λ. Finally, to make life difficult the wave number $\bar{\nu}$ was defined through

$$\bar{\nu} = \frac{1}{\lambda} \tag{9.14}$$

Eqs 9.9, 9.10, and 9.14 connect E, ν, $\bar{\nu}$, and λ to each other. You can specify the energy of the photon by giving E or ν or $\bar{\nu}$ or λ.

§8. *How to Convert from One Unit to Another.* Now I can show you how the system of units works. In Table 9.1 the vacuum UV range is given as 200 nm. The letters nm stands for nanometer.

Before I go on I will remind you some of the prefixes used with units, such as nano, tera, femto, etc. They are listed in Table 9.2. By using this table I can tell that 200 nm means 200×10^{-9} m. Since this quantity is a length, it is the wavelength λ of the photon. From Eq. 9.13 I can calculate the "old" frequency ν:

$$\nu = \frac{c}{\lambda} = \frac{2.9979458 \times 10^8 \text{ m/s}}{200 \times 10^{-9} \text{ m}}$$

$$= 1.49897 \times 10^{15} \text{ s}^{-1} = 1.4987 \text{ PHz}$$

Table 9.2 Prefixes used with units. For example, nm is a nanometer, fs is a femtosecond, and THz is terahertz.

Fraction	Prefix	Symbol	Multiple	Prefix	Symbol
10^{-2}	centi	c	10^3	kilo	k
10^{-3}	milli	m	10^6	mega	M
10^{-6}	micro	μ	10^9	giga	G
10^{-9}	nano	n	10^{12}	tera	T
10^{-12}	pico	p	10^{15}	peta	P
10^{-15}	femto	f	10^{18}	exa	E
10^{-18}	atto	a			

This is 1.49897 petahertz (see Table 9.2 for the meaning of "peta").

The wave vector k (Eq. 9.8) can also be calculated in units of cm^{-1}:

$$k = \frac{1}{\lambda} = \frac{1}{200 \times 10^{-9} \text{ m}}$$

$$= \frac{1}{200 \times 10^{-7} \text{ cm}}$$

$$= 5 \times 10^{4} \text{ cm}^{-1}$$

This is the same as the number given in Table 9.1.

Finally, let us calculate the energy in joules. From Eqs 9.11 and 9.5, I have

$$E = h\nu = 2\pi\,\hbar\nu$$

$$= 2\pi \times 1.054 \times 10^{-34} \text{ J} \times \text{s} \times 1.5 \times 10^{15} \text{ s}^{-1}$$

$$= 9.93 \times 10^{-19} \text{ J}$$

All these different, but equivalent, ways of giving the energy of the photon are ungainly. For this energy range we prefer to use the electron volt (eV) as a unit of energy. You have learned about the eV in your physics class and I will not repeat that material here. The conversion factor from eV to joule is

$$1 \text{ J} = 6.2415 \times 10^{18} \text{ eV} \tag{9.15}$$

In this unit, the energy that we have calculated above in a variety of units becomes $E = 6.198$eV. This is an easier number to work with.

I was telling you that various fields favor different units. In X-ray spectroscopy the energy of the photon is given in eV. In vacuum UV and visible spectroscopy, either eV or nanometer are used. In near IR and mid-IR, either cm^{-1} or THz (terahertz) are used. In far IR, cm^{-1} or GHz (gigahertz) are preferred. Microwave spectroscopists use cm^{-1} and GHz; and radio-wave spectroscopists use MHz. Often when you study a specific system, information from all these spectroscopies is used. Sooner or later you will have to learn how to use all these units.

This seems crazy and the rational approach would be to use the joule with prefixes for all ranges. However tradition, practical convenience, and the fact that so much information is given in the "irrational" units, keeps these units alive.

§9. *Why Do We Need to Know the Laws of Light Absorption and Emission?*
Light is everywhere around us. We see because light of certain frequencies excites molecules in our retina, causing a radiationless process that sends signals to the brain. The surfaces of the objects around us are colored because they absorb photons of a certain frequency from the light falling on them. As a result, the light reflected by the surface has an excess of photons of the color complementary to the one absorbed; a surface that absorbs green photons is colored red.

A rose is red whether or not we understand why; knowing that red roses absorb green light does not increase our enjoyment of them, nor does it help us improve them. There are, however, situations where understanding light absorption has been very fertile. You may have spent some time surfing the Internet today. You sent a request to a remote server and a stream of data came back. This information was carried by light pulses, through thousands of miles of optical fiber.

Optical fibers are not found in the garden, like roses. Someone had to develop the concept and implement the thought. Consider the task. A small pulse of light has to travel very long distances with almost no loss. This can happen only if the material does not absorb or scatter light. As you know, light travels in a straight line but when sent through a fiber it must follow every twist and turn. As if this is not hard enough, the pulses of light must maintain their shape. We cannot allow some photons in a pulse to travel faster than others, making the pulse spread. All these conditions must be satisfied by a fiber that is mechanically stable, has the same radius for miles, and is not affected by cold, heat, oxygen or water. That such a material was made, is the closest humans come to performing miracles. The success relied on an intimate knowledge of light propagation, absorption, and scattering.

You may decide, during a cruel southern California winter day, to spend a leisurely afternoon in front of a fireplace, amusing yourself with some light reading (say, a book on physical chemistry). The wood burning in the fire produces CO_2, water, and minerals (ash). This is what the tree was made from: CO_2 from the air and water and minerals from the ground. Burning returned the tree to its "initial state" and it also *produced a lot of energy*. Where does this energy come from? It is sunlight. A poet might be pleased to discover that burning wood in a stove releases the sunlight stored in it or that coal is fossil CO_2, water, and sunlight.

How is the light energy accumulated and stored in the plant? In the first step the chlorophyll molecules in the leaves absorb sunlight. Three things can happen after a photon is absorbed. (1) The molecule re-emits the photon, in which case the photon energy is lost to the plant. (2) A radiationless transition takes place and some heat is produced. The surrounding air cools the leaf removing the energy brought by the photon. (3) Some photochemistry occurs and the photon energy is turned into chemical energy. This stays in the leaf and it can be used to make the plant live and grow.

The light-absorbing and processing system in plants is exquisitely designed to absorb sunlight efficiently and cause a photochemical reaction *before* photon re-emission or any radiationless process can take place; not only is light absorbed efficiently but most of it is converted to chemical energy. By extremely complex processes this chemical energy is used to turn CO_2, water and minerals into a tree. The process is thermodynamically "uphill": cellulose and the other organic matter in the plant are less stable than CO_2 and water. Such a synthesis requires energy and this is provided by the absorbed photons.

Since plants are at the bottom of the food chain, they feed animals that feed us, so much of the energy in our bodies is the chemical equivalent of sunlight.

There are people who think it might be useful to develop molecular systems that imitate photosynthesis, to use them to harvest the energy of the sun. To create them, they will have to find molecules that absorb sunlight well and turn the photon energy efficiently into usable chemical energy. This means that they must be engineered so that photochemistry is faster than any of the energy losing processes. This cannot be done without a good understanding of photon absorption and emission and the chemistry of the excited molecules.

You have heard of global warming and probably worry about it. The Sun sends a gigantic amount of energy to Earth. There would be no life on Earth without it, but there will be no life if we retain most of it. Our planet is "designed" to retain the right amount of sunlight energy (people living at the equator, in deserts or near the poles may disagree with this statement).

If we had no atmosphere, light that is not absorbed by the ground would be reflected back into the cosmos. But we do have an atmosphere and this plays an essential role. First, it absorbs harmful UV radiation coming from the sun. This is a good thing, since UV radiation is very energetic and would cause quite a bit of harmful photochemistry. The ozone molecules, high in the atmosphere, are very important UV absorbers. The lack of ozone above parts of Australia is held responsible for a high incidence of skin cancer. This is why we have banned the use of many gases that destroy ozone.

The same absorption process that protects us from the UV radiation, works against us when it comes to infrared photons. Some IR photons are absorbed and turned into heat and some are sent back where they came from. There must be a fine balance between the amount of energy we keep and the amount we return. Global warming is a speculation (I use the word theory for statements that are proven beyond doubt) that says that we produce gases that will cause Earth to retain too much heat.

Not all infrared radiation coming to us is absorbed and turned into heat. Some is reflected by the ground and some is absorbed and re-emitted. Infrared photons are also produced on earth. A molecule that has absorbed a UV or visible photon may lose energy by emitting the absorbed energy in bits of infrared radiation. Some molecules in a warm body have enough energy to get excited when they collide and then emit infrared photons. This is why night vision glasses are able to see humans when their temperatures differ from that of the environment.

Consider now the fate of all this IR light that is not turned into heat. Most of it is directed towards the cosmos but to get there it must pass through the atmosphere. There some of it is absorbed and turned into heat. Some is absorbed and re-emitted. But the direction of the emitted photon is random and instead of going towards the universe, the photon may get back towards Earth or move parallel to it in the atmosphere. Such photons have another chance of being absorbed and turned into heat. Thus photons that started moving straight towards the cosmos end up moving on a circuitous route on which they have a substantial chance of being turned into heat. The absorbing gases trap some of the outward bound photons and heat the atmosphere. Carbon dioxide, which is the parent of all plant life, is also one of the worst offenders on the list of the global warming gases, because it is fairly abundant and has a very high IR absorption rate. It is produced by driving cars that run on petroleum products, by burning organic molecules or coal in power plants, and by decomposition of dead plants.

Understanding all aspects of photon emission and absorption is essential for deciding whether global warming is a real threat to us. Important and far-reaching political and economic consequences rest on the conclusion reached by scientists.

I will end this lengthy advertisement for photon absorption and emission by describing one more process in which photon–matter interactions play an important role. A film of TiO_2 can absorb a photon and this will excite an electron inside. If we use the film as an electrode, immersed in a water solution of an acid, the excited electron can be captured by a proton to make hydrogen. The TiO_2 film, made electron-deficient by the formation of a hydrogen atom, can take an electron from OH^- in solution and form water and oxygen. Light energy is used to make hydrogen and oxygen from water. These can be used in a fuel cell, to make electricity and water. A battery of fuel cells can run your car. If you do the energy balance you find that you have taken water and sunlight and got a car to move and produce water.

The net process uses sunlight to move the car. The water is recovered, which is important since we may run out of drinking water before we run out of oil. Not even an environmentalist can complain about this process.

This scheme is not sufficiently efficient to be put in use right now, but people are working to improve it. Among the goals: an increase in the photon absorption rate and a decrease of the rate with which the excited electron emits a photon and goes back to being a sedate electron that will not react with the proton.

I would not say that photon absorption and emission rule that world. But they are likely to play a role in shaping our future.

10

LIGHT EMISSION AND ABSORPTION: EINSTEIN'S PHENOMENOLOGICAL THEORY

Introduction

§1. You may have noticed that some of the discussions in the previous section use the language of chemical kinetics: we were concerned with how fast the photons were absorbed or emitted. Einstein was the first to examine light emission and absorption as if it was a "chemical reaction"

$$A + \text{photon} \; \rightleftharpoons A^*$$

Here A represents the molecule in the initial state (prior to photon absorption) and A^* is the molecule after photon absorption (i.e. the excited molecule). The "left-to-right" reaction is photon absorption, the "right-to-left" reaction is photon emission. By adopting this assumption Einstein could use the machinery of chemical kinetics to make statements about the rate of absorption and emission.

If that was all, this work would not have been worthy of Einstein. He made an additional observation, which is not surprising now, but was a novel approach in the time of his youth. He said: assume that the light and the gas of molecule are enclosed in a black body box, in which the radiation and the molecule are in equilibrium.

The radiation at equilibrium must satisfy Planck's law and the molecules at equilibrium must satisfy Boltzmann's law. These two laws will specify how many photons of a given frequency, how many excited molecules, and how many molecules in the ground state are present in the box (when the system is in equilibrium).

The kinetic equations can also be used to study equilibrium, because when equilibrium is reached, the number of photons and the number of molecules must be constant. Therefore, the rate of change of these quantities must be zero.

Einstein had at his disposal two approaches to the same problem, one kinetic and one thermodynamic. This allowed him to calculate the *same quantities* (the equilibrium "populations") in two ways. The results of the two calculations must be the same if they are both correct. By using this equality he expressed the rate of emission and absorption in terms of the molecular quantities contained in the Planck and Boltzmann equations. The analysis based on this idea forced Einstein to suggest that a new process, which he called *stimulated emission*, must take place; otherwise the two approaches would not give the same result. Moreover, he managed to derive a relationship between the rate of photon emission and that of photon absorption.

The stimulated emission is the process indispensable for making lasers, but even Einstein did not foresee that the new process could have such interesting applications.

Photon Absorption and Emission: the Model

§2. *Photon Energy and Energy Conservation.* Einstein postulated that light of frequency Ω consists of "particles," called photons, that have the energy $\hbar\Omega$. Light absorption is the disappearance of a photon simultaneously with the excitation of the molecule to a state of higher energy.

Let us assume molecules are in a state ψ_i with energy E_i before photon absorption and reach the state ψ_f having energy E_f after photon absorption. Here ψ_i and ψ_f are energy eigenfunctions of the molecule, and E_i and E_f are the corresponding eigenvalues. In principle these quantities can be calculated by solving the Schrödinger equation of the molecule.

Energy conservation says that the photon can be absorbed only if

$$\hbar\Omega = \hbar\omega_{fi} \equiv E_f - E_i; \tag{10.1}$$

the energy $\hbar\Omega$ of the disappearing photon is entirely given to the molecule.

To describe this process we use the following nomenclature: a molecule that absorbs a photon undergoes a *transition* from the *initial state* ψ_i of energy E_i to the final state of ψ_f of energy E_f. The quantity ω_{fi} is called the *transition frequency*.

If the energy eigenvalues E_n, $n = 0, 1, 2, \ldots$, of the molecules in a gas or a liquid are such that no transition frequencies matches the photon frequency (i.e. there are no i and f for which $\hbar\Omega = E_f - E_i$), the photon passes through the sample unmolested; no absorption takes place.

§3. *How do we Reconcile this Energy Conservation with the Existence of a Line-width?* At this point you should wonder what has happened to the line-width mentioned in the previous chapter. We said there (see §3 of Chapter 9) that the absorption spectrum has peaks at certain frequencies, but absorption also takes place at frequencies close to those of the peaks. Eq. 10.1 does not allow this: it says that absorption takes place *only* when $\Omega = \omega_{fi}$.

It turns out that Eq. 10.1 is approximate. It is the result of a theory that oversimplifies the interaction of the molecule with the radiation and with the neighboring molecules. A better theory is available, which shows that the peaks in the absorption spectrum have a width.

If you paid attention you should have been suspicious of Eq. 10.1 when you first saw it. It tells you that you will observe light absorption only if you prepare a light beam whose frequency matches $(E_f - E_i)/\hbar$ *exactly*. Well, if you are going to do this experiment for the first time, you would want to know how accurately you need to match these two numbers. To ten decimal places? To twenty? You will soon come to think that if Eq. 10.1 is correct no-one would be able to observe light absorption: the two numbers will always differ at some decimal place. On the other hand we see light absorption everywhere around us: this is why we see color and why sunlight feels warm, etc. Einstein's genius consisted, among others, in the ability to overlook such objections, which would have paralyzed a lesser physicist. He probably decided that this contradiction was a detail that could be taken care of later, when a better theory was developed.

§4. *The Model.* Einstein considered a gas of molecules that have only two energy eigenstates, E_0 and E_1, in contact with photons of frequencies $\Omega = (E_1 - E_0)/\hbar$. Several processes can take place when the photons interact with the molecules.

(a)   Photons are absorbed by the molecule and become excited (absorption).

(b)   The excited molecules emit photons spontaneously (spontaneous emission).

(c)   The presence of the photons speeds up the rate of photon emission (stimulated emission).

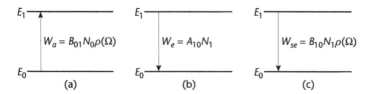

Figure 10.1 Energy-level diagrams representing the processes caused by the inter-action of the molecules with light. (a) Photon absorption; (b) spontaneous photon emission; (c) stimulated photon emission. In each case, the photon frequency is $\Omega = (E_1 - E_0)/\hbar$. The equations give the rate of each transition.

The introduction of stimulated emission was a novelty and I will tell you soon why Einstein decided to postulate its existence. The transitions corresponding to these processes are shown schematically in Fig. 10.1.

To describe the evolution of such a system we need to specify the concentration N_0 of the molecules in the ground state, the concentration N_1 of the molecules in the excited state, and the "concentration" $\rho(\Omega)$ of the photons whose frequency Ω equals the transition frequency $(E_1 - E_0)/\hbar$.

Einstein assumed that the evolution of these quantities is described by rate equations of the kind used in chemical kinetics. To complete the theory we need to:

• Write down the rate equations.

• Solve them.

• Analyze the meaning of the results.

Photon Absorption and Emission: the Rate Equations

§5. *The Rate of Photon Absorption.* Einstein proposed that light absorption changes the concentration of the molecules in the ground state with the rate:

$$W_a \equiv \left\{ \frac{dN_0(t)}{dt} \right\}_{abs} = -B_{01}\rho(\Omega)N_0(t) \tag{10.2}$$

This definition is in keeping with what Newton (and Leibnitz) taught us when they invented the derivative: if you want to know the rate of change of a quantity $x(t)$, calculate the derivative dx/dt.

Absorption takes place only if a photon of frequency $\Omega = (E_1 - E_0)/\hbar$ encounters a molecule in the state ψ_0. The probability that this happens ought to be proportional to the concentration $\rho(\Omega)$ of photons of frequency Ω, times the concentration N_0 of the molecules in the state ψ_0 (which are the only molecules that can absorb the photon).

The quantity $\rho(\Omega)$ is defined so that $\rho(\Omega)dVd\Omega$ gives the number of photons in the volume dV having frequency between Ω and $\Omega + d\Omega$. One can calculate this quantity by using Maxwell's equations, but the result depends on the nature of the light beam used in each particular experiment (we could use a pulse, or we could place the molecules in a laser cavity, or we could use polarized light moving in one direction, etc.). Fortunately, most of the results obtained by Einstein can be understood without knowing the specific formula for $\rho(\Omega)$. For what we do here it is important to realize that $\rho(\Omega)$ increases when we increase the intensity of light.

B_{01} is called the *Einstein B coefficient*. It is a proportionality constant, analogous to the rate constant in the ordinary chemical kinetics. Its role is to account for the fact that not all encounters between photons and molecules result in proton absorption; there is a certain probability that the two particles pass by each other with no damage done.

The index 01 reminds us that B_{01} is specific to the transition from the state ψ_0 to the state ψ_1. The value of B for a transition between another pair of states will be different. B also changes from molecule to molecule. All the intricate changes taking place in a molecule when it absorbs a photon are hidden in the magnitude of B.

§6. *The Rate of Spontaneous Photon Emission*. Once excited in the state ψ_1, the molecule emits photons spontaneously, with the rate

$$W_e \equiv \left\{ \frac{dN_0(t)}{dt} \right\}_{\text{spem}} = -\left\{ \frac{dN_1(t)}{dt} \right\}_{\text{spem}} = A_{10}N_1(t) \tag{10.3}$$

The first equality takes place because

$$N = N_0(t) + N_1(t) \tag{10.4}$$

is the total number of molecules, which does not change in the course of the experiment. The last equality says that the rate of photon emission is proportional to the number of excited molecules; a sample having more excited molecules in the unit volume will emit more photons per unit time.

The coefficient A_{10} is called *Einstein's A coefficient*. It depends on the particular molecule and the particular transition studied. The subscript 10 indicates that the coefficient refers to the transition from the state ψ_1 to ψ_0.

§7. *The Rate of Stimulated Emission.* So far we have used ordinary chemical kinetics to describe the rate of photon absorption and emission. However, a person familiar with freshman chemistry would have stopped here and would soon have discovered that the theory does not work properly. To fix it, Einstein invoked a new process, called *stimulated photon emission*. He postulated that the interaction between the photons of frequency $\Omega = (E_1 - E_0)/\hbar$ and the molecules excited to the state ψ_1 will "stimulate" the latter to emit photons of frequency Ω faster than when the radiation was absent. The rate of this process is proportional to the concentration $\rho(\Omega)$ of the photons of frequency Ω, and the concentration of the molecules in the excited state ψ_1:

$$W_{se} \equiv \left\{ \frac{dN_0(t)}{dt} \right\}_{\text{stem}} = - \left\{ \frac{dN_1(t)}{dt} \right\}_{\text{stem}} = B_{10}\rho(\Omega)N_1(t) \qquad (10.5)$$

B_{10} is the proportionality constant.

§8. *The Total Rate of Change of N_0 (or N_1).* We have discussed these processes separately, but the net change of N_0 is caused by their joint action. Therefore

$$\frac{dN_0(t)}{dt} = -\frac{dN_1(t)}{dt}$$

$$= \left\{ \frac{N_0(t)}{dt} \right\}_{\text{abs}} + \left\{ \frac{N_0(t)}{dt} \right\}_{\text{spem}} + \left\{ \frac{N_0(t)}{dt} \right\}_{\text{stem}}$$

$$= -B_{01}\rho(\Omega)N_0(t) + A_{10}N_1(t) + B_{10}\rho(\Omega)N_1(t) \qquad (10.6)$$

The equation tells us that the number of photons in the ground state is decreased by photon absorption and it is increased by spontaneous and stimulated photon emission by the excited molecules.

At this point I need to warn you of the limitations of these equations. Normally in such an experiment the molecules are inside a container with glass walls and a beam of light enters the container on one side and exits on the opposite side. The number of photons in the container is increased by the flux coming in and is decreased by the flux going out. The rate equation should contain the effect of these two fluxes and it does not. Moreover, the number of photons in the container should decrease along the container. This effect is not represented in the equations either.

Strictly speaking this particular model is valid only when both the photons and the molecules are enclosed inside a container. It has to be modified when dealing with experiments in which light comes in and goes out. Why do we bother with it then? Because it teaches us some important lessons about the Einstein coefficients and light absorption and emission processes. It can also be improved to deal more realistically with the system used in the laboratory, but we do not have time to go into that here.

Photon Absorption and Emission: the Detailed Balance

§9. *Molecules in Thermal Equilibrium with Radiation.* After writing the rate equations, Einstein took a next step, by using a beautiful argument, which was very original at the time he made it. He realized that if the system is closed (i.e. no molecules or photons go in or out of the container) and it is held at constant temperature and pressure, it would have to reach equilibrium. This meant that the concentrations obtained from Eq. 10.6 after a long time must be the same as the equilibrium concentration calculated from the equilibrium theory of light and matter. Since the equilibrium theory was already developed at that time, Einstein could use this equivalence to learn something new about his coefficients.

The concept of thermal equilibrium between radiation and the molecules is subtle and requires a bit of clarification. Before Einstein's time people did experiments and used theory to study the so-called black-body radiation. Roughly, this is obtained by taking a box and coating the inside walls with a black substance (carbon black would do). Pump the air out of the box and then heat it up. Because of the high temperature, the atoms in the black substance move fast and collide with each other and some of them are excited and emit radiation. The emitted photons are not allowed to leave the box and a certain photon concentration builds up inside. However, the walls also absorb photons and in time an equilibrium is reached: the rate of photon emission by the walls equals the rate of photon absorption by the walls; when this happens, the concentration of the photons in the box becomes constant. We say that the radiation (the photons) in the box is in thermal equilibrium with the atoms in the black walls. The reason for using a black wall is to ensure that the wall emits and absorbs radiation of all frequencies.

The equilibrium between the radiation and a black body has given a lot of grief to a large number of distinguished physicists. Well-constructed theories based on classical theory of radiation, thermodynamics, and statistical mechanics gave results that disagreed with the measurements. Faced with this failure, Planck made the bold assumption that the energy of the emitted and absorbed light is discrete. Using that, he showed that the number of photons of frequency Ω per unit volume, per unit frequency at equilibrium, inside a black box is

$$\rho_{eq}(\Omega) = \left(\frac{2\hbar}{\pi}\right)\left(\frac{\Omega}{c}\right)^3 \frac{1}{\exp\left[\frac{\hbar\Omega}{k_B T}\right] - 1} \tag{10.7}$$

Here c is the speed of light, $k_B = 1.38010^{-23}$ J/K (joule per kelvin) is Boltzmann's constant, T is the temperature in kelvin, and $\hbar$ is Planck's constant.

This formula fitted the data perfectly for all light frequencies and for all temperatures. It showed that the principle of continuity of energy, so fundamental to classical physics, does not hold for light. If Planck allowed the light energy to take continuous values he obtained the erroneous equations produced by his predecessors.

This theory introduced a new fundamental constant (the Planck constant) and allowed the first accurate determination of the magnitude of the Boltzmann constant, of the Planck constant, and of Avogadro's number.

Exercise 10.1

Use Eq. 10.7 to do the following.

(a)   Plot of the dependence of $\rho_{eq}(\Omega)$ on Ω for $T = 300$ K and $T = 3000$ K.

(b)   Explain the physical implications of these plots. What kind of radiation are you exposed to at room temperature? What kind when you are close to a very hot metal bar (e.g. a toaster)?

(c)   A glowing metal bar is sometimes used as a source of infrared radiation. At what IR frequencies is the glow-bar most efficient?

(d)   Would a glow-bar be a practical source of UV radiation?

If we put molecules in a black-body box whose walls are kept at a constant temperature T, the molecules will reach thermal equilibrium. Boltzmann showed that when this happens

$$\left\{\frac{N_1}{N_0}\right\}_{\text{eq}} = \exp\left[-\frac{E_1 - E_0}{k_B T}\right] = \exp\left[-\frac{\hbar\Omega}{k_B T}\right] \tag{10.8}$$

To obtain the last equality I used Eq. 10.1. This form of Boltzmann's formula is valid when the states ψ_0 and ψ_1 are not degenerate.

Einstein asked himself what will happen if he puts the molecules in a *black-body box* held at a constant temperature T. In such an experiment we are not bringing light from outside: it is generated by the hot molecules and the hot black walls of the box. After a while the molecules and the radiation will all be in thermal equilibrium. When this happens Planck's formula, Eq. 10.7, for the density of radiation and Boltzmann's formula, Eq. 10.8, for the ratio N_1/N_0 must hold.

§10. *The Detailed Balance.* Furthermore, when equilibrium is reached N_0 no longer varies in time and $dN_0(t)/dt = 0$. Using this condition in the rate equation (Eq. 10.6) gives

$$\rho_{\text{eq}}(\Omega) = \frac{A_{10}}{\left\{\dfrac{N_0}{N_1}\right\}_{\text{eq}} B_{01} - B_{10}} \tag{10.9}$$

This equation is valid only if the system is in equilibrium. This means that in Eq. 10.9, $\rho_{\text{eq}}(\Omega)$ must be given by Planck's equation (Eq. 10.7) and N_0/N_1 by Boltzmann's equation (Eq. 10.8). Using these relations in Eq. 10.9 leads to:

$$\left(\frac{2\hbar}{\pi}\right)\left(\frac{\Omega}{c}\right)^3 \frac{1}{\exp\left[\frac{\hbar\Omega}{k_B T}\right] - 1} = \frac{A_{10}}{\exp\left[\frac{\hbar\Omega}{k_B T}\right] B_{01} - B_{10}} \tag{10.10}$$

This equality is possible only if

$$B_{01} = B_{10} \equiv B \tag{10.11}$$

and

$$A_{10} = B\left(\frac{2\hbar}{\pi}\right)\left(\frac{\Omega}{c}\right)^3 \equiv CB \tag{10.12}$$

The constant

$$C = \left(\frac{2\hbar}{\pi}\right)\left(\frac{\Omega}{c}\right)^3 \tag{10.13}$$

depends on light frequency but it is independent of the nature of the molecule. Due to these two equations, B is the only molecule-dependent quantity left in the theory. Within Einstein's approach B has to be determined from experiments.

This procedure, which demands that the rate equation describes equilibrium in a way consistent with thermodynamics, is quite general and is called the *principle of detailed balance*. Its use imposes certain limitations on the rate constants. In ordinary chemical kinetics, the principle connects the ratio of the forward and reverse rate constants of a reversible reaction to the equilibrium constant of the reaction.

The Solution of the Rate Equations

§11. *Introduction.* All differential equations have an infinite number of solutions. This happens because the solution contains constants whose values are not specified by the equation. The number of such constants in the solution is equal to the order of the equation. Our kinetic equation is first order and we expect its solution to contain one unspecified constant. When differential equations are used to describe physical processes the value of the constant is determined by initial conditions or by boundary condition. In general, these conditions describe some particular features of the experiment being analyzed.

Here we consider two physical situations, resulting in two initial conditions. First we assume that at $t = 0$ all molecules are in the ground state and they are suddenly exposed to light. We use the rate equation to calculate how the concentration of the molecules in the ground state varies in time. In doing this we discover an interesting phenomenon, called saturation: no matter how hard we try, the quantity $N_1(t)/N$ cannot be greater than 1/2.

We also consider the case when a pulse of light created a certain concentration of excited molecule and ask how this concentration will evolve in time if the system is not exposed to light (i.e. no stimulated emission takes place). We discover that the concentration decays exponentially in a time of order A^{-1}.

These conclusions are valid only when the rate equation is valid, and this means relatively low light intensity.

§12. *Use the Detailed Balance Results to Simplify the Rate Equation.* Using Eqs 10.11 and 10.12, which connect A to B, simplifies the rate equation Eq. 10.6 to

$$\frac{dN_0(t)}{dt} = -B\rho(\Omega)N_0(t) + B\rho(\Omega)N_1(t) + BCN_1(t) \tag{10.14}$$

This differential equation appears to contain two unknown quantities, N_0 and N_1. However, we also know that

$$N_0(t) + N_1(t) = N$$

where N is the total number of molecules. We use this relation to eliminate N_1 from the rate equation, Eq. 10.14. This leads to

$$\frac{dN_0(t)}{dt} = -BN_0(t)\rho + B\left[\rho(\Omega) + C\right]\left[N - N_0(t)\right] \tag{10.15}$$

Now we have one differential equation, containing one unknown function, namely $N_0(t)$.

But this is not enough. The solution of this first-order differential equation will contain an unknown constant C_1. Indeed, if we ask **Mathematica** to solve this equation we obtain (see Workbook QM10.1, Cell 1)

$$N_0(t) = \frac{N(C + \rho)}{C + 2\rho} + C_1 \exp\left[B(C + 2\rho)t\right] \tag{10.16}$$

This expression satisfies the differential equation regardless of the value of the constant C_1. This is not unexpected: from the elementary theory of differential equations we know that all general solutions of a first-order differential equation contain a constant whose value is arbitrary.

Exercise 10.2

Show that Eq. 10.16 satisfies the differential equation for any value of the constant C_1.

To determine what this constant is, we must use the physical situation to impose additional conditions on the solution.

§13. *The Initial Conditions.* In what follows I solve this equation for two physical situations.

- At time $t = 0$, all molecules are in the ground state and therefore

$$N_0(t = 0) = N \tag{10.17}$$

where N is the total number of molecules. In the laboratory this situation is achieved approximately if we start exposing the molecules to light at time $t = 0$ and measure the ground-state concentration (or the excited-state concentration) at a time t.

- At time zero, the concentration of the molecules in the excited state has a known value N_1 and the molecules are no longer in contact with the light. We can achieve this situation approximately if we expose the molecules to a pulse of light and ask what happens to the excited state concentration after the light pulse no longer interacts with the molecules. The initial condition in this case is

$$\frac{N_1(t = 0)}{N} = x_1 \tag{10.18}$$

where x_1 is a known number.

§14. *The Ground State Population when the Molecules are Continuously Exposed to Light.* In this case we use the initial condition Eq. 10.17 to determine the constant C_1 in Eq. 10.16:

$$N_0(t = 0) = N = \frac{N(C + \rho)}{C + 2\rho} + C_1 \tag{10.19}$$

The first equation is the initial condition and the second is obtained by setting $t = 0$ in Eq. 10.16.

Solving Eq. 10.19 for C_1 gives

$$C_1 = -\frac{(C + \rho)N}{C + 2\rho} \tag{10.20}$$

Now we know the value of C_1. Introducing it in the expression (Eq. 10.16) for the ground-state concentration leads to

$$\frac{N_0(t)}{N} = \frac{C + \rho + \rho \exp\left[-Bt(C + 2\rho)\right]}{C + 2\rho} \qquad (10.21)$$

Exercise 10.3

(a) Test that $N_0(t)$ given by Eq. 10.21 satisfies the differential rate equation Eq. 10.15 and the initial condition Eq. 10.17. (b) Try to solve the differential equation by using the elementary methods you learned in calculus.

§15. *Analysis of the Result: Saturation.* Let us first use our physical intuition to establish how we expect $N_0(t)$ to behave. Then we can use the mathematical results to see if our expectations were justified (or, if you prefer, to test whether the mathematics gives reasonable results).

At $t = 0$ we have $N_0(t = 0)/N = 1$; no radiation acts on the system and the molecules are in the ground state. Upon contact with the radiation the molecules absorb light and $N_0(t)$ starts to decrease, while $N_1(t)$ increases. The increase of $N_1(t)$ brings about an increase of photon emission rate (which is proportional to $N_1(t)$) and a decrease of the absorption rate (which is proportional to $N_0(t)$). At some point the emission rate will become equal to the absorption rate and $N_1(t)$ and $N_0(t)$ will stop changing. When this happens we say that the system has reached a *steady state*.

We expect that steady-state concentration to depend on the intensity of light, hence on ρ; when ρ is high, there should be fewer molecules in the ground state, when the steady state is reached.

A simple way of testing whether Eq. 10.21 behaves in this way is to plot $N_0(t)$ as a function of time, for three values of ρ. Such a plot (made in Workbook QM10.1, Cell 4) is shown in Fig. 10.2. We see that the equation behaves as expected: $N_1(t)$ increases with time and reaches a constant value.

The steady state is reached faster when photon density is higher (i.e. the light is more intense). This statement can be made more precise by examining Eq. 10.21, which tells us that $N_0(t)$ becomes independent of time when the exponent in $B(2\rho + C)t$ becomes larger than 1. Therefore, the system reaches a steady state when

$$t \gg \frac{1}{B(2\rho + C)} \equiv \frac{1}{2\rho B + A} \tag{10.22}$$

To obtain the last equality, I used $BC = A$. The steady state is reached faster when ρ, A, and B are larger.

Exercise 10.4

Can you explain why A is present in Eq. 10.22, and why the time to reach the steady state could be very short if $B\rho$ is very small and A is large?

There is also a hint in Fig. 10.2 that the steady-state value of $N_1(t)$ does not increase indefinitely with ρ. We can use Eq. 10.21 to show that the highest value $N_1(t)$ can reach is 0.5, regardless of how high ρ is. Indeed, when ρ is very large the exponential in Eq. 10.21 is negligible and

$$\left(\frac{N_0}{N}\right)_{ss} = \frac{C + \rho(\Omega)}{C + 2\rho(\Omega)} \tag{10.23}$$

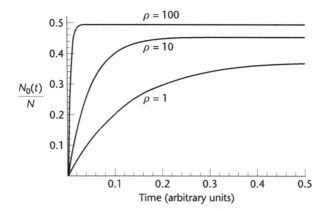

Figure 10.2 The dependence of the excited state concentration $N_1(t)$ on time, for three values of the photon density ρ. Note that the concentration $N_1(t)$ reaches the steady state and that the steady-state value increases with ρ. It seems that the steady-state value has ceiling at 0.5.

The subscript 'ss' indicates that this is the ratio N_0/N when the system has reached steady state. Note that $(N_0/N)_{ss}$ depends only on the properties of light (through ρ) and some universal constants contained in C. The maximum value that can be reached, when $\rho \gg C$ (or equivalently $B\rho \gg BC = A$) is $N_0/N = 1/2$. At most half the molecules in the sample can be excited, no matter how powerful the laser is.

This limitation is called the saturation of absorption. It happens because of the existence of the stimulated emission: as we increase the intensity of the light we increase both the rate of absorption and the rate of stimulated emission. The value of $\frac{1}{2}$, for the steady-state concentration, is reached when the stimulated emission is much faster than the spontaneous emission.

Exercise 10.5

Examine the solution of the rate equation in a fictitious system in which stimulated emission does not exist. Can we overcome saturation, in this case, by increasing the intensity of the laser?

§16. *Warning.* This theory worked very well in Einstein's time because no powerful light sources were available. As quantum theory developed, and as more and more powerful lasers became available, it was shown that new phenomena take place, as light intensity is increased, which are not covered by Einstein's analysis. As an extreme example consider the following phenomenon: if you expose xenon or a piece of glass to a strong laser of frequency Ω the sample emits intense light of frequency $\Omega, 2\Omega, 3\Omega \ldots$ all the way to rather high multiples of the incident frequency. The strange phenomena, taking place when matter is exposed to strong lasers, are studied by *nonlinear optics* or *nonlinear spectroscopy*.

§17. *Why did Einstein Introduce Stimulated Emission?* Einstein must have realized that a new emission process had to be postulated when he made the detailed balance analysis of his rate equation. If the stimulated emission is not present, then rate equation Eq. 10.6 gives the equilibrium condition (use $dN_0(dt)/dt = 0$):

$$0 = \left\{ \frac{N_0}{dt} \right\}_{abs} + \left\{ \frac{N_0}{dt} \right\}_{spem}$$

$$= -B_{01}\{\rho(\Omega)\}_{eq}\{N_0(t)\}_{eq} + A_{10}\{N_1(t)\}_{eq} \qquad (10.24)$$

This leads to

$$\frac{\{N_1\}_{eq}}{\{N_0\}_{eq}} = \frac{B_{01}}{A_{10}} \{\rho(\Omega)\}_{eq} \tag{10.25}$$

The left-hand side must satisfy Boltzmann's equation Eq. 10.8, and in the right-hand side $\{\rho(\Omega)\}_{eq}$ (with $\Omega = (E_1 - E_0)/\hbar$) must satisfy Planck's formula Eq. 10.7. You can see that this is impossible if Eq. 10.25 is correct. At this point Einstein must have realized that unless the rate equation has an additional term (the stimulated emission), it will violate the detailed balance principle.

§18. *Stimulated Emission and the Laser.* At the time of their discovery neither stimulated emission nor the phenomenon of saturation seemed very important. They were interesting discoveries but they did not seem to affect us: the light sources used at the time had low intensity, the rate of stimulated emission was very small and saturation could not be reached.

Things changed when lasers were designed. To have a working laser it is necessary to have (and maintain) a *population inversion*, which means that $N_1 \gg N_0$. Because of saturation this cannot be achieved by shining light on the system: the population inversion had to be created by another process. Different types of laser have different ways of achieving this.

But population inversion is not enough. In a laser, the excited molecules are placed inside an optical cavity, which is a space enclosed by mirrors. The mirrors keep the photons, emitted by the excited molecules, inside the cavity. There they cause stimulated emission, which creates more photons, which creates more stimulated emission.... At some point the number of photons in the cavity is so high that some escape and this produces the laser light. The photons emitted from the cavity have special properties because an overwhelming majority were created by stimulated emission. They differ from the photons created by ordinary light sources, where each excited molecule emits on its own and there is no correlation between the properties of the emitted photons.

A laser is a much more complicated device than I have described above and so is its history. The idea of stimulated emission is, however, at its foundation. It is doubtful that we would have discovered the laser without Einstein's work.

Exercise 10.6

Assume that you have a process, other than light absorption, that excites the molecules with the rate $dN_1(t)/dt = kN_0$. Add this rate to Einstein's equation and determine whether in this case you can obtain a steady state in which $N_1(t)/N > \frac{1}{2}$.

§19. *The Rate of Population Relaxation.* As our last inquiry into Einstein theory we will solve the rate equation with the initial condition $N_1(t = 0)/N = x_1$ (x_1 is the fraction of molecules in excited state) and $\rho = 0$. These conditions can be realized experimentally as follows. Enclose the molecules in a container and expose them to a pulse of light which raises the concentration of the molecules in the excited state so that $N_1(t = 0)/N \equiv x_1$. After the pulse has left the sample, the molecules will be emitting light. Assume that the intensity of the emitted light is small enough to neglect its influence on the molecules (i.e. we can set $\rho = 0$). We want to know how the concentration of the excited molecules decays in time.

We are interested in the concentration $N_1(t)$ of the excited molecules which is given by $N_1(t) = N - N_0(t)$. Since $\rho = 0$ the rate equation Eq. 10.6 becomes

$$\frac{dN_1(t)}{dt} = -AN_1(t) \tag{10.26}$$

This has the solution (see Workbook QM10.2)

$$N_1(t) = C_2 \exp[-At] \tag{10.27}$$

where C_2 is an unspecified constant.

Using the initial condition $N_1(t = 0) = Nx_1$ leads to

$$Nx_1 = C_2 \tag{10.28}$$

We can now use Eq. 10.28 to eliminate C_2 from Eq. 10.27 to obtain

$$\frac{N_1(t)}{N} = x_1 \exp[-At] \tag{10.29}$$

The concentration of the excited molecules decays exponentially in a time of order of A^{-1}.

Exercise 10.7

Solve the rate equation for the case when at time $t = 0$ the fraction of the excited molecules is $N_1(0)/N = x_1$ and $\rho \neq 0$. Try to guess the behavior of the solution before you perform any calculations. What is the steady-state concentration? What happens if $N_1(0)$ is higher than the steady-state concentration? What happens if $N_1(0)$ is smaller than the steady-state concentration?

11

LIGHT ABSORPTION: THE QUANTUM THEORY

Introduction

§1. Current experimental studies of the interaction between light and matter use light pulses produced by a variety of lasers. Even the continuous light emitted by some sources (lasers or lamps) consists of a superposition of pulses. Because of this, the elementary event in photon absorption is the interaction of a molecule with a pulse of light. The theory determines the probability that a molecule in a state ψ_i, before the arrival of the pulse, is in the state ψ_f after the pulse has departed. In almost all cases of practical or scientific interest, the states ψ_i and ψ_f are eigenstates of the energy operator.

The derivation of this probability is not usually presented in introductory courses (you can find it in the book by Merzbacher[a]), and I will follow this tradition. I present here some of the results of quantum theory and explain under what physical conditions the theory is reliable. Light absorption by a quantum dot and a simplified model of a vibrating diatomic molecule, are examined in Chapter 12.

The equation for the probability contains four elements: a group of constants, the polarization vector $\mathbf{e}$ of light, a quantity $N(\omega)$ related to the amount of

---

[a] E. Merzbacher, *Quantum Mechanics*, 2nd edition, John Wiley & Sons, New York, 1961, p. 458.

electromagnetic energy carried by the pulse, and a property of the molecule, $\langle \psi_f | \mathbf{m} | \psi_i \rangle$, called the *transition dipole moment*. The main body of the chapter explains what these quantities are and why they appear in the theory.

The formula given here is valid only when the light has low intensity and its wavelength is substantially larger than the size of the molecule (the latter condition is not satisfied for X-rays). Molecules exposed to intense light produce a large number of spectacular effects, which are the subject of nonlinear optics and nonlinear spectroscopy. These phenomena are not discussed here, even though they are extensively used in research.

Of all the features present in the formula, the transition dipole moment is the most informative. For some transitions $\psi_i \to \psi_f$, the transition dipole moment is zero, which means that even though the frequency of light Ω satisfies the energy conservation conditions $\Omega = (E_f - E_i)/\hbar$, the molecule does not absorb a photon. In such cases we say that the transition is *forbidden*. Sometimes forbidden transitions take place, with very low probability, because the approximations made to derive the formula for the absorption probability are not strictly valid.

Absorption spectroscopy is one of the most extensively used methods for determining the energies and the states a molecule can have. Being aware that some transitions are forbidden is important. If we were to assume, because we did not know the theory, that the molecule has only the states that contribute to the absorption spectrum, we could misinterpret other molecular phenomena which are influenced by the states that do not show up in the spectrum (sometimes called dark states). Once we suspect that such states exist we can hunt for them by using other kinds of spectroscopies.

Quantum Theory of Light Emission and Absorption

§2. *The Absorption Probability.* In the experiment we examine here we expose a molecule in the state ψ_i, of energy E_i, to a pulse of light. The probability that the molecule absorbs a photon and is excited to a state ψ_f, of energy E_f, is:

$$P_{i \to f} = \frac{\pi}{\epsilon_0 c \hbar^2} |\langle \psi_f \, | \, \hat{\mathbf{m}} \cdot \mathbf{e} \, | \, \psi_i \rangle|^2 N(\Omega) L(\Omega) \tag{11.1}$$

This expression is quite a handful and we will spend some time examining its parts to determine their physical significance.

§3. *Electrodynamic Quantities: Light Pulses.* Maxwell's theory of electrodynamics shows that light is an electromagnetic wave, consisting of oscillating electric and magnetic fields. It is very likely that in your physics course you have learned that the variation in space of an electromagnetic wave is of the form $\exp[-i\mathbf{k}.\mathbf{r}]$ and its time evolution is given by $\exp[i\Omega t]$. Here $\mathbf{r}$ is the position at which the wave is observed, $\mathbf{k}$ is the wave vector, Ω is the frequency of light, t is the time when the observation is made, and $i \equiv \sqrt{-1}$. This monochromatic plane wave propagates in the direction of the wave vector $\mathbf{k}$. If light propagates through vacuum, the frequency ω and the magnitude k of the wave vector $\mathbf{k}$ are connected through the relation $\omega = kc$.

A monochromatic plane wave is a simple and helpful theoretical concept, but unfortunately the light described by such a wave cannot be made in the laboratory. Let us start with the requirement that the light is monochromatic (i.e. has one frequency). If you take any source of light, no matter how sophisticated, and measure and plot its intensity versus its frequency, you will obtain a curve like the one shown in Fig. 11.1; the photons in the beam of light have more than one frequency. This is not due to the limitations of our instruments. In any light source the photons are emitted by excited atoms or molecules, or by excited electrons in a solid. This light may be "processed" in a laser cavity or in a variety of optical devices and its properties may be modified. However, one fact remains: this photon emission is a random process; we do not know the exact time when the atom was excited or the exact time when the photon will be emitted. This uncertainty in the time between absorption and emission causes an uncertainty in the energy of the emitted photon, due to a principle of quantum mechanics similar to Heisenberg's uncertainty principle. The emitted light is not monochromatic and

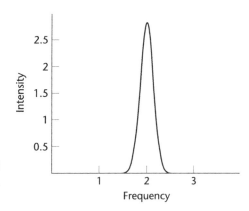

Figure 11.1 The dependence of the intensity of light, emitted by a source, on frequency.

the probability that the emitted photon has a certain frequency looks like the curve shown in Fig. 11.1.

Maxwell's theory tells us that the wave vector and the frequency are connected. Therefore, if we do not know for sure the frequency of the wave, we do not know for sure its wave vector. This means that the electromagnetic wave produced by a laboratory source is not a plane wave.

If light is not a monochromatic plane wave then what is it? One can show that light source emit pulses of light. Mathematically, these can be represented as a superposition of monochromatic, plane waves, having the frequency distribution shown in Fig. 11.1.

This seems to be in contradiction with the fact that many sources (such as some lasers or the bulbs in the lamps in your room) emit light continously. There is, however, no contradiction: the continuous light is a superposition of many light pulses emitted by the huge number of excited atoms or electrons in the source. These pulses overlap so well that the light appears to be emitted continuously. You can see that this is true if you measure the intensity of light as a function of frequency: the curve you get is like that shown in Fig. 11.1. If the light were a plane wave, then the curve would have no width.

It is worth pointing out that many lasers used in the laboratory are designed to produce pulses and that most experiments in spectroscopy and photochemistry are done intentionally with pulses of light.

We are led therefore to examine the interaction of a sample with a pulse of light. If the molecules in the sample do not interact strongly with one another, we can calculate the probability that one molecule is excited (by using Eq. 11.1). Then, the number of molecules excited by the pulse is the probability that one molecule is excited times the number of molecules exposed to the pulse. If the light consists of a superposition of pulses then we need to know the number of pulses arriving at the sample per unit time. Multiplying the probability that one molecule is excited by interacting with one pulse, by the number of pulses interacting with the molecule per second, we can calculate the excitation rate (the probability that the molecule is excited, per unit time).

Thus, we can calculate all the quantities of interest to the experimentalists from Eq. 11.1 for the excitation probability.

§4. *Electromagnetic Quantities: the Polarization of Light.* As already mentioned, a pulse is a superposition of electromagnetic, monochromatic plane waves. Every such wave consists of an electric field $\mathbf{E}$ and a magnetic field $\mathbf{H}$. The magnetic field interacts with the molecules much more weakly than the electric field, and its presence can be neglected in most spectroscopic studies. There are spectroscopic methods that rely on the interaction of the magnetic field of light with the molecule, but we will not study them here.

A laser produces *polarized* light, which means that the electric field associated with the pulse has a known direction. The polarization vector **e** (which appears in Eq. 11.1) is a vector of unit length, that has the same direction as the electric field of the polarized light. The manufacturer gives the orientation of **e** for the light produced by the laser. There are instruments that can measure the polarization of light.

The light produced by sun or by a light bulb is not polarized. In this case we only know the probability that **e** has a certain direction. Thus, light could be partially polarized (e.g. the sunlight reflected by the ocean) or it could be such that every orientation of **e** is equally probable (e.g. the light in a black-body box). If the light is not polarized, we calculate the excitation probability (from Eq. 11.1) for a given polarization and then average the result over all polarizations.

There is a lot more to say about polarization but we confine ourselves here to *linearly polarized* light. If I say that the light is (linearly) polarized in the OX direction, I mean that its polarization vector **e** is oriented along the OX axis and has the components $\mathbf{e} = \{1, 0, 0\}$. Elliptically or circularly polarized light is frequently used in practice, but we do not discuss such experiments here.

§5. *Electromagnetic Properties: the Direction of Propagation.* A pulse of light moves in a certain direction. Maxwell's theory shows that this is the direction towards which the wave vector **k** points. Thus, by directing a laser pulse along the direction OZ of some coordinate system, we specify that the wave vector points along the OZ axis and therefore it has the components $\mathbf{k} = \{0, 0, \omega/c\}$. Maxwell's theory also requires that, for light propagating in vacuum, **k** must be perpendicular to **e**. If a pulse propagates along the OZ axis (the direction of **k**) then the polarization vector **e** must lie in the OXY plane.

§6. *Electromagnetic Quantities: the Energy and the Intensity of the Pulse.* In most measurements performed on light, we measure its energy. For this reason an important characteristic of a light pulse is the amount of energy it carries. This is defined as follows. First you place an imaginary plane perpendicular to the direction of propagation of the pulse. Then, you determine the total amount of electromagnetic energy, with a frequency between Ω and $\Omega + d\Omega$, that passes through the unit area of this plane, from the time the pulse arrived at the plane's location until it passed through it. This amount of energy is, by definition, equal to $N(\Omega)d\Omega$. Note that $N(\Omega)$ appears in Eq. 11.1.

Another way of describing this quantity is to count all the photons having frequency between Ω and $\Omega + d\Omega$ that pass through the unit area of the plane and multiply that number by the photon energy $\hbar\Omega$. The result of this measurement is $N(\Omega)d\Omega$.

$N(\Omega)$ is related to the intensity $I(\Omega, t)$ of the light in the pulse, which is the energy of the light with frequency between Ω and $\Omega + d\Omega$, passing through the plane at time t, per unit area and unit time. $N(\Omega)$ is the integral of $I(\Omega, t)$ over time. Eq. 11.1 contains $N(\Omega)$ and not $I(\Omega, t)$ because we asked what is the probability that the molecule is excited *after* the pulse departed from the region where the molecule is located. Had we asked what the probability is that the molecule is excited at a time t, while the pulse still overlaps with the molecule, we would have had to know $\int_{-\infty}^{t} I(\omega, t)dt$ and use a different equation for the probability.

We have described light by using classical Maxwell electrodynamics. This gives correct results in many, but not all, cases. A logically consistent and correct theory treats light quantum mechanically, by using quantum electrodynamics. This topic is beyond this introductory text.

§7. *The Properties of the Molecule: the Transition Dipole Moment* $\langle \psi_f | \hat{\mathbf{m}} | \psi_i \rangle$. The quantity $\hat{\mathbf{m}}$ is the *dipole moment operator* defined by

$$\hat{\mathbf{m}} = \sum_{\alpha=1}^{n} q\mathbf{r}(\alpha) - \sum_{\beta=1}^{p} qZ_\beta \mathbf{R}(\beta) \tag{11.2}$$

In this expression $\mathbf{r}(1), \ldots, \mathbf{r}(n)$ are the coordinates of the n electrons in the molecule, $\mathbf{R}(1), \ldots, \mathbf{R}(p)$ are the coordinates of the p nuclei, q is the electron charge (a negative number), and Z_β is the number of protons in the nucleus β (for example, for the helium atom $Z = 2$). The negative sign in front of the second sum appears because the charge of a proton is $-q$.

To calculate $\langle \psi_f | \hat{\mathbf{m}} | \psi_i \rangle$ we must know the energy eigenstates ψ_i and ψ_f of the molecule. We obtain them by solving numerically (or analytically, when possible) the Schrödinger equation for the molecule, in the absence of light.

In Chapter 2 you encountered the notation $\langle \psi_f \mid \hat{O} \mid \psi_i \rangle$ which in our case translates to

$$\mathbf{m}_{if} \equiv \langle \psi_f | \, \hat{\mathbf{m}} \, | \psi_i \rangle$$

$$\equiv \int \psi_f(\mathbf{r}(1),\dots,\mathbf{r}(n),\mathbf{R}(1),\dots,\mathbf{R}(p))^* \, \hat{\mathbf{m}} \, \psi_i(\mathbf{r}(1),\dots,\mathbf{r}(n),\mathbf{R}(1),\dots,\mathbf{R}(p))$$

$$\times \, d\mathbf{r}(1)\cdots d\mathbf{r}(n) \, d\mathbf{R}(1) \cdots d\mathbf{R}(p) \tag{11.3}$$

For a molecule having n electrons and p nuclei, the integral is over $d\mathbf{r}_1,\dots,d\mathbf{r}_n$ positions of the electrons and over $d\mathbf{R}_1,\dots,d\mathbf{R}_p$ positions of the nuclei. Since the position of each particle is described by three coordinates, this is a $(3n + 3p)$-dimensional integral.

The transition dipole $\mathbf{m}_{if}$ is a vector and Eq. 11.1 contains its dot product with the polarization vector $\mathbf{e}$. This is equal to:

$$\langle \psi_f | \hat{\mathbf{m}} | \psi_i \rangle \cdot \mathbf{e} = \langle \psi_f | \hat{m}_x | \psi_i \rangle \cdot e_x + \langle \psi_f | \hat{m}_y | \psi_i \rangle \cdot e_y + \langle \psi_f | \hat{m}_z | \psi_i \rangle \cdot e_z \tag{11.4}$$

The quantities $\hat{m}_x$, $\hat{m}_y$, and $\hat{m}_z$ are the components of the vector $\hat{\mathbf{m}}$ defined by Eq. 11.2.

If the light is linearly polarized, we can choose the coordinate system so that the light propagates along the OZ axis, and $\mathbf{e}$ is parallel to the OX axis. This means that $\mathbf{e}$ has the components $\{e_x, e_y, e_z\} = \{1, 0, 0\}$. Consequently, Eq. 11.4 becomes

$$\langle \psi_f | \hat{\mathbf{m}} | \psi_i \rangle \cdot \mathbf{e} = \langle \psi_f | \hat{m}_x | \psi_i \rangle \tag{11.5}$$

As you will see in the next chapter, for a given initial state, the choice of the orientation of the polarization vector with respect to the absorbing system determines, in many cases, the nature of the final state that is excited by light absorption, and/or the magnitude of the absorption probability.

§8. *The Properties of the Molecule: the Line Shape.* The quantity

$$L(\Omega) = \frac{1}{\pi} \frac{\Gamma/2}{(\Omega - \omega_{if})^2 - (\Gamma/2)^2} \tag{11.6}$$

is called the *line-shape* of the transition $\psi_i \rightarrow \psi_f$. The *Lorentzian* function given by Eq. 11.6 is most commonly observed in experiments. The shape of this function is shown in Fig. 11.1. In this expression Ω is the frequency of a photon in the pulse of light and

$$\omega_{if} \equiv \frac{E_f - E_i}{\hbar} \tag{11.7}$$

is called the *transition frequency*.

Viewed as a function of the photon frequency Ω, the Lorentzian function has a peak when

$$\Omega = \omega_{if} \equiv \frac{E_f - E_i}{\hbar} \tag{11.8}$$

This condition is energy conservation.

There is a very peculiar thing about this function: it is not zero when

$$|\Omega - \omega_{if}| \leq \Gamma \tag{11.9}$$

Using this information in Eq. 11.1 indicates that there is a finite probability that the molecule will absorb a photon that does not satisfy the energy conservation condition, Eq. 11.8. Photons satisfying the conservation condition are absorbed with the highest probability, but there is a fair chance that photons whose frequency is not very different from ω_{if} are also absorbed. The parameter Γ, called the *line width*, controls by how much the energy conservation condition can be violated.

While we do not study line shapes here, it is necessary to explain why all spectra have one, and how quantum mechanics deals with this apparent violation of energy conservation. When we solved the Schrödinger equation in Chapter 8 sharply defined eigenvalues were obtained. We have also stated that when energy is measured, only values equal to the eigenvalues of the energy operator can be obtained. Now I am telling you that spectroscopic measurements show the existence of energies that are close to an eigenvalue, but can deviate from it by an amount of order Γ. How do we reconcile these contradictory statements, and what is the meaning of Γ?

If we examine a molecule that does not interact with anything (it is truly alone in the world), we obtain the well-defined energy eigenvalues that we have discussed so far. However, a molecule is never alone. Because it consists of charged particles, it generates an electromagnetic field and interacts with it. If we quantize this field and allow for the interaction of the molecular charges with the field, we find that the joint molecule–field system has many more eigenvalues than the isolated molecule. Some of these eigenvalues are equal to the energies $E_i, i = 1, 2, 3 \ldots$

of the isolated molecule (very slightly shifted by the interaction with the electromagnetic field). The others are very close to the E_i-s and they are involved in the adsorption processes that give the line-shape. These eigenvalues have been computed with an extraordinary accuracy, for the hydrogen atom, and they agree well with the experiments. There is no doubt that, in those cases that are simple enough that we can perform accurate calculations, this theory explains the most minute detail of the observed spectra, with greater accuracy than is needed in any practical application.

If a molecule is in contact with other molecules (i.e. it is in a gas, or dissolved in a liquid, or embedded in a solid), the interaction with them also "broadens" the absorption spectrum. In such cases the line shape can be complicated: it depends on temperature and contains information about the interaction between the molecule and its surroundings, and it is not necessarily a Lorentzian.

A different, but equivalent, interpretation of the line shape is provided by the fact that a molecule in an excited state has a finite lifetime; sooner or later it will have to emit a photon or give its energy to the medium in which it is embedded. One can show that in this situation a kind of *uncertainty principle* is valid: if an energy eigenstate has a life-time τ, then its energy is uncertain by an amount $\hbar/\tau$. This uncertainty accounts for the line width.

These results may seem strange but they are well anchored in both theory and experiments.

§9. *Put it all Together and Examine Eq. 11.1.* We can now go back to Eq. 11.1 and see how these quantities influence the transition probability P_{if}. Let us look first at the dot product between the transition dipole matrix element $\langle \psi_f \mid \hat{\mathbf{m}} \mid \psi_i \rangle$ and the polarization vector $\mathbf{e}$. If this product is zero, the probability of the transition $\psi_i \to \psi_f$ is zero. When this happens we say that the transition is forbidden. Note that this quantity depends on light polarization and it is possible that a transition that is forbidden for a certain polarization (e.g. linear polarization) may be allowed when the polarization is changed (e.g. to circular polarization).

The line-shape function $L(\Omega)$ makes sure that the transition $\psi_i \to \psi_f$ takes place only if the photon frequency Ω is very close to the transition frequency ω_{if}. 'Very close' means within the range allowed by the magnitude of the line-width Γ. $L(\Omega)$ is multiplied by $N(\Omega)$, which has a peak around a frequency Ω_0, called the *pulse frequency*. The transition probability is proportional to the product $L(\Omega)N(\Omega)$ and it differs from zero only if the pulse frequency Ω_0 is very close to the transition frequency ω_{if}. Another way to put it is to say that the transition probability differs from zero only if the line-shape $L(\Omega)$ overlaps with $N(\Omega)$. This overlapping condition is a generalization of the simple energy conservation rule that required that $\Omega = \omega_{\text{if}}$.

§10. *Trouble with Units Again!* The equation for the probability of photon absorption involves electrodynamic quantities, such as the charge of the particles and the amount of energy carried by the pulse of light. You must be aware that, as far as units are concerned, electrodynamics is worse than spectroscopy: the formulae are different when different units are used.

Eq. 11.1 is given in the rationalized MKSA system of units in which the electron charge is $q = -1.60 \times 10^{-19}$ C, the length is given in meters, the permittivity of vacuum is $\epsilon_0 = 10^7/c^2$ f/m, the speed of light is $c = 2.997 \times 10^8$ m/s, Planck's constant is $\hbar = 1.054 \times 10^{-34}$ J s, and $N(\omega)$ is given in J s/m^2.

Many standard textbooks give the formula for the excitation probability in the Gaussian system of units, in which the equation is:

$$P_{\text{i} \to \text{f}} = \frac{4\pi^2}{c\hbar^2} \, |\langle \psi_\text{f} \,|\, \hat{\mathbf{m}} \cdot \mathbf{e} | \psi_\text{i} \rangle|^2 N(\omega_{\text{if}}) \tag{11.10}$$

In this system the speed of light is $c = 2.997 \times 10^{10}$ cm/s, Planck's constant is $\hbar = 1.054 \times 10^{-27}$ erg s, the electron charge is $q = -4.80 \times 10^{-10}$ esu (electrostatic units of charge), and $N(\omega)$ has units of erg s/cm^2.

§11. *How is the Transition Probability Formula Used?* It would seem that we should be able to use the equation Eq. 11.1 to calculate the absorption probability and compare it to the peaks in the absorption spectrum. However, such comparisons are rarely made because the absolute intensity of a spectrum is difficult to measure. This would require the experimentalist to "count" all the photons absorbed by the sample. This can be done only if the number of photons interacting with the molecules is known accurately, if the efficiency of the light detector is carefully calibrated, and if the geometry of the apparatus is accounted for. It is rare that the results of such analysis are worth the trouble. The most common use of the equation for the probability is to analyze the ratios of the intensities of any two absorption peaks, measured in the same apparatus, with the same light source and source–sample–detector geometry. This simplifies the analysis greatly, because the properties of the light source and the geometry of the apparatus are the same for any two peaks in the absorption spectrum, and these quantities drop out when the ratio is calculated.

§12. *The Probability of Stimulated Emission.* Eq. 11.1 has an interesting symmetry. If the molecule is in a high energy state ψ_f, before interacting with the pulse of light, then exposing it to light of frequency $(E_f - E_i)/\hbar$ will stimulate the emission of a photon of the same frequency, by causing the transition $\psi_f \rightarrow \psi_i$ to the lower energy state ψ_i. The probability $P_{f \rightarrow i}$ of this transition is equal to the absorption probability $P_{i \rightarrow f}$. This follows from the fact that

$$|\langle \psi_f \,|\, \hat{\mathbf{m}} \,|\, \psi_i \rangle|^2 = |\langle \psi_i \,|\, \hat{\mathbf{m}} \,|\, \psi_f \rangle|^2$$

You can verify easily that this equation is correct by using the definition of the symbols involved in it (see Chapter 2).

This conclusion is in agreement with Einstein's theory that says that the Einstein coefficient B_{fi} of stimulated photon emission is equal to the coefficient B_{if} for absorption (see Chapter 10, §10, Eq. 10.11).

§13. *Validity Conditions.* Eq. 11.1 is valid under the following conditions:

(a)   The light intensity is sufficiently low. One criterion for this is that the probability $P_{i \rightarrow f}$ should be at most 0.2. Any value higher than that makes the use of this equation questionable. A practical test in the laboratory is to check whether the absorption efficiency is proportional to the intensity of the laser.

(b)   The size of the molecule is much smaller (e.g. by a factor of 5 or more) than the wavelength of light. This condition is satisfied for microwave, infrared, visible and near-UV light, but not for X-rays.

(c)   The light in the pulse is linearly polarized and is streaming in one direction. However, no matter what kind of light is used, the absorption probability is proportional to $|\langle \psi_f \,|\, \hat{\mathbf{m}} \cdot \mathbf{e} \,|\, \psi_i \rangle|^2$ and to the intensity of light as long as condition (b) is satisfied.

The Connection to Einstein's B Coefficient

§14. Einstein's theory of absorption showed that the emission and absorption rate are described by the B coefficient. This is the absorption rate coefficient in the equation (see Chapter 10, §5)

$$\left\{ \frac{dN_0(t)}{dt} \right\}_{abs} = -B\rho(\Omega)N_0(t) \tag{11.11}$$

Here $N_0(t)$ is the concentration of ground-state molecules and $\rho(\Omega)dVd\Omega$ is the number of photons in the volume dV, having frequency between Ω and $\Omega + d\Omega$. In Chapter 10 I used the notation B_{01} but here I drop the subscript.

It should be clear that there must be a connection between Eq. 11.1 and the B coefficient: they both describe the ability of a molecule to absorb a photon of a certain frequency. However, $P_{i \rightarrow f}$ is a probability that the molecule is in the state f, after interacting with a pulse of polarized light, moving in a given direction; B, on the other hand, is the rate of the transition i $\rightarrow$ f when the molecule is continously exposed to light moving in all directions and having a random polarization.

It is possible to start with $P_{i \rightarrow f}$ and derive a formula for B. Here I outline how such a calculation proceeds. Since the B coefficient is less useful than $P_{i \rightarrow f}$, I will not give the detailed derivation.

A continuous source of light consists of a superposition of overlapping pulses. Let us denote by n the number of pulses passing through the location of the molecule per second. The absorption rate by one molecule bathed in continuous light is then (in the Gaussian system of units):

$$R_{i \rightarrow f} = nP_{i \rightarrow f} = \frac{4\pi^2}{c\hbar^2} |\langle \psi_f | \hat{\mathbf{m}} | \psi_i \rangle \cdot \mathbf{e}|^2 N(\omega_{if})n \qquad (11.12)$$

To conform with Einstein's model, I took the line-shape function $L(\Omega)$ to be equal to 1 when $\Omega = \omega_{if}$ and zero otherwise. The quantity $N(\omega_{if})n\,d\omega_{if}$ is the total amount of energy passing through the unit area (of a plane perpendicular to the direction of propagation of the pulse), per unit time. In other words this is the intensity of light integrated over the time the molecule is in contact with the pulse. One can show that

$$N(\omega_{if})n = \rho(\omega_{if})c \qquad (11.13)$$

where $\rho(\omega_{if})d\omega$ is the energy, per unit volume, of the photons having a frequency between ω_{if} and $\omega_{if} + d\omega_{if}$. Combining this with Eq. 11.12 leads to

$$R_{i \rightarrow f} = nP_{i \rightarrow f} = \frac{4\pi^2}{\hbar^2} |\langle \psi_f | \hat{\mathbf{m}} \cdot \mathbf{e} | \psi_i \rangle|^2 \rho(\omega_{if}) \qquad (11.14)$$

This formula assumes that the light moves in one direction and it is polarized. We need to average this over all directions of propagation and over all polarizations, because this corresponds to the black body box considered by Einstein. After this averaging the rate is

$$\overline{R}_{i \to f} = \frac{4\pi^2}{3\hbar^2} |\langle \psi_f \,|\, \hat{m}_x^2 + \hat{m}_y^2 + \hat{m}_z^2 \,|\, \psi_i \rangle|^2 \rho(\omega_{if}) \qquad (11.15)$$

The absorption rate by N molecules in a unit volume is $\overline{R}_{i \to f} N$ and this is equal to $BN\rho(\omega_{if})$. Comparing these expressions to each other gives a formula for the B coefficient.

Single Molecule Spectroscopy and the Spectroscopy of an Ensemble of Molecules

§15. There is an important difference between the absorption rate per molecule (i.e. R) and the rate of absorption in a single-molecule experiment. R is an average quantity: to obtain it we must measure the rate of absorption (i.e. the number of photons absorbed per unit time) by an ensemble of N molecules (where N is very large) and divide it by N. In such an experiment a large number of photons stream through the sample and the rate tells me how many are lost per unit time. Since the number of photons is very large the measured intensity of the light emerging from the sample is a continuous quantity.

This is not what is observed in a single-molecule experiment. To do such an experiment, I would spread on a glass slide an extremely dilute solution of the molecule I want to study. Then I will shine light on the slide through a microscope objective, to create a very small spot. Since the spot is small and the solution is very dilute, the spot will contain either one molecule or none.

To do the experiment choose a molecule that absorbs a photon of frequency Ω, undergoes a radiationless transition to a new state, and then emits a photon of frequency Ω_1. Use a continuous source of light of frequency Ω and detect light of frequency Ω_1 coming from the illuminated spot. Even though the molecule is bathed by a continuous stream of incident photons, the emission is discrete: I detect individual photons of frequency Ω_1, arriving at the detector at discrete times. For a while nothing happens, then a photon of frequency Ω_1 arrives at the detector, at time t_1, then nothing happens, then a photon arrives at time t_2, etc. If you repeat the same experiment, with the same molecule, with the same experimental set-up, you get a different time series $t_1, t_2, \ldots$. However, the average time $(t_1 + t_2 + \cdots + t_m)/m$ obtained in the two experiments is *the same*, if m is large enough (about 1000).

Because of what you learned in Chapter 5, this will not surprise you. Quantum mechanics tells us that we do not know the exact time when a photon is absorbed or emitted, but only the probability that this happens at a given time. Emission and absorption are games of chance.

The emission times reflect several processes. If the molecule is in the ground state at time zero, a photon is absorbed at time τ_1, then the molecule undergoes a radiationless transition a time τ_2 later, and emits a photon after a time τ_3. We have $t_1 = \tau_1 + \tau_2 + \tau_3$.

This is very different from what we observe when we perform an absorption or an emission experiment on an ensemble of N molecules. But these observations must be connected in some way. Consider, for simplicity, an ensemble of absorbing molecules that do not interact with each other (i.e. a dilute solution or a gas). Each molecule interacts with light on its own: it does not matter whether the other molecules are excited or not. Therefore, what we see when we shine light on the ensemble is the sum of the events undergone by the individual molecules. Photon absorption and emission occur at "random" times; but, because the experiment involves so many molecules and so many photons, the time difference between the arrival of two photons at the detector is so small that we cannot resolve it with the existing devices. It appears to us that the emission and the absorption are continuous processes. Moreover, they are reproducible since we measure the average number of photons emitted per unit time.

Single-molecule experiments give very interesting results when used to study biological systems. For example, in favorable cases we can determine the times at which a molecule enters or leaves the pocket of an enzyme if the enzyme has a fluorescent residue whose emission frequency changes when a molecule enters the pocket.

Exercise 11.1

The energy eigenstates of a harmonic oscillator, which represents a vibrating diatomic molecule, are

$$\psi_0(x) = \left(\frac{m\omega}{\hbar\pi}\right)^{\frac{1}{4}} \exp\left[-\frac{m\omega x^2}{2\hbar}\right]$$

$$\psi_1(x) = \left(\frac{4m^3\omega^3}{\hbar^3\pi}\right)^{\frac{1}{4}} x \exp\left[-\frac{m\omega x^2}{2\hbar}\right]$$

$$\psi_2(x) = \left(\frac{m\omega}{\hbar\pi}\right)^{\frac{1}{4}} \left(\frac{2mx^2\omega}{\hbar} - 1\right) \exp\left[-\frac{m\omega x^2}{2\hbar}\right]$$

Here, $x = r - r_0$ is the deviation of the bond length r from the equilibrium bond length r_0.

(a) Calculate the transition dipole moment for a diatomic molecule exposed to light polarized along the molecular axis (**e** has the same direction as the axis of the molecule).

(b) Calculate the transition dipole in the case when **e** is perpendicular to the molecular axis.

(c) Calculate the transition dipole in the case when **e** makes an angle of 30 degrees with the molecular axis.

12

LIGHT EMISSION AND ABSORPTION BY A PARTICLE IN A BOX AND A HARMONIC OSCILLATOR

Introduction

§1. In this chapter I use Eq. 11.1 to examine light absorption by two systems: a quantum dot and a harmonic oscillator.

A quantum dot is a small parallelepiped made of a doped semiconductor (e.g. germanium with trace amounts of arsenic). The few electrons produced by doping move freely (without interacting with the ions or with each other) in the box formed by the semiconductor. The energy eigenstates and eigenvalues for such a system were derived in Chapter 8.

If a diatomic molecule is in the gas phase, its infrared spectrum is determined by its vibrations *and* its rotation. However, if the molecule is embedded in an ice of argon or it is bound to a solid surface, it will no longer rotate. In this case the vibrational motion alone determines the absorption spectrum. Because the bond in a diatomic molecule is very stiff, its vibrational motion is similar to that of two

169

atoms connected by an elastic spring. This is called a *harmonic oscillator* and we know its energy eigenvalues and eigenstates.

We will use Eq. 11.1 to calculate, for both systems, the probability that the system absorbs a photon when it is exposed to a pulse of light. These calculations reveal two general features.

Firstly, if we are guided by energy conservation only, then we might believe that whenever the system is in the state ψ_i and the frequency of light Ω is very close to $(E_f - E_i)/\hbar$, the molecule has a finite probability $P_{i \to f}$ to absorb a photon and undergo the transition $\psi_i \to \psi_f$. Here ψ_i and ψ_f are eigenstates of the Hamiltonian operator and E_i and E_f are the corresponding energies.

You will learn here that the statement above is not always true. When we use Eq. 11.1 to evaluate the absorption probability, we find that for some pairs of states ψ_i and ψ_f, the transition dipole $\langle \psi_f | \hat{\mathbf{m}} | \psi_i \rangle$ is equal to zero. This means that the absorption of a photon of frequency $\Omega = (E_f - E_i)/\hbar$ is impossible. When this happens we say that the transition $\psi_i \to \psi_f$ is forbidden.

Secondly, Eq. 11.1 indicates that the probability of photon absorption depends on the polarization of light (i.e. on the orientation of $\mathbf{e}$, which gives the direction of the electric field of light). If the molecule does not rotate, we can determine its orientation by studying how its ability to absorb photons changes with the direction of $\mathbf{e}$.

Light Absorption by a Particle in a Box

§2. *Quantum Dots.* The initial justification of the particle-in-a-box model was purely pedagogical: it is the simplest model one can imagine. It was created for displaying how quantum mechanics works, without the smokescreen created by tedious mathematics.

In the past few decades, people have managed to make systems whose behavior is well approximated by the particle-in-a-box model. An example is shown in Fig. 12.1. This shows a pillar whose base is made of AlAs and whose top (gray in Fig. 12.1) is made from GaAs. The pillar is grown on a flat surface, not shown in the figure. With modern processing technology, one can cover a flat surface with a periodic array of such pillars. The base of the pillars could be as small as $100 \, \text{Å} \times 100 \, \text{Å}$ and they could have an even smaller height. This is the system that we will study here.

The technology used to make such a structures is rather spectacular and it is worth knowing, at least as an example of the marvels that chemistry can produce. One starts with a very flat surface, which will serve as a support. This is placed in an ultra-high-vacuum chamber (a pressure of 10^{-10} Torr) where it is cleaned thoroughly. Then, a beam of Al and As vapor is sent towards the surface. These atoms bind to the hot surface, move around, and arrange themselves into a very flat film, of

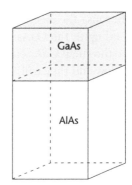

Figure 12.1 A pillar made of AlAs and GaAs. One usually makes millions of such structures supported on a flat, solid surface.

a perfect AlAs crystal. The technology is so advanced that one can control how many atomic layers the film will have. Once the film reached the desired thickness, one sends towards the surface a beam of Ga and As atoms, which form a perfectly crystalline, very flat film of GaAs. The end result is a solid support with a film of AlAs on it, covered by a film of GaAs. Because of the high vacuum and the high purity of the vapor, these films are extremely pure chemically, which is essential for what follows.

This sandwich is the starting point in the fabrication of the pillars that interest us. The GaAs film can be covered with a photosensitive polymer and then light or X-rays or an electron beam can be used to make periodic holes in the polymer. Then, this surface can be exposed to metal vapor to deposit metal in the holes, on top of the GaAs, and on the polymer. Next, we dissolve the polymer in a solvent and are left with a periodic array of thin metal "islands" on top of GaAs (the islands are located where the holes in the photoresist were). Next, we etch the GaAs and the AlAs films, all the way down to the support. The metal islands protect the material underneath them from etching. The result is a periodic array of pillars like the ones shown in Fig. 12.1, but with a metal island on top. We end the fabrication by dissolving the metal with the appropriate reagent.

In what follows, I concentrate on the GaAs parallelepiped. The electrons in a very pure GaAs or AlAs crystal form the chemical bonds and are tied tightly to the atoms. Since they cannot move freely through the material the two crystals have no electrical conductivity.

A very neat trick, however, allows us to add mobile electrons in the GaAs sector of the pillar. When we make the pillar, a few Si atoms can be placed in the AlAs. It so happens that Si embedded in AlAs will ionize and donate an electron to the AlAs solid. Since all the bonds of the atoms in the AlAs are already formed, the additional electrons do not bind to any of the atoms and move freely through the AlAs.

Because the electron energy in GaAs is lower than in AlAs, the additional electrons end up in the GaAs layer. Thus, by adding Si to AlAs, we add additional electrons to the small GaAs cube. If I add only a few free electrons, their density in the GaAs is low and they will interact very weakly with each other; they will behave as if each is alone inside the GaAs box.

In the theory of a particle in a box, developed in Chapter 8, we assumed that the particle cannot get out of the box. This is not true for the electrons in the GaAs box. If the electrons absorb a high-energy photon, they aquire enough energy to get out into the vacuum. However, the energy of the electrons exposed to low frequency light is small compared to the energy needed to get out of the box. We can therefore assume that the electron cannot get out at all.

This is not science fiction, though it sounds more complicated than anything Rube Goldberg managed to invent. If we perform light absorption experiments on this system we find that the frequency of the absorbed photons matches very well the values calculated by assuming that the absorbing particle is a free electron in a box. Moreover, there are sound theoretical reasons for this free-electron behavior.

Devices like these are made in many laboratories in the world. The lasers in your CD player or in a supermarket scanner are made by a simpler version of such systems.

So here we are: we have created an array of very small GaAs boxes containing freely moving electrons, which behave as if each is alone in the box. In this section we examine an experiment in which we shine light at the surface (with the pillars on it) and measure how much light is reflected. When the intensity of reflected light is plotted, as a function of light frequency, the graph has dips at certain frequencies (at which photon absorption occurs). The depth of these dips is proportional with the absorption probability given by Eq. 11.1.

§3. *Photon Absorption Probability.* To calculate the absorption probability $P_{i \to f}$, I will use Eqs 11.1 and 11.2, which I repeat here. In the MKSA system of units:

$$P_{i \to f} = \frac{\pi}{\epsilon_0 c \hbar^2} |\langle \psi_f \,|\, \hat{\mathbf{m}} \cdot \mathbf{e} \,|\, \psi_i \rangle|^2 N(\omega_{if}) \qquad (12.1)$$

and $\hat{\mathbf{m}}$ is the *dipole moment operator* defined by

$$\hat{\mathbf{m}} = \sum_{\alpha=1}^{n} q\mathbf{r}(\alpha) - \sum_{\beta=1}^{p} qZ_\beta \mathbf{R}(\beta) \qquad (12.2)$$

For simplicity, I have taken the line shape $L(\Omega)$ to be equal to 1 if $\Omega = \omega_{if}$ and zero otherwise.

§4. *The Amount of Light and its Frequency.* Recall that $N(\Omega)$ is the amount of electromagnetic energy, having the frequency Ω, that passes through a unit area perpendicular to the direction of propagation of the pulse of light. The magnitude of $N(\Omega)$ depends on the light source used in the experiment and the position of the sample with respect to the source. Absorption takes place only if the pulse of light is such that $N(\omega)$ has a large value for one of the transition frequencies $\omega_{if} = (E_f - E_i)/\hbar$.

In what follows I will not attempt to calculate $N(\Omega)$, since it depends on the specific experimental set-up. This means that I will not be able to calculate the absolute value of the absorption probability, but I can understand the relative magnitude of different peaks in the absorption spectrum (or dips in the reflection spectrum). Because of practical difficulties in determining precisely the fraction of light absorbed, most analysis of spectra is confined to attempts to understand the relative magnitude of the peaks (i.e. to understand why some peaks are higher than others).

§5. *The Energies and the Absorption Frequencies.* In Chapter 8, I derived the following equation (see Eq. 8.33) for the energy eigenstates of a particle in a box:

$$E(n,j,k) = E_x(n) + E_y(j) + E_z(k)$$

$$= \frac{\hbar^2 \pi^2}{2\mu} \left[\left(\frac{n}{L_x} \right)^2 + \left(\frac{j}{L_y} \right)^2 + \left(\frac{k}{L_z} \right)^2 \right] \tag{12.3}$$

Here n, j, and k are the *quantum numbers* and they take the values

$$\left. \begin{array}{l} n = 1, 2, 3, \ldots \\ j = 1, 2, 3, \ldots \\ k = 1, 2, 3, \ldots \end{array} \right\} \tag{12.4}$$

The mass of the electron is μ, and L_x, L_y, and L_z are the lengths of the sides of the parallelepiped. The coordinate system has its origin at a corner of the parallelepiped and the axes are parallel with the edges of the parallelepiped (see Fig. 8.1) in which the electron is trapped.

I will assume in what follows that the initial state corresponds to the quantum numbers $\{n,j,k\} = \{1,1,1\}$. Therefore, the box will only absorb photons having one of the frequencies

$$\omega_{if} = \frac{E_f - E_i}{\hbar} \equiv \frac{E_{n,j,k} - E_{1,1,1}}{\hbar} \tag{12.5}$$

which can be calculated by using Eqs 12.3 and 12.4. Some of the possible absorption frequencies are shown in Table 8.1.

Exercise 12.1

An electron is trapped in a parallelepiped of whose sides are equal to $L_x = 100$ Å, $L_y = 200$ Å, and $L_z = 1000$ Å. Calculate the frequencies of the transitions from $\{n,j,k\} = \{1,1,1\}$ to $\{2,1,1\}$, $\{1,2,1\}$, $\{1,1,2\}$, $\{3,1,1\}$, $\{1,3,1\}$, $\{1,1,3\}$, $\{2,2,1\}$, $\{1,2,2\}$, and $\{2,1,2\}$. Try first to predict which transitions have higher frequencies and which have the smallest frequency, without doing any computations.

§6. *The Transition Dipole.* To make further progress in understanding light absorption by an electron in a box, I have to calculate the transition dipole moment for this system. I will do this in stages.

1. I collect the expressions for the initial and final wave functions, from Chapter 8.

2. I establish the correct expression for the dipole operator $\hat{\mathbf{m}}$.

3. I use this information to calculate the magnitude of the transition dipole.

§7. *The Energy Eigenfunctions for a Particle in a Box.* The energy eigenfunctions for a particle in a box are given by Eq. 8.31 as:

$$\phi_{n,j,k}(x,y,z) = \left(\frac{8}{L_x L_y L_z}\right)^{1/2} \sin\left(\frac{\pi n x}{L_x}\right) \sin\left(\frac{\pi j y}{L_y}\right) \sin\left(\frac{\pi k z}{L_z}\right) \tag{12.6}$$

The quantum numbers n, j, and k can take only the values allowed by Eq. 12.4. I have shown in Chapter 8 that this is a set of orthonormal functions, which means that

$$\langle a,b,c \mid n,j,k \rangle \equiv \langle \phi_{a,b,c} \mid \phi_{n,j,k} \rangle \equiv \int \phi^*_{a,b,c} \phi_{n,j,k}\, dx\, dy\, dz = \delta_{a,n} \delta_{b,j} \delta_{c,k} \tag{12.7}$$

with the Kronecker delta given by:

$$\delta_{m,h} = \begin{cases} 0 & \text{if } m \neq h \\ 1 & \text{if } m = h \end{cases} \qquad (12.8)$$

In the equations above, a, b, c, n, j, k, m, and h are integers.

§8. *The Dipole Operator* $\hat{\mathbf{m}}$. The GaAs quantum dot, which is the physical realization of the particle-in-a-box model, contains a large number of electrons and nuclei. However, most electrons are tied up in chemical bonds and absorb only photons having high energy. If we work with light of low frequency, which we do, we need not worry about them. The only electrons of interest to us are the ones introduced by doping. These are few and because the dot is fairly roomy (for the electrons), they interact weakly with each other. We can therefore assume that each electron absorbs light as if it is alone in the dot. We can then calculate the probability that *one* such electron will absorb a photon, and obtain the number of photons absorbed by the dot by multiplying this probability with the number of "free" electrons in the dot. The dipole operator $\hat{\mathbf{m}}$ that appears in Eq. 12.1 is therefore (see Eq. 12.2)

$$\hat{\mathbf{m}} = -q\hat{\mathbf{r}} \qquad (12.9)$$

where $\hat{\mathbf{r}}$ is the position operator of the electron and q is the absolute value of the electron charge. According to Eq. 12.2, we must add to this the dipole operator of the charged nuclei. However, since only light absorption by electrons was considered here we will ignore this contribution to the absorption probability. This is an approximation made mostly because of the introductory nature of this text. Had we been serious about the results, we would have included the nuclear contribution to account for the possibility that photon absorption may take place with the simultaneous excitation of the electronic and the nuclear motion.

The outcome of this argument is that Eq. 12.9 will be used for $\hat{\mathbf{m}}$ in the equation for $P_{i \to f}$. Therefore, the transition dipole that we want to calculate is

$$\langle \psi_f | \hat{\mathbf{m}} | \psi_i \rangle = -\langle \psi_f | q\hat{\mathbf{r}} | \psi_i \rangle \equiv -q\langle \phi_{n,j,k} | \hat{\mathbf{r}} | \phi_{1,1,1} \rangle \qquad (12.10)$$

In this equation we use the symbols ψ_i and ψ_f for the initial and final state of a general system and $\phi_{n,j,k}$ and $\phi_{1,1,1}$ for the initial and final states in the case of the particle in a box.

§9. *The Role of Light Polarization.* We are making progress, but are not done yet. The transition dipole is a vector which appears, in Eq. 12.1, in a dot product with the polarization vector **e** of the light. I will assume in what follows that the experimentalist has mounted the sample so that the polarization of light is perpendicular to the surface of the support, and therefore parallel to the pillars shown in Fig. 12.1. This means that $\mathbf{e} = \{1, 0, 0\}$ and we have

$$\langle \psi_\mathrm{f} \,|\, \hat{\mathbf{m}} \,|\, \psi_\mathrm{i} \rangle \cdot \mathbf{e} = q \langle \psi_\mathrm{f} \,|\, \hat{\mathbf{r}} \,|\, \psi_\mathrm{i} \rangle \cdot \mathbf{e} = q \langle \psi_\mathrm{f} \,|\, \hat{x} \,|\, \psi_\mathrm{i} \rangle$$

$$= q \langle \phi_{n,j,k} \,|\, \hat{x} \,|\, \phi_{1,1,1} \rangle \tag{12.11}$$

§10. *The Evaluation of the Transition Moment for a Particle in a Box.* As explained in Chapter 3, the symbol $\langle \phi_{n,j,k} \,|\, \hat{x} \,|\, \phi_{1,1,1} \rangle$ is a shorthand for a triple integral over the position of the electron:

$$\langle n, j, k \,|\, \hat{x} \,|\, 1, 1, 1 \rangle \equiv \langle \phi_{n,j,k} \,|\, \hat{x} \,|\, \phi_{1,1,1} \rangle$$

$$= \int_{-\infty}^{\infty} dx \int_{-\infty}^{\infty} dy \int_{-\infty}^{\infty} dz \, \phi_{n,j,k}(x,y,z)^*$$

$$\times x \, \phi_{1,1,1}(x,y,z) \tag{12.12}$$

We can now use Eq. 12.6 for the wave functions $\phi_{n,j,k}(x,y,z)$ and $\phi_{1,1,1}(x,y,z)$ in Eq. 12.12 to obtain:

$$\langle n, j, k \,|\, \hat{x} \,|\, 1, 1, 1 \rangle \equiv \int_{-\infty}^{+\infty} dx \, dy \, dz \, \phi_{n,j,k}(x,y,z)^* \, x \, \phi_{1,1,1}(x,y,z)$$

$$= \left[\frac{2}{L_x} \int_0^{L_x} dx \, \sin\left(\frac{n\pi x}{L_x}\right) x \, \sin\left(\frac{1\pi x}{L_x}\right) \right]$$

$$\times \left[\frac{2}{L_y} \int_0^{L_y} dy \, \sin\left(\frac{j\pi y}{L_y}\right) \sin\left(\frac{1\pi y}{L_y}\right) \right]$$

$$\times \left[\frac{2}{L_z} \int_0^{L_z} dz \, \sin\left(\frac{k\pi z}{L_z}\right) \sin\left(\frac{1\pi z}{L_z}\right) \right]$$

$$\equiv \langle n \,|\, \hat{x} \,|\, 1 \rangle \, \langle j \,|\, 1 \rangle \, \langle k \,|\, 1 \rangle \tag{12.13}$$

The last equality contains the usual shorthand symbols for the one-dimensional integrals shown below:

$$\langle n \mid \hat{x} \mid 1 \rangle = \frac{2}{L_x} \int_0^{L_x} dx \, \sin\left(\frac{n\pi x}{L_x}\right) x \sin\left(\frac{\pi x}{L_x}\right) \tag{12.14}$$

$$\langle j \mid 1 \rangle = \frac{2}{L_y} \int_0^{L_y} dy \, \sin\left(\frac{j\pi y}{L_y}\right) \sin\left(\frac{\pi y}{L_y}\right) \tag{12.15}$$

$$\langle k \mid 1 \rangle = \frac{2}{L_z} \int_0^{L_z} dz \, \sin\left(\frac{k\pi z}{L_z}\right) \sin\left(\frac{\pi z}{L_z}\right) \tag{12.16}$$

The integrals can be evaluated by using the methods you have learned in calculus or a symbolic manipulation program (see Workbook QM12.1). The results are:

Workbook

$$\langle n \mid \hat{x} \mid 1 \rangle = -\frac{4L_x n \left(1 + \cos\left(n\pi\right)\right)}{\pi^2 \left(n^2 - 1\right)} \tag{12.17}$$

$$\langle j \mid 1 \rangle = \delta_{j1} \tag{12.18}$$

$$\langle k \mid 1 \rangle = \delta_{k1} \tag{12.19}$$

Combining these results as shown by the last equality in Eq. 12.13 leads to our final result for the integral:

$$\langle n, j, k \mid \hat{x} \mid 1, 1, 1 \rangle = -\delta_{j,1}\delta_{k,1}\frac{4L_x n \left(1 + \cos\left(n\pi\right)\right)}{\pi^2 \left(n^2 - 1\right)} \tag{12.20}$$

Insert this in the formula for the absorption probability, Eq. 12.1, and obtain

$$P_{i\to f} = \frac{\pi}{\epsilon_0 c \hbar^2} |\langle \psi_f \mid \hat{\mathbf{m}} \cdot \mathbf{e} \mid \psi_i \rangle|^2 N(\omega_{if})$$

$$= \frac{q^2 \pi}{\epsilon_0 c \hbar^2} \delta_{j,1}\delta_{k,1}\left(\frac{4L_x n \left(1 + \cos\left(n\pi\right)\right)}{\pi^2 \left(n^2 - 1\right)}\right)^2 N(\omega_{if}) \tag{12.21}$$

I have used here the fact that $\delta_{ij}^2 = \delta_{ij}$, which you can easily verify from the definition of δ_{ij} (Eq. 12.8).

This equation is valid only if the light polarization is oriented in the OX direction, which is perpendicular to the support and parallel to the pillars shown in Fig. 12.1.

§11. *The First Selection Rule.* The peak height in the absorption is proportional to the probability of photon absorption given by Eq. 12.21. I remind you that this equation is valid for light whose polarization is such that the electric field vector points in the OX direction, which is parallel to the pillars shown in Fig. 12.1. The presence of the Kronecker deltas in this formula tells us that if the initial state is $\{1, 1, 1\}$ then the final state must be of the form $\{m, 1, 1\}$, where m is an unspecified integer. In other words *if* e *is oriented in the OX direction, photon absorption cannot change the quantum numbers j and k associated with the OY and OZ directions.*

The fact that the transitions $\{1, 1, 1\} \rightarrow \{n, j \neq 1, k \neq 1\}$ are forbidden means that Einstein's B coefficients for such transitions are zero. But the A coefficients, which control the rate of photon emission in the transitions $\{n, j \neq 1, k \neq 1\} \rightarrow \{1, 1, 1\}$, are proportional to the B coefficients for those transitions; therefore, the A coefficients for the $\{n, j \neq 1, k \neq 1\} \rightarrow \{1, 1, 1\}$ transitions are also equal to zero. This means that the emission of photons *polarized in the OX direction* in the transitions $\{n, j \neq 1, k \neq 1\} \rightarrow \{1, 1, 1\}$ is forbidden.

We have already remarked that $P_{i \rightarrow f} = P_{f \rightarrow i}$, which immediately tells us that if a molecule does not absorb a photon in a transition i→f, then it will not emit a photon in a transition f→i. The particular conclusions reached for the particle in a box are in agreement with this general statement.

§12. *Physical Interpretation.* To understand why this happens and what it means, I remind you that in a state $\{n, j, k\}$ the first quantum number (n in this case) describes the *kinetic energy along the x axis* (in classical physics this would be $mv_x^2/2$). The fact that the absorption of a photon polarized in the OX direction can only change n means that the interaction with light can only increase the kinetic energy of the electron motion in the OX direction; the kinetic energy in the OY and OZ directions will remain unchanged. This is reasonable. The electric field is the force with which light acts on the electron. If the force is in the direction OX, then it could not increase the kinetic energy in the direction OY or OZ. This interpretation is correct only for systems like the particle in a box, where the only force acting, on the particle is the electric force due to light.

§13. *The Second Selection Rule.* So far we have said nothing about the transitions $\{1, 1, 1\} \rightarrow \{n, 1, 1\}$. Are these transitions possible regardless of the value of the integer n?. The answer is no. In Eq. 12.21, the expression $1 + \cos(n\pi)$ has the property

$$
\left.\begin{aligned}
1 + \cos(n\pi) = 0 \quad &\text{if } n \text{ is odd} \\
1 + \cos(n\pi) = 2 \quad &\text{if } n \text{ is even}
\end{aligned}\right\} \tag{12.22}
$$

This means that the probability of transition $\{1,1,1\} \to \{n,1,1\}$ is

$$
\left.\begin{aligned}
P_{\{1,1,1\}\to\{n,1,1\}} &= 0 & &\text{if } n \text{ is odd} \\
P_{\{1,1,1\}\to\{n,1,1\}} &= \frac{q^2\pi}{\epsilon_0 c\hbar^2}\left(\frac{8L_x n}{\pi^2(n^2-1)}\right)^2 & &\text{if } n \text{ is even}
\end{aligned}\right\} \tag{12.23}
$$

We have now an additional *selection rule*: if the initial state of the particle is $\{n = 1, j = 1, k = 1\}$, and the light is polarized in the x-direction, then transitions to states with n even are allowed and those to states with n odd are forbidden.

I can now summarize all selection rules for the particle in a box: if the initial state of the system is $i = \{n = 1, j = 1, k = 1\}$, and the light is polarized in the x-direction, then only transitions to the final states $f = \{n, j = 1, k = 1\}$ with even n are allowed; all other transitions are forbidden. Furthermore, if the particle is excited to a state $\{n, 1, 1\}$ then it can emit light polarized in the x-direction, in a transition to the state $\{1, 1, 1\}$, only if n is even.

Here are a few parting remarks concerning the transition moment and the selection rules. Keep in mind that these particular rules are strictly valid for the model employed in the calculations. Here it is assumed that the particle-in-a-box model describes the system accurately and that the electrons in the box do not interact with each other. If one of these approximations does not hold, the selection rules can be violated. If the model describes reality closely but not exactly, then the forbidden transitions will be observed in the laboratory but they will have small probabilities and the peaks corresponding to them in the absorption spectrum will be small. Similar comments are valid for the rules concerning emission.

Finally, I must warn you of a hidden approximation: I assumed that the polarization of the light can be controled and, in particular, that I can take it along the x-axis (for example). Life is not that simple. The polarization of the light coming out of the laser can be controled, but to calculate the absorption probability I need the electric field **E** *at the position of the electron inside the box*. In the example used here, an array of quantum dots supported on a flat surface, the presence of the quantum dots and of the supporting surface *modifies the electric field of the light*. We can calculate this modification by solving Maxwell's equations for the electromagnetic radiation. This will tell us that even though the electric field vector **E** of the light coming out of the laser is oriented along the x-direction, the electric field inside the box is not. Moreover, $\mathbf{E}$ has different values at different locations $\mathbf{r}$ in the box. The function $\mathbf{E}(\mathbf{r})$ must be used to calculate the term $\langle \psi_f \,|\, \mathbf{r} \cdot \mathbf{E}(\mathbf{r}) \,|\, \psi_i \rangle$. This transition dipole depends on the polarization of the light coming out of the laser but the dependence is rather complicated. The selection rules are no longer strictly valid.

Exercise 12.2

(a) Calculate the absorption probability and derive the selection rules, for a particle in a box, for the general transition $\{n_i, j_i, k_i\} \rightarrow \{n_f, j_f, k_f\}$, where $n_i, j_i, k_i, n_f, j_f, k_f$ can take any of the values $1, 2, \ldots$. If you need help, look in Workbook QM12.2.
(b) Solve Problem (a) for the polarization vector $\mathbf{e} = \{0.6, 0.8, 0\}$.

§14. *A Calculation of the Spectrum in Arbitrary Units.* As already mentioned, it is customary to measure the relative absorption spectrum. This means that height of all the peaks is divided by the height of one of the peaks, chosen as a reference. This is equivalent to saying that the height of reference peak is chosen as a unit, or that the spectrum is given in arbitrary units chosen to make the height of the reference peak equal to one.

Here I calculate the absorption peak frequencies and the peak heights for an electron, in a box with $L_x = 100$ Å, when the light is polarized in the x-direction. I choose as a reference the probability of the transition $\{1, 1, 1\} \rightarrow \{2, 1, 1\}$. This means that I calculate

$$S(1, 1, 1 \rightarrow n, 1, 1) \equiv P_{\{1,1,1\} \rightarrow \{n,1,1\}} / P_{\{1,1,1\} \rightarrow \{2,1,1\}} \tag{12.24}$$

I will also assume that $N(\omega)$ depends weakly on frequency and it is therefore the same for all transitions; this means that $N(\omega)$ is no longer present in S.

It is customary in spectroscopy to plot the spectrum versus the energy of the absorbed photon, which is given by (use Eqs 12.3 and 12.4)

$$\hbar \Omega_{n,1} \equiv E_{n,1,1} - E_{1,1,1} = \frac{\hbar^2 \pi^2}{2 m L_x^2} (n^2 - 1) \tag{12.25}$$

The computations were performed in Workbook QM12.2. The frequencies and the relative peak intensities are given in Table 12.1.

Table 12.1 Absorption frequencies in eV and relative absorption intensity (see Eq. 12.24) for some values of the quantum number n, calculated in Cell 4 of Workbook QM12.1. Note that the relative intensity is independent of box size.

n	$(E_{n,1,1} - E_{1,1,1})/\hbar$	$S(1,1,1 \to n,1,1)$
2	0.0188	1
3	0.0376	0
4	0.0639	0.16
5	0.0978	0
6	0.1390	0.0661
7	0.1880	0
8	0.2440	0.0363
9	0.3080	0
10	0.3800	0.0230

Exercise 12.3

I have an ensemble of parallelepipeds on a surface (as in Fig. 12.1) of lengths $L_1 = 100$, $L_2 = 200$, $L_3 = 600$, but I do not know how the boxes are oriented.

(a)   How would you use polarized light to determine the orientation of the boxes?

(b)   If you do not know the lengths L_1, L_2, L_3 but do know that the boxes are all parallel to each other, how could spectroscopy be used to determine the orientation and the lengths of the sides of the boxes?

(c)   Describe what will happen to the spectrum if the boxes are perpendicular to the surface but they are randomly oriented with respect to the OX and OY directions (the coordinate system is pinned to the box as in Fig. 8.1.)

Light Absorption by a Harmonic Oscillator

§15. *Introduction.* I discuss here light absorption and emission by a large number of diatomic molecules that *vibrate but do not rotate*. We can obtain such a system in the laboratory by trapping diatomic molecules in an ice of noble gas, or by binding the molecules chemically to a flat solid surface. While in the noble gas matrix the molecules are randomly oriented, the ones bound to a surface tend, in most cases, to be oriented perpendicular to it. If the surface is very flat this means that the molecules are parallel to each other.

We want to calculate the probability that a molecule in such a system has absorbed a photon after being exposed to a pulse of light. The starting point of the theory is Eq. 11.1, combined with Eq. 11.2, which gives:

$$P_{i \to f} = \frac{4\pi}{\epsilon_0 c \hbar^2} |\langle \Psi_f | \hat{\mathbf{m}} \cdot \mathbf{e} | \Psi_i \rangle|^2 N(\omega_{if})$$ (12.26)

with the dipole operator $\hat{\mathbf{m}}$ given by:

$$\hat{\mathbf{m}} = q \left(\sum_{\alpha=1}^{n} \hat{\mathbf{r}}(\alpha) - \sum_{\beta=1}^{2} Z_\beta \hat{\mathbf{R}}(\beta) \right)$$ (12.27)

These equations are in rationalized MKSA units. Ψ_i and Ψ_f are the initial and final wave functions of the electrons and the nuclei.

In applying this equation there is a fundamental difference from the case of an electron in a box. There it was assumed that the nuclear degrees of freedom are not excited by photon absorption. Here this assumption cannot be made, since it is the excitation of the vibrational motion that we want to study. Therefore, we are forced to consider the response of both nuclei and electrons to the interaction with the light. This will make the theory a little different from that used when we studied light absorption by an electron in a quantum dot. The details of the theory are explained in Chapter 17. Here I use the result, which says that $\hat{\mathbf{m}} \cdot \mathbf{e}$ in Eq. 12.26 becomes

$$\left(\frac{d\mathbf{m}(x)}{dx} \right)_{x=0} \cdot \mathbf{e} \, x$$ (12.28)

where $\mathbf{m}$ is the dipole moment of the molecule and $x = r - r_0$ is the difference between the interatomic distance r and the interatomic distance r_0 for which the atoms have minimum energy. Using this in Eq. 12.26 gives

$$P_{i \to f} = \frac{4\pi q^2}{\epsilon_0 c \hbar^2} \left\{ \mathbf{e} \cdot \left(\frac{d\mathbf{m}(x)}{dx} \right)_{x=0} \right\}^2 |\langle \phi_n | \hat{x} | \phi_0 \rangle|^2 N(\omega_{if})$$ (12.29)

The wave functions ϕ_n and ϕ_0 describe the states of the oscillator. The wave function describing the electrons enters into the calculation of $d\mathbf{m}(x)/dx$.

To evaluate the absorption probability $P_{i \to f}$ we need to know the eigenstates $\phi_n(x)$ of the harmonic oscillator.

§16. *The Eigenstates and Eigenvalues of a Harmonic Oscillator.* The energy eigenvalues of a harmonic oscillator are

$$E_n = \hbar\omega \left(n + \frac{1}{2} \right), \quad n = 0, 1, 2, \ldots \tag{12.30}$$

with the eigenfunctions

$$\phi_{n=0} \equiv \phi_0 = \left(\frac{\mu\omega}{\hbar\pi} \right)^{\frac{1}{4}} \exp\left[-\frac{\mu\omega x^2}{2\hbar} \right] \tag{12.31}$$

$$\phi_{n=1} \equiv \phi_1 = \left(\frac{4\mu^3\omega^3}{\hbar^3\pi} \right)^{\frac{1}{4}} x \, \exp\left[-\frac{\mu\omega x^2}{2\hbar} \right] \tag{12.32}$$

$$\phi_{n=2} \equiv \phi_2 = \left(\frac{\mu\omega}{\hbar\pi} \right)^{\frac{1}{4}} \left(\frac{2\mu\omega x^2}{\hbar} - 1 \right) \exp\left[-\frac{\mu\omega x^2}{2\hbar} \right] \tag{12.33}$$

I have stopped at $n = 2$ even though we know these functions for all values of n. Here μ is the reduced mass of the oscillator:

$$\mu = \frac{m_1 m_2}{m_1 + m_2} \tag{12.34}$$

where m_1 and m_2 are the masses of the two atoms making up the diatomic molecule.

§17. *The Transition Dipole.* It is now a matter of simple mathematics to calculate the transition dipole $\langle \phi_n \,|\, \hat{x} \,|\, \phi_0 \rangle$, since all we have to do is evaluate some integrals. This is done in Workbook QM12.3 and the results are:

Workbook

$$\langle \phi_1 \,|\, \hat{x} \,|\, \phi_0 \rangle \equiv \int_{-\infty}^{\infty} \phi_1(x) x \phi_0(x) dx = \sqrt{\frac{\hbar}{2\mu\omega}} \tag{12.35}$$

and

$$\langle \phi_n \,|\, \hat{x} \,|\, \phi_0 \rangle \equiv \int_{-\infty}^{\infty} \phi_n(x) x \phi_0(x) dx = 0 \quad \text{for } n \geq 2 \tag{12.36}$$

The result is rather surprising. If the harmonic oscillator is in the ground state (which is the case for the majority of the diatomic molecules in a sample) then we can excite it only to the state $n = 1$ through photon absorption, even though the harmonic oscillator has an infinite number of states. This is a powerful selection rule! Of course, its reliability depends on the accuracy with which the harmonic oscillator model describes the vibrations of a diatomic molecule. It turns out that the model works rather well, and that, although the transitions $0 \to 2$ and $0 \to 3$ can be observed, the number of photons absorbed in these transitions is much smaller than the number absorbed by $0 \to 1$ transitions.

We will study the absorption of light by a diatomic molecule more thoroughly in Chapter 17, where we will also take into account the role of rotations.

Exercise 12.4

Calculate the transition dipole $\langle \phi_2 \, | \, \hat{x} \, | \, \phi_1 \rangle$.

§18. *Molecular Orientation and Polarization.* Let us take a further step and calculate the absorption probability by a harmonic oscillator when the light polarization makes an angle θ with the direction of the oscillation (the axis of the diatomic molecule). Since in most molecules the dipole moment is oriented along the molecular axes, $\mathbf{m} \cdot \mathbf{e} = m(x) \cos(\theta)$. Using this in Eq. 12.29 for the absorption probability we obtain:

$$P_{0 \to 1} = \frac{4\pi q^2}{\epsilon_0 c \hbar^2} \left\{ \left(\frac{dm(x)}{dx} \right)_{x=0} \right\}^2 \frac{\hbar}{2\mu\omega} \cos^2(\theta) N(\omega) \qquad (12.37)$$

The symbol $P_{0 \to 1}$ is used to remind us that only transitions from the ground vibrational state ϕ_0 to the first excited vibrational state ϕ_1 are allowed. If the polarization of the light is perpendicular to the molecular axis, $\theta = \pi/2$ and the photon is not absorbed. If the diatomic molecules are frozen in an argon ice, then only those whose axis is almost parallel with $\mathbf{e}$ will adsorb light.

Exercise 12.5

Some diatomic molecules are absorbed on solid surfaces. When their concentration on the surface is low, they bind parallel to the surface. When the concentration is high, they will start to stand up and bind perpendicular to the surface. Explain how absorption spectroscopy can be used to determine when the transition from recumbent to erect takes place.

Exercise 12.6

Derive a formula like Eq. 12.37 that is valid for the transition from $n = 1$ to $n = 2$.

Exercise 12.7

The diatomic molecules "dissolved" in an Ar ice are randomly oriented. Describe how Eq. 12.29 can be used to calculate the absorption probability of linearly polarized light. *Hint.* Take the OZ axis parallel to the electric field of the light and average Eq. 12.29 over all possible orientations. Each orientation is equally probable.

13

TWO-PARTICLE SYSTEMS

Introduction

§1. *Overview.* In Chapters 13–18 we study systems consisting of two particles that interact through a force directed along the line joining them (a *central force*). In this chapter you will learn how to derive the Schrödinger equation for such a system.

Analysis based on classical mechanics suggests that one of the terms in the Schrödinger equation ought to be connected to the angular momentum. For this reason we take a detour, in Chapter 14, to study how the angular momentum is treated in quantum mechanics and determine the operator representing it and its eigenvalues and eigenfunctions. This allows the Schrödinger equation for two particles to be written in a convenient form and will help us interpret its solutions physically.

In Chapters 15–18 we solve the Schrödinger equation for two examples of two-particle systems, a diatomic molecule and the hydrogen atom, and use the results to interpret their absorption and emission spectra.

The behavior of a diatomic molecule is radically different from that of a hydrogen atom, even though both systems consist of two particles bound by a central force. Their energies and spectra show no similarity whatsoever. Nevertheless, up to a certain point the mathematics describing these two systems is identical.

This happens because in both cases the potential energy depends only on the distance between the particles. This confers the system a *spherical symmetry*, which dominates the mathematics of the problem.

The properties of a diatomic molecule differ qualitatively from those of a hydrogen atom because the forces between the particles in these systems are very different. The interaction between the atoms in a diatomic molecule is similar to a stiff spring. As a result, the atoms vibrate with a very small amplitude, around a fixed position (the bond length); to a good approximation, the molecule rotates as a rigid rod that is not affected by the vibrational motion. The Coulomb interaction between the electron and the proton, in the hydrogen atom, is soft and long-ranged; it is impossible to describe the motion of the electron as a superposition of a small amplitude vibration and a rigid-body rotation, decoupled from each other.

These two examples display the tremendous unifying powers of mathematics and the power of physics to create very diverse behavior within the same mathematical framework.

§2. *A Brief Outline.* Before we plunge into mathematics, let us use our intuition to define some goals. After all, two particles bound by a force form a familiar object. Think of it as a barbell whose balls are connected by a spring, flying through a space devoid of gravity. The particles exert a force on each other but no force acts on the system from outside; there is an *internal*, but no *external*, force. This means that the two particles (i.e. the diatomic or the hydrogen atom) move, as a whole, with *uniform velocity*. This suggests that there might be a point in the system that is unaffected by the change in the orientation or in the distance between the two particles; oblivious to the agitation of the particles, this point moves serenely with *constant velocity*. Mechanics shows that such a point, called the *center of mass*, exists and its coordinates can be identified.

Imagine now that you are weightless, and sit on the center of mass and move with it, but do not rotate with the molecule (or H atom). When observing the motion of the particles from this position, you are no longer aware of the translational motion of the system: you only see the *internal motion,* the change in the distance between the particles and of the orientation in space of the line joining them. This is very similar to watching a ping-pong game inside a car of a moving train. The ball moves with the train, but you cannot detect this motion. Because of this, the motion of the ball appears to you simpler than to a person watching it from the ground. This suggests that it might be convenient to choose a coordinate system pinned to the center of mass and translating with it; in this coordinate system, the motion of the two particles should be simpler.

When describing the internal motion of the particles you are very likely to talk about the change of the distance r between them. We call this the *vibrational* motion when studying a diatomic molecule, or the *radial* motion when studying the hydrogen atom. You will also observe how the orientation of the line joining the particles is changed by the *rotational* motion of the two particles. This can be described by two angles, θ and ϕ. It seems that, if we want to decompose the motion into a rotation and a vibration, we should describe it by using the variables r, θ, and ϕ. Technically this means that we will change variables from Cartesian to *spherical coordinates*.

An essential characteristic of our system is that the two particles interact with each other through a *central force*. Because the force acts along an imaginary line joining the particles, no *torque* acts on the system. When you studied physics you learned that if there is no torque, the *angular momentum* of the system does not change. Thus any two-particle system twirling in a space devoid of gravity (or any other external forces) will have a constant angular momentum. We expect therefore that specifying the magnitude and the orientation of the angular-momentum vector will be an essential part of specifying how the system moves.

In this chapter we put some mathematical muscle behind these statements and use them to write the Schrödinger equation in a form that reveals the radial and the rotational components of the motion and displays explicitly the role of angular momentum. We thus take advantage of the spherical symmetry of the system (i.e. of the fact that the force is central).

The procedure for deriving the Schrödinger equation is that outlined in the previous chapters: write the equation for the total energy of the system in classical mechanics and replace the potential and the kinetic energy with the appropriate operators. This gives the Schrödinger equation in Cartesian coordinates. But, as noted above, the equation ought to be simpler if we change the coordinate system twice: first to place the origin of the Cartesian system in the center of mass and then to go to spherical coordinates.

The Schrödinger Equation for the Internal Motion

§3. *The Laboratory Coordinate System.* To describe the position of the particles, I use a *laboratory coordinate system*, which is fixed in space (Fig. 13.1). The location of particle 1 is given by the vector $\mathbf{r}(1)$ and that of particle 2 by $\mathbf{r}(2)$. The vector

$$\mathbf{r} = \mathbf{r}(2) - \mathbf{r}(1) \tag{13.1}$$

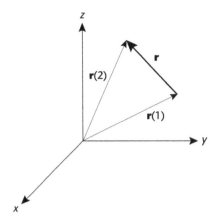

Figure 13.1 The coordinate system used for defining the positions **r**(1) and **r**(2) of the two particles and the vector **r** between them.

points from particle 1 to particle 2, and its length r is the distance between the particles. The variation of the interparticle distance r describes the *radial motion* of the system (i.e. the "vibration" of the two-particle system) and the change in the orientation of the vector **r** describes the rotational motion.

§4. *The Energy.* In classical mechanics, the total energy is

$$H = m(1)\frac{\mathbf{v}(1) \cdot \mathbf{v}(1)}{2} + m(2)\frac{\mathbf{v}(2) \cdot \mathbf{v}(2)}{2} + V(r)$$

$$\equiv K + V(r) \tag{13.2}$$

$\mathbf{v}(1)$ and $\mathbf{v}(2)$ are the velocities of the particles

$$\mathbf{v}(1) \equiv \frac{d\mathbf{r}(1)}{dt} \tag{13.3}$$

$$\mathbf{v}(2) \equiv \frac{d\mathbf{r}(2)}{dt} \tag{13.4}$$

The first term in Eq. 13.2 is the kinetic energy of particle 1 and the second is the kinetic energy of particle 2; their sum is the total kinetic energy K.

The function $V(r)$ is the potential energy due to the interaction between particles. It is important to note that V depends only on the length (magnitude) r of the vector **r**; it is a function of the distance between the particles, and it is independent of the way in which the line joining the particles is oriented in space.

If we know the potential energy of the particles we can calculate the components of the force they exert on each other from the equations

$$F_x = -\frac{\partial V}{\partial x}; \quad F_y = -\frac{\partial V}{\partial y}; \quad F_z = -\frac{\partial V}{\partial z} \tag{13.5}$$

It is not difficult to show that because the potential energy depends only on r, the force between the particles is oriented along the line joining them.

Exercise 13.1

Show that if the potential energy V depends only on the length of the vector $\mathbf{r} = \mathbf{r}(2) - \mathbf{r}(1)$, then the force acts along the line joining the particles.

§5. *Because There is no External Force, the System Moves with Constant Velocity.* In what follows we assume that we know the potential energy of the two-particle system. This means that we can calculate the forces between the particles and use Newton's equations

$$m(1)\frac{d^2\mathbf{r}(1)}{dt^2} = \mathbf{F}_{21} \tag{13.6}$$

$$m(2)\frac{d^2\mathbf{r}(2)}{dt^2} = \mathbf{F}_{12} \tag{13.7}$$

to study their classical motion. In these equations $\mathbf{F}_{21}$ is the force exerted by particle 2 on particle 1 and $\mathbf{F}_{12}$ is the force exerted by particle 1 on particle 2. These equations are valid only if *no external force* acts on the particles.

Add Eqs 13.6 and 13.7 to obtain

$$\frac{d^2}{dt^2}[m(1)\mathbf{r}(1) + m(2)\mathbf{r}(2)] = \mathbf{F}_{21} + \mathbf{F}_{12} \tag{13.8}$$

Newton's third law says that

$$\mathbf{F}_{21} + \mathbf{F}_{12} = 0 \tag{13.9}$$

The force exerted by particle 1 on particle 2 is equal and opposite in sign to the force exerted by particle 2 on particle 1. If I push you, the force you feel on your chest is equal in magnitude to and opposite in sign from the force I feel on my hand.

Using Eq. 13.9 in Eq. 13.8 leads to

$$\frac{d^2}{dt^2}[m(1)\mathbf{r}(1) + m(2)\mathbf{r}(2)] = 0$$

By writing this as

$$\frac{d}{dt}\left[\frac{d}{dt}[m(1)\mathbf{r}(1) + m(2)\mathbf{r}(2)]\right] = 0$$

I conclude that

$$\frac{d}{dt}[m(1)\mathbf{r}(1) + m(2)\mathbf{r}(2)] = \text{a constant}$$

For later convenience, I write this as

$$\frac{d}{dt}\mathbf{R} = \text{a constant} \tag{13.10}$$

$$\mathbf{R} \equiv \frac{m(1)\mathbf{r}(1) + m(2)\mathbf{r}(2)}{m(1) + m(2)} \tag{13.11}$$

The vector $\mathbf{R}$ is the position of the *center of mass*.

Equation 13.10 tells me that the velocity of the center of mass is constant. This is remarkable! The particles undergo a rather complicated motion (they rotate and vibrate), but the point of coordinates $\mathbf{R}$ moves with constant velocity. I emphasize that this happens because no external force acts on the system; the particles only interact with each other. An external force would accelerate (or decelerate) the center of mass.

That internal forces do not affect the translational motion of the body as a whole is not surprising. Imagine that you are in free fall in outer space, where there is no gravity. Can you change your motion by pulling at your hair (which is an internal force)? I think not. If you do not believe me, sit on a scale and try to change your weight by pulling your hair upwards. The reading of the balance will not change. The force exerted by your hand on your hair is canceled by the force exerted by your hair on your hand. Now ask a friend to pull upwards on your hair. *Surprise!* You lost some weight. His hand exerts an external force.

A mild surprise in our result is that there is a point in the system that moves with constant velocity, unaffected by the internal motion (rotation or vibration) of the particles forming the system. This is not a property specific to two-particle systems: it is also true if the system has a thousand particles.

Exercise 13.2

(a)   Define $\mathbf{s}(1)$ and $\mathbf{s}(2)$ through

$$\mathbf{R} + \mathbf{s}(1) = \mathbf{r}(1)$$

$$\mathbf{R} + \mathbf{s}(2) = \mathbf{r}(2)$$

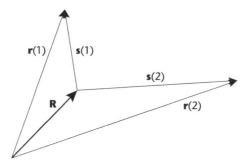

Show that $\mathbf{s}(1)$, $\mathbf{s}(2)$, and $\mathbf{r} = \mathbf{r}(2) - \mathbf{r}(1)$ are collinear (they lie on the same line). Therefore the picture must be redrawn as shown below.

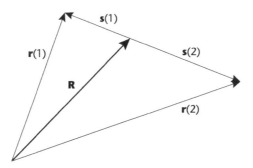

(b)   Show that the tip of the vector $\mathbf{R}$ (i.e. the location of the center of mass) is on the line joining particle 1 to particle 2. Moreover, if $m(1) \gg m(2)$ then $\mathbf{R}$ is very close to $\mathbf{r}(1)$. (*Hint.* Three vectors $\mathbf{a}$, $\mathbf{b}$, and $\mathbf{c}$ are collinear if $\mathbf{a} = \text{constant} \times \mathbf{c}$ and $\mathbf{b} = \text{constant} \times \mathbf{c}$.)

§6. *New Variables.* From now on we describe the motion by following the evolution of **R** and **r**, rather than **r**(1) and **r**(2). It is clear that **R** is a very convenient variable: its evolution is so simple.

But why would I use **r**? I like **r** because its length r tells me how the distance between the particles changes and its orientation tells me how the line joining the particles rotates.

To express the total energy of the system in terms of the new variables, I have to express **r**(1) and **r**(2) in terms of **R** and **r**. It is not hard to show that Eqs 13.1 and 13.11 can be solved for **r**(1) and **r**(2) to give (see Workbook QM13.1)

$$\mathbf{r}(1) = \mathbf{R} - \frac{m(2)}{m(1) + m(2)}\,\mathbf{r} \tag{13.12}$$

$$\mathbf{r}(2) = \mathbf{R} + \frac{m(1)}{m(1) + m(2)}\,\mathbf{r} \tag{13.13}$$

The velocities **v**(1) and **v**(2) of the particles are

$$\mathbf{v}(1) \equiv \frac{d\mathbf{r}(1)}{dt} = \frac{d\mathbf{R}}{dt} - \frac{m(2)}{m(1) + m(2)}\frac{d\mathbf{r}}{dt} \tag{13.14}$$

$$\mathbf{v}(2) \equiv \frac{d\mathbf{r}(2)}{dt} = \frac{d\mathbf{R}}{dt} + \frac{m(1)}{m(1) + m(2)}\frac{d\mathbf{r}}{dt} \tag{13.15}$$

§7. *A Coordinate System with the Origin in the Center of Mass.* Now let us inject some physics into the analysis. I am interested in the internal motion of the system. If I stand on the laboratory floor and look at the two particles flying through space, the motion I see is made more complicated by the translational motion of the whole system. Now imagine that I am pinned to the center of mass and translate with the two-particle system, but I *do not rotate with it.* I no longer perceive the translational motion of the system, and from this vantage point, the motion of the two particles is simpler.

This tells me that to get rid of the translational motion, I should use a coordinate system whose *origin is at the center of mass* and that translates but does not rotate with the two-particle system. In this system of coordinates, the velocity of the center of mass is zero, and the velocities of the particles become (set $d\mathbf{R}/dt = 0$ in

Eqs 13.14 and 13.15)

$$\mathbf{v}(1) = -\frac{m(2)}{m(1) + m(2)} \frac{d\mathbf{r}}{dt} \tag{13.16}$$

$$\mathbf{v}(2) = \frac{m(1)}{m(1) + m(2)} \frac{d\mathbf{r}}{dt} \tag{13.17}$$

Inserting the expressions for $\mathbf{v}(1)$ and $\mathbf{v}(2)$, given by Eqs 13.16 and 13.17, in the kinetic energy

$$K = \frac{m(1)}{2}\mathbf{v}(1) \cdot \mathbf{v}(1) + \frac{m(2)}{2}\mathbf{v}(2) \cdot \mathbf{v}(2)$$

gives

$$K = \frac{\mu}{2}\, \mathbf{v} \cdot \mathbf{v} \tag{13.18}$$

with

$$\mu = \frac{m(1)m(2)}{m(1) + m(2)} \tag{13.19}$$

and

$$\mathbf{v} = \frac{d\mathbf{r}}{dt} \tag{13.20}$$

The quantity μ is the *reduced mass* and $\mathbf{v}$ is the *relative velocity* of the system. It tells me how fast one particle moves towards or away from the other particle and how the system rotates.

§8. *The Total Energy in Terms of Momentum.* I am interested in the total energy (the Hamiltonian) because I intend to *quantize it* by using the rules described in Chapter 2 to derive the Hamiltonian operator of the system. To do this the classical quantities in the total energy need to be replaced with operators. Unfortunately, I do not know the operator corresponding to the velocity. This is not a disaster: the velocity can be expressed in terms of momentum and then the classical momentum can be replaced with the corresponding operator. This is what I will do next.

In classical mechanics the momentum associated with a coordinate x is obtained by taking the derivative of the kinetic energy with respect to the velocity $v_x \equiv dx/dt$.

Following this rule, the momentum component p_x associated with the velocity v_x is (use Eq. 13.18)

$$p_x \equiv \frac{\partial K}{\partial v_x} = \frac{\partial}{\partial v_x}\left[\frac{\mu}{2}\left(v_x^2 + v_y^2 + v_z^2\right)\right] = \mu v_x$$

Similarly, $p_y = \mu v_y$ and $p_z = \mu v_z$, or $\mathbf{p} = \mu \mathbf{v}$. With this definition of momentum, Eq. 13.18 for the kinetic energy becomes

$$K = \frac{\mathbf{p}\cdot\mathbf{p}}{2\mu} \tag{13.21}$$

and the total energy is

$$H = \frac{\mathbf{p}\cdot\mathbf{p}}{2\mu} + V(r) \tag{13.22}$$

This is the energy of the internal motion, as perceived by an observer riding on the center of mass.

§9. *The Hamiltonian Operator.* From Eq. 13.22 we can obtain the Hamiltonian operator by replacing the classical quantities with the appropriate operators. In our case the momentum p_x is replaced by the operator

$$\hat{p}_x \equiv \frac{\hbar}{i}\frac{\partial}{\partial x} \tag{13.23}$$

Similar equations are used for $\hat{p}_y$ and $\hat{p}_z$.

Replacing the classical momenta in Eq. 13.22 with momentum operators leads to the Hamiltonian operator

$$\hat{H} = -\frac{\hbar^2}{2\mu}\left[\frac{\partial^2}{\partial x^2} + \frac{\partial^2}{\partial y^2} + \frac{\partial^2}{\partial z^2}\right] + V(x,y,z) \tag{13.24}$$

Here x, y, and z are the coordinates of the vector $\mathbf{r}$. As explained in Chapter 3, the potential energy operator applied to a function is a multiplication of the function by the classical potential energy $V(x,y,z)$.

The eigenvalue problem for this Hamiltonian operator

$$\hat{H}\psi(x,y,z) = -\frac{\hbar^2}{2\mu}\left[\frac{\partial^2\psi}{\partial x^2} + \frac{\partial^2\psi}{\partial y^2} + \frac{\partial^2\psi}{\partial z^2}\right] + V(x,y,z)\psi(x,y,z)$$

$$= E\,\psi(x,y,z) \tag{13.25}$$

is the Schrödinger equation for the two-particle system, in a coordinate system with the origin at the center of mass.

This equation tells us nothing about the translation of the system as a whole. Fortunately, that motion plays a minor role in spectroscopy. The frequency of a photon absorbed by a moving molecule (or atom) is slightly shifted, as compared to the frequency of a photon absorbed by a stationary molecule. A similar shift is observed in the frequency of the emitted photons. This *Doppler shift* is small and it can be ignored when analyzing spectra.

There are, however, cases in which the Doppler shift is useful. Since its magnitude depends on the velocity of the molecule, it has been used to determine the velocity of the fragments produced by the photo-decomposition of a molecule.

§10.   *The Concept of Quasi-particle.* We started this analysis with the classical energy, Eq. 13.2, which can be rewritten (by using $m(1)\mathbf{v}(1) = \mathbf{p}(1)$ and $m(2)\mathbf{v}(2) = \mathbf{p}(2)$) as

$$H = \frac{\mathbf{p}(1)\cdot\mathbf{p}(1)}{2m(1)} + \frac{\mathbf{p}(2)\cdot\mathbf{p}(2)}{2m(2)} + V(r) \tag{13.26}$$

Had we replaced the classical momenta in Eq. 13.26 with the appropriate operators we would have obtained the Schrödinger equation:

$$\hat{H}\psi(x,y,z) = -\frac{\hbar^2}{2m(1)}\left[\frac{\partial^2\psi}{\partial x(1)^2} + \frac{\partial^2\psi}{\partial y(1)^2} + \frac{\partial^2\psi}{\partial z(1)^2}\right]$$

$$- \frac{\hbar^2}{2m(2)}\left[\frac{\partial^2\psi}{\partial x(2)^2} + \frac{\partial^2\psi}{\partial y(2)^2} + \frac{\partial^2\psi}{\partial z(2)^2}\right]$$

$$+ V(r)\psi(\mathbf{r}(1),\mathbf{r}(2))$$

$$= E\,\psi(\mathbf{r}(1),\mathbf{r}(2)) \tag{13.27}$$

Eq. 13.27 is quite different from Eq. 13.25, even though they both represent the same two-particle system. Note that Eq. 13.27 is a two-particle equation: its wave function $\psi(\mathbf{r}(1), \mathbf{r}(2))$ depends on the coordinate of both particles and the kinetic energy of both particles is represented in the equation. On the other hand, Eq. 13.25 looks as if it is the Schrödinger equation of one particle. The eigenfunction is a function of only three coordinates (instead of six) and there is only one kinetic energy term. This metamorphosis took place when we switched to a coordinate system pinned to the center of mass. In such a system we are unable to "see" the translational motion of the two-particle system as a whole. To remind ourselves of this situation, we sometimes say that the behavior of the two-particle system is described by that of a *quasi-particle*, whose Schrödinger equation is Eq. 13.25.

Exercise 13.3

Start with the Schrödinger equation Eq. 13.27, where the wave function ψ is $\psi(x(1), y(1), z(1), x(2), y(2), z(2))$. Show that if you use

$$\mathbf{r}(1) = \{x(1), y(1), z(1)\} = \mathbf{R} - \frac{m(2)}{m(1) + m(2)} \mathbf{r}$$

and

$$\mathbf{r}(2) = \{x(2), y(2), z(2)\} = \mathbf{R} + \frac{m(1)}{m(1) + m(2)} \mathbf{r}$$

to change variables to $\mathbf{R}$ and $\mathbf{r}$, then, for $\mathbf{R} = \{X, Y, Z\}$ and $\mathbf{r} = \{x, y, z\}$, the Schrödinger equation becomes

$$-\frac{\hbar^2}{2M} \left[\frac{\partial^2 \psi}{\partial X^2} + \frac{\partial^2 \psi}{\partial Y^2} + \frac{\partial^2 \psi}{\partial Z^2} \right] - \frac{\hbar^2}{2\mu} \left[\frac{\partial^2 \psi}{\partial x^2} + \frac{\partial^2 \psi}{\partial y^2} + \frac{\partial^2 \psi}{\partial z^2} \right]$$
$$+ V(r)\psi = E\psi \tag{13.28}$$

where

$$M = m(1) + m(2)$$

and

$$\mu = \frac{m(1)m(2)}{m(1) + m(2)}$$

In the center-of-mass system of coordinates, the kinetic energy of the center of mass is zero and therefore we set

$$-\frac{\hbar^2}{2M}\left[\frac{\partial^2\psi}{\partial X^2} + \frac{\partial^2\psi}{\partial Y^2} + \frac{\partial^2\psi}{\partial Z^2}\right]$$

to zero. As a result we obtain the quasi-particle Hamiltonian Eq. 13.25.

Hint. First transform the part

$$-\frac{\hbar^2}{2m(1)}\frac{\partial^2\psi}{\partial x(1)^2} - \frac{\hbar^2}{2m(2)}\frac{\partial^2\psi}{\partial x(2)^2}$$

by using $x(1) = X - \frac{m(2)}{M}x$ and $x(2) = X + \frac{m(1)}{M}x$. The transform of the parts involving y and z can then be written down by analogy.

Also show that the solution of Eq. 13.28 is of the form $\phi(X, Y, Z)\psi(x, y, z)$ and find the solution for $\phi(X, Y, Z)$. *Hint.* Use the method of separation of variables.

The Hamiltonian of the Quasi-particle in Spherical Coordinates

§11. *Why Spherical Coordinates?* The variables x, y, and z in the Schrödinger equation, Eq. 13.25, are the Cartesian coordinates of the vector $\mathbf{r}$. The length r of this vector is the distance between particles and its evolution describes the radial (vibrational) motion. The orientation of $\mathbf{r}$ tells us how the two-particle system rotates. It would therefore seem advantageous to cast the Schrödinger equation in a form in which the variables are r and the two angles describing the orientation of the system. For example, we can use the angles θ and ϕ defined in Fig. 13.2. I assume that you have already encountered this *spherical coordinate system*, so I will not dwell on it.

From Fig. 13.2, you can see, with a bit of trigonometry, that

$$\left.\begin{array}{l} x = r\,\sin\theta\,\cos\phi \\ y = r\,\sin\theta\,\sin\phi \\ z = r\,\cos\theta \end{array}\right\} \tag{13.29}$$

In these equations,

$$0 \le r \le \infty,\ \ 0 \le \theta \le \pi,\ \ 0 \le \phi \le 2\pi \tag{13.30}$$

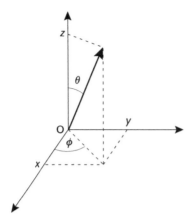

Figure 13.2 The spherical coordinates θ, ϕ. The length of the vector defined by the arrow is r.

Solving these equations for r, θ, and ϕ, allows us to write the spherical coordinates in terms of x, y, and z:

$$\left. \begin{aligned} r &= \sqrt{x^2 + y^2 + z^2} \\ \cos\theta &= \frac{z}{\sqrt{x^2 + y^2 + z^2}} \\ \tan\phi &= \frac{y}{x} \end{aligned} \right\} \tag{13.31}$$

Next, I want to change variables in the Schrödinger equation, Eq. 13.25, from $x, y,\ z$ to r, θ, ϕ. In principle, this is easy: use Eq. 13.31 and the chain rule to replace the derivatives with respect to x, y, and z with derivatives with respect to r, θ, and ϕ. A lot of work and a bit of patience will get the job done.

However, I believe that you can find a better use for your time and energy. The book by Morse and Feshbach[a] gives general methods for performing such changes of variables and a table with the form of many operators, in a large number of coordinate systems. On page 116 of this book, I found the following expression for the Laplace operator

$$\nabla^2\psi \equiv \frac{\partial^2\psi}{\partial x^2} + \frac{\partial^2\psi}{\partial y^2} + \frac{\partial^2\psi}{\partial z^2}$$

---

[a] P. M. Morse and H. Feshbach, *Methods of Theoretical Physics, Vol. 1*, McGraw Hill, New York, 1953, pp. 1–117.

$$= \frac{1}{r^2} \frac{\partial}{\partial r} \left(r^2 \frac{\partial \psi}{\partial r} \right) + \frac{1}{r^2 \sin \theta} \frac{\partial}{\partial \theta} \left(\sin \theta \frac{\partial \psi}{\partial \theta} \right) + \frac{1}{r^2 \sin^2 \theta} \frac{\partial^2 \psi}{\partial \phi^2}$$

Using this formula in the Schrödinger equation Eq. 13.25 gives:

$$-\frac{\hbar^2}{2\mu} \left[\frac{1}{r^2} \frac{\partial}{\partial r} \left(r^2 \frac{\partial \psi}{\partial r} \right) + \frac{1}{r^2 \sin \theta} \frac{\partial}{\partial \theta} \left(\sin \theta \frac{\partial \psi}{\partial \theta} \right) + \frac{1}{r^2 \sin^2 \theta} \frac{\partial^2 \psi}{\partial \phi^2} \right]$$

$$+ V(r)\psi = E\psi \tag{13.32}$$

Here ψ is a function of r, θ, and ϕ.

To determine the physical meaning of the terms in this equation, I make a detour and examine the classical motion of the system, to highlight the role of the angular momentum.

The Role of Angular Momentum in the Motion of the Two-particle System

§12. *Introduction.* Since no external force acts on the particles, the external torque applied to them is zero. When you studied physics you were told that in the absence of an external torque the angular momentum of the system is *constant*. Because of this, the magnitude and the direction of the angular momentum is an important descriptor of the motion.

In this section, I review a few statements classical mechanics makes about the angular momentum. This will help us describe and understand the motion in both the classical and the quantum systems. In Supplement 13.1 you will find a derivation of the classical results mentioned below.

§13. *Angular Momentum in Classical Mechanics.* The orbital angular momentum in classical mechanics is defined by

$$\mathbf{L} = \mathbf{r} \times \mathbf{p} \tag{13.33}$$

Here $\mathbf{r}$ is the position of the quasi-particle and $\mathbf{p} = \mu\, d\mathbf{r}/dt$ is its momentum. This definition uses the cross-product between $\mathbf{p}$ and $\mathbf{r}$. Some properties of this operation are reviewed in Supplement 13.1.

§14. *The Properties of the Angular Momentum.* Starting from the definition of the angular momentum in Eq. 13.33, we can use Newton's equation (see Supplement 13.1) to prove the following statements. If no external force acts on the two-particle system then:

1. the angular momentum vector does not change in time;
2. the motion of the two particles is confined to a plane;
3. the angular momentum vector **L** is perpendicular to the plane in which the particles move;
4. the magnitude L of the angular momentum vector **L** is

$$L = \mu r^2 \frac{d\phi}{dt} \tag{13.34}$$

   where $d\phi/dt$ is the angular velocity of the molecular axis;
5. the energy of the two particles is

$$H = \frac{1}{2}\mu \left(\frac{dr}{dt}\right)^2 + \frac{L^2}{2\mu r^2} + V(r) \tag{13.35}$$

   where $L^2 = \mathbf{L} \cdot \mathbf{L}$ is the length of the angular momentum vector, squared.

To explain the meaning of these statements, I will use the model of a barbell whose balls are connected by a spring. The barbell moves because it was given an initial impulse (someone threw it in the air). After the throw, no force acts on the barbell (we ignore gravity since that force is too weak to play a role in quantum mechanics). The evolution of the two particles, as observed from the center of mass, is a complicated superposition of their rotational and vibrational (radial) motions. However, according to item 1 above, they *must* move so that neither the *direction* nor the *magnitude* of the vector **L** change. The direction and the magnitude of the angular momentum are determined by the initial toss of the barbell, but after that they do not change. Moreover, items 2 and 3 tell us that the barbell does not wobble: the two balls will forever be contained in a plane and this plane is perpendicular to the immovable angular momentum vector.

The magnitude of the angular momentum is proportional (see item 4) to the *angular velocity $d\phi/dt$*. The angle ϕ is defined in Fig. 13.3. In it, I use a coordinate system whose origin is pinned to the center of mass and whose xy-plane coincides with the plane in which the two particles (and therefore the vector **r**) move (see item 2). The z-axis is taken along the angular momentum (which is perpendicular to the plane (see item 3)). In this coordinate system ϕ is the angle between the vector **r**

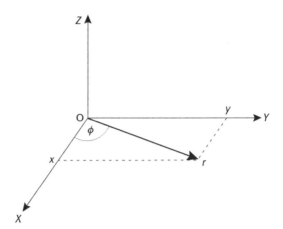

Figure 13.3 Polar coordinates.

and the x-axis of the coordinate system. The time evolution of this angle describes the rotation of the barbell and $d\phi/dt$ is the velocity with which the angle changes.

The energy of motion, Eq. 13.35, can be decomposed into three terms (item 5). The first term

$$K_v \equiv \frac{1}{2}\mu \left(\frac{dr}{dt}\right)^2$$

differs from zero only if the motion causes the distance r between the particles to vary with time. This term is the kinetic energy of the vibrational (radial) motion.

The last term in Eq. 13.35 is the potential energy $V(r)$ of the interaction between the particles. It depends only on the distance r between the particles and therefore it affects only the radial (vibrational) motion.

The second term in Eq. 13.35

$$K_{rv} \equiv \frac{L^2}{2\mu r^2} \tag{13.36}$$

is more interesting.

This term differs from zero *only if the two-particles rotate*; if they do not, the angular velocity is zero, hence L is zero (see Eq. 13.34), which makes K_{rv} zero. However, it would be a mistake to think that K_{rv} is exclusively rotational energy, since it also depends on the distance r between particles. When this is changed (by the radial motion), K_{rv} changes. This means that K_{rv} is modified by both rotation and vibration; this term *couples* the two kinds of motion.

The role of this term can be understood better if a limiting case is examined. If a diatomic molecule is given a small internal energy (by the initial impulse mentioned earlier), the distance between atoms undergoes small-amplitude oscillations around a constant value r_0 (the bond length). This means that we can make an approximation and replace r in Eq. 13.36 with r_0. After this replacement, K_{rv} *depends only on the rotational motion*, and the total energy H has one term, $K_v + V(r)$, which depends only on the radial (vibrational) coordinate, and another, K_{rv}, which depends only or the rotational variable ϕ. When this happens we say that the rotational and the vibrational (radial) motions are *decoupled*. They evolve independently of each other: the rotational motion is not affected by the manner in which the molecule vibrates and vice versa.

If we give the molecule a lot of energy, then the amplitude of the vibrational motion can be large and we can no longer replace r with r_0 in K_{rv}. Now the two kinds of motions are coupled. This means that if we start the molecule with a large vibrational motion this will affect the manner in which the molecule rotates. When r is large, the *moment of inertia* μr^2 is large; because L is constant and equal to $\mu r^2 \, d\phi/dt$, an increase in μr^2 requires a decrease in the angular velocity $d\phi/dt$. When the molecule is stretched by the radial motion its angular velocity is diminished. This is what we mean when we say that the two motions are coupled: the coupling term forces the energy to flow from rotation into vibration, and vice versa. It is also possible to give the molecule so much rotation that the bond will be stretched until it breaks.

You have seen effects of this coupling in everyday experiences. In a cylindrical rotating chamber, common at county fairs, you can walk from the wall to the axis of rotation fairly easily, if the angular velocity of the chamber is small. In that case the small amount of rotational energy makes the coupling term K_{rv} small and its effect on the radial motion is negligible. However, once the angular velocity picks up, you are "glued" to the wall and can no longer walk back to the center of the chamber. The rotational motion strongly affects your ability to move radially.

The decoupling of rotation from radial motion, performed above, was possible because the chemical bond in the diatomic is very stiff and the radial motion is a small-amplitude vibration. In a hydrogen atom the Coulomb interaction between the electron and the proton is much softer, and it is no longer possible to describe the radial motion as a small-amplitude oscillation. Because of this, replacing r in K_{rv} with r_0 is no longer a reasonable approximation. The hydrogen atom is fundamentally different from a diatomic molecule because of this difference in the interaction between the constituent particles. The masses are also very different but, had the interactions been very similar, this mass difference would only result in a scaling of the energy of the system and not in a qualitative difference in the spectrum.

§15. *The Schrödinger Equation of the Quasi-particle and the Square of the Angular Momentum.* The classical formula for the energy of the quasi-particle is (see Eq. 13.35)

$$H = \frac{\mu}{2} \left(\frac{dr}{dt}\right)^2 + \frac{L^2}{2\mu r^2} + V(r) \tag{13.37}$$

The Schrödinger equation for the same system is (see Eq. 13.32)

$$-\frac{\hbar^2}{2\mu} \left[\frac{1}{r^2} \frac{\partial}{\partial r} \left(r^2 \frac{\partial \psi}{\partial r}\right) + \frac{1}{r^2 \sin\theta} \frac{\partial}{\partial \theta} \left(\sin\theta \frac{\partial \psi}{\partial \theta}\right) + \frac{1}{r^2 \sin^2\theta} \frac{\partial^2 \psi}{\partial \phi^2} \right]$$

$$+ V(r)\psi = E\psi \tag{13.38}$$

It is reasonable to assume that there is a connection between these two equations. To find it, I will examine a limiting case when the distance r between the particles is fixed at some value $r = r_0$ and no longer changes in time. In the classical theory this means that dr/dt becomes zero and

$$H = \frac{L^2}{2\mu r_0^2} + V(r_0)$$

The interaction energy $V(r_0)$ is a constant and such constants can always be taken to be zero: the physical behavior of a system does not change if I add (or subtract) a constant to (or from) the total energy. Then, for this *rigid-rotor* model, the classical Hamiltonian is

$$H = \frac{L^2}{2\mu r_0^2} \tag{13.39}$$

Now take the same limit in the Schrödinger equation, Eq. 13.38. Since $r = r_0$, the wave function no longer depends on r and its derivative with respect to r is zero.

In this case the Schrödinger equation becomes

$$-\frac{\hbar^2}{2\mu}\left[\frac{1}{r_0^2 \sin\theta}\frac{\partial}{\partial\theta}\left(\sin\theta\frac{\partial\psi}{\partial\theta}\right) + \frac{1}{r_0^2 \sin^2\theta}\frac{\partial^2\psi}{\partial\phi^2}\right] = E\psi \qquad (13.40)$$

A comparison of the classical expression, Eq. 13.39, with the quantum expression, Eq. 13.40, suggests that

$$\frac{1}{\sin\theta}\frac{\partial}{\partial\theta}\left(\sin\theta\frac{\partial\psi}{\partial\theta}\right) + \frac{1}{\sin^2\theta}\frac{\partial^2\psi}{\partial\phi^2} = \frac{\hat{L}^2\psi}{2\mu} \qquad (13.41)$$

where $\hat{L}^2$ is the operator corresponding to the angular momentum squared. Once we accept Eq. 13.41, the Schrödinger equation for the quasi-particle can be rewritten as (use Eq. 13.41 in Eq. 13.38)

$$-\frac{\hbar^2}{2\mu}\frac{1}{r^2}\frac{\partial}{\partial r}\left(r^2\frac{\partial\psi}{\partial r}\right) + \frac{1}{2\mu r^2}\hat{L}^2\psi + V(r)\psi = E\psi \qquad (13.42)$$

In Chapter 14, I show that this equation is correct. It is this form that we will use, in the next few chapters, in our studies of the diatomic molecule and the hydrogen atom.

§16. *Why Go Through so Much Trouble?* You may wonder why we spend so much time rewriting the Schrödinger equation. With all the computer power available to us, why not just solve the two-particle Schrödinger equation, Eq. 13.27, numerically and be done with it? Such a solution can be obtained, but there is a price to pay. If the equation is solved by brute force, a six-dimensional wave function $\psi_n(x(1), y(1), z(1), x(2), y(2), z(2))$ and energy eigenvalues E_n are obtained. This is all we need for calculating various measurable quantities. However, given this information it is difficult to determine how the total energy breaks up into translational, rotational, and radial components. While the numbers produced by this procedure are correct, our understanding of their meaning suffers. After eliminating the translation of the center of mass to obtain the Schrödinger equation of the quasi-particle, and then rewriting the equation in spherical coordinates, we can separate the energy into radial and rotational components. Moreover this form of the equation is much simpler and the solution can be expressed in terms of special functions (spherical harmonics, Laguerre polynomials and Hermite polynomials) that can be easily evaluated. Furthermore, because the physical meaning of various terms in the equations is understood, we are able to make good approximations, such as the rigid-rotor and the harmonic approximations, which are useful in dealing with the vibrational and rotational motion of the molecules.

Supplement 13.1 The Role of Angular Momentum in the Motion of the Two-particle System

§17. The angular momentum in classical mechanics is defined by

$$\mathbf{L} = \mathbf{r} \times \mathbf{p} \tag{13.43}$$

Here $\mathbf{r}$ is the position of the quasi-particle and $\mathbf{p} = \mu \, d\mathbf{r}/dt$ is its momentum.

This definition makes use of the *cross-product* $\mathbf{r} \times \mathbf{p}$. Recall some of its properties.

If $\mathbf{a} = a_x\mathbf{i} + a_y\mathbf{j} + a_z\mathbf{k}$ and $\mathbf{b} = b_x\mathbf{i} + b_y\mathbf{j} + b_z\mathbf{k}$, then

$$\mathbf{a} \times \mathbf{b} = (a_yb_z - a_zb_y)\mathbf{i} + (a_zb_x - a_xb_z)\mathbf{j} + (a_xb_y - a_yb_x)\mathbf{k} \tag{13.44}$$

The x-component of the vector $\mathbf{a} \times \mathbf{b}$ is $(a_yb_z - a_zb_y)$, the y-component is $(a_zb_x - a_xb_z)$, and the z-component is $(a_xb_y - a_yb_x)$.

The vector

$$\mathbf{c} = \mathbf{a} \times \mathbf{b}$$

is perpendicular to the plane that contains the vectors $\mathbf{a}$ and $\mathbf{b}$ ($\mathbf{c}$ is perpendicular to both $\mathbf{a}$ and $\mathbf{b}$). Furthermore, we know that if $\mathbf{a}$ and $\mathbf{b}$ are parallel, then $\mathbf{a} \times \mathbf{b} = 0$.

Exercise 13.4

Prove that if $\mathbf{a}$ and $\mathbf{b}$ are parallel, then $\mathbf{a} \times \mathbf{b} = 0$. *Hint.* $\mathbf{a}$ is parallel to $\mathbf{b}$ if $\mathbf{a} = \alpha\mathbf{b}$ where α is a constant.

§18. *The Evolution of the Angular Momentum.* To find out how $\mathbf{L}$ changes in time, I take the derivative of Eq. 13.43:

$$\frac{d\mathbf{L}}{dt} = \frac{d\mathbf{r}}{dt} \times \mathbf{p} + \mathbf{r} \times \frac{d\mathbf{p}}{dt} \tag{13.45}$$

Newton's equation gives

$$\frac{d\mathbf{p}}{dt} = \mathbf{F} \tag{13.46}$$

where $\mathbf{F}$ is the force. Also $\mathbf{p} = \mu \frac{d\mathbf{r}}{dt}$ is parallel to $\frac{d\mathbf{r}}{dt}$, which means that their cross-product is zero:

$$\frac{d\mathbf{r}}{dt} \times \mathbf{p} = 0 \tag{13.47}$$

Using Eqs 13.46 and 13.47 in Eq. 13.45 gives

$$\frac{d\mathbf{L}}{dt} = \mathbf{r} \times \mathbf{F} \tag{13.48}$$

But the force in our system is oriented along the vector $\mathbf{r}$ connecting the particles. Therefore, $\mathbf{r}$ and $\mathbf{F}$ are parallel and $\mathbf{r} \times \mathbf{F} = 0$. This gives

$$\frac{d\mathbf{L}}{dt} = 0 \tag{13.49}$$

This equation tells us that the motion of the two particles must be such that the vector $\mathbf{L} = \mathbf{r} \times \mathbf{p}$ does not change in time. This means that neither the *direction* nor the *magnitude* of $\mathbf{L}$ changes. Now remember that because $\mathbf{L} = \mathbf{r} \times \mathbf{p}$, both $\mathbf{r}$ and $\mathbf{p}$ must be in a plane perpendicular to $\mathbf{L}$. Since $\mathbf{L}$ never changes direction, $\mathbf{r}$ and $\mathbf{p}$ must be forever in this plane. Therefore, the motion of the quasi-particle is planar.

§19. *A Polar Coordinate System.* We can take advantage of the planarity of the motion and take the coordinate system to have the x and y axes in the plane of motion (see Fig. 13.3). This means that $\mathbf{r} = \{x, y, 0\}$; the OZ component of $\mathbf{r}$ is zero. The angular momentum is perpendicular to this plane and therefore $\mathbf{L} = \{0, 0, L_z\}$.

The length r of $\mathbf{r}$ describes the radial (vibrational) motion of the particles (the change of the distance between them) and the change in the angle ϕ characterizes the rotation of the particles. Since we want to express the energy (and the particle motion) in terms of rotation and vibration, it is useful to use r

and ϕ as variables, instead of x and y. From Figure 13.3, it can be seen that

$$x = r \cos \phi \tag{13.50}$$

$$y = r \sin \phi \tag{13.51}$$

The components of momentum are (use the preceding equations; see Workbook QM13.2)

Workbook

$$p_x = \mu \frac{dx}{dt} = \mu \cos \phi \frac{dr}{dt} - \mu r \sin \phi \frac{d\phi}{dt} \tag{13.52}$$

$$p_y = \mu \frac{dy}{dt} = \mu \sin \phi \frac{dr}{dt} + \mu r \cos \phi \frac{d\phi}{dt} \tag{13.53}$$

Here dr/dt is a radial velocity, telling me how fast the distance between the particles changes, and $d\phi/dt$ is an angular velocity, telling me how fast the particle rotates in the plane (how fast the angle ϕ in Fig. 13.3 changes).

Exercise 13.5

Determine the equation of motion for r and ϕ. *Hint*. Newton's equations are

$$\mu \frac{d^2x}{dt^2} = \frac{\partial p_x}{\partial t} = F_x$$

$$\mu \frac{d^2y}{dt^2} = \frac{\partial p_y}{\partial t} = F_y$$

with

$$F_x = -\frac{\partial V}{\partial x} = -\frac{\partial V}{\partial r}\frac{\partial r}{\partial x} \quad \text{and} \quad F_y = -\frac{\partial V}{\partial r}\frac{\partial r}{\partial y}$$

The final result should connect $\partial V/\partial r$ to the derivatives of r and ϕ with respect to time.

Exercise 13.6

If you feel brave, solve Newton's equations obtained in the previous exercise. Use the initial conditions for ϕ, $d\phi/dt$, r, and dr/dt, and calculate how ϕ and r

change in time. If you use **Mathematica**, **DSolve** or **NDSolve** will give you the solution. Use $V(r)$ in eV and r in Å. Look at two examples. For one, take $V(r)$ given by Coulomb interaction, and use the masses of a proton and an electron. For the other, take

$$V(r) = 1.6 \left[\left(\frac{1.7}{r} \right)^{12} - \left(\frac{1.7}{r} \right)^{6} \right] \text{ eV, with } r \text{ in Å}$$

and use the masses of HCl. Examine several cases where the system has high and low energies in the radial and the rotational motion. See what happens if you take $r = r_0$ in the coupling term, where r_0 is the solution of $dV/dr = 0$.

---

§20.   *The Angular Momentum in Polar Coordinates.* I can now calculate how the components of the angular momentum depend on the polar coordinates r and ϕ. I know that **L** has only one component L_z (L_x and L_y are zero because of the choice of coordinate system) given by Eqs 13.43 and 13.44:

$$L_z = (\mathbf{r} \times \mathbf{p})_z = x p_y - y p_x \tag{13.54}$$

Inserting the expressions for x, y, p_x and p_y given by Eqs 13.50–13.53 in this expression gives

$$L_z = r \cos \phi \left(\mu \sin \phi \frac{dr}{dt} + \mu r \cos \phi \frac{d\phi}{dt} \right) - r \sin \phi \left(\mu \cos \phi \frac{dr}{dt} - \mu r \sin \phi \frac{d\phi}{dt} \right)$$

$$= \mu r^2 \frac{d\phi}{dt} \tag{13.55}$$

You can see in Workbook QM13.2 the algebra used to arrive at this expression.

§21.   *The Energy in Polar Coordinates.* The energy of the quasi-particle is given by

$$H = \frac{\mathbf{p} \cdot \mathbf{p}}{2\mu} + V(r) = \frac{p_x^2}{2\mu} + \frac{p_y^2}{2\mu} + V(r) \tag{13.56}$$

Remember that our coordinate system was chosen to make $p_z = 0$; therefore $\mathbf{p} \cdot \mathbf{p} = p_x^2 + p_y^2$.

By using Eqs 13.52 and 13.53 for p_x and p_y, we have

$$H = \frac{1}{2\mu}\left(\mu\cos\phi\frac{dr}{dt} - \mu r\sin\phi\frac{d\phi}{dt}\right)^2 + \frac{1}{2\mu}\left(\mu\sin\phi\frac{dr}{dt} + \mu r\cos\phi\frac{d\phi}{dt}\right)^2 + V(r)$$

Some algebra (see Workbook QM13.4) gives

$$H = \frac{1}{2}\mu\left(\frac{dr}{dt}\right)^2 + \frac{\mu r^2}{2}\left(\frac{d\phi}{dt}\right)^2 + V(r) \qquad (13.57)$$

The first term is the kinetic energy of the radial motion of the quasi-particle. For the two-particle system, r is the distance between the particles, and $\mu(dr/dt)^2/2$ is the kinetic energy associated with the change in that distance.

The second term depends on both angular velocity $d\phi/dt$ and the distance r. I will rewrite this term by expressing $(d\phi/dt)^2$ in terms of L^2. I start from Eq. 13.55 to obtain

$$L^2 = L_z^2 = \left(\mu r^2 \frac{d\phi}{dt}\right)^2 \qquad (13.58)$$

Solving for $(d\phi/dt)^2$ gives

$$\left(\frac{d\phi}{dt}\right)^2 = \frac{L^2}{\mu^2 r^4}$$

By inserting this expression for $(d\phi/dt)^2$ into Eq. 13.57, I get

$$H = \frac{1}{2}\mu\left(\frac{dr}{dt}\right)^2 + \frac{L^2}{2\mu r^2} + V(r) \qquad (13.59)$$

This is Eq. 13.35, which I wanted to prove. It expresses the energy in terms of L^2, which is a constant of motion. The term $L^2/2\mu r^2$ can be viewed as the potential energy associated with the centrifugal force. It disappears if $L^2 = 0$. From Eq. 13.58, we see that $L^2 = 0$ means that $d\phi/dt = 0$ and therefore ϕ is constant in time. There is no centrifugal energy when the angle does not change, i.e. when the particle does not rotate.

14

ANGULAR MOMENTUM IN QUANTUM MECHANICS

Introduction

§1. Angular momentum appeared naturally when we analyzed the classical mechanics of a two-particle system (see Chapter 13). To understand the rotational motion of a molecule, we need to know how quantum mechanics treats angular momentum.

The usefulness of the angular momentum is not limited to systems that have two particles. Any molecule that is not subjected to an external force will rotate without impediment and its angular momentum must be conserved. This happens no matter how many atoms the molecule has.

Why would anyone be interested in the rotational motion of a molecule? It turns out that rotational energy is quantized and a rotating molecule can absorb or emit photons, by undergoing transitions between rotational states. The magnitude of the rotational energies is such that the frequency of these photons corresponds to microwave radiation. Microwave spectroscopy has developed complex instruments and procedures for determining the rotational energies of various molecules and for using these data to extract information about molecular structure. The bond lengths and the shapes of many molecules have been determined very accurately by this method.

You may have heard at a seminar or read in a magazine that various molecules have been detected in interstellar space. The detection was not made by sending a graduate student to a remote galaxy to do some analytical chemistry there. It was done by measuring the radiation emitted by these molecules.

To emit radiation a molecule must be in an excited state. This excitation occurs when molecules in interstellar space collide with each other. Because the temperature of interstellar space is very low, the molecules located there have very low kinetic energy (when you study statistical mechanics you will learn why). When two of them collide, neither can gain or lose much energy (a collision between two slow-moving cars does not cause much damage). There is enough energy to excite the rotational motion, but not enough to excite the vibrational motion or the motion of the electrons. Because of these collisions, some of the molecules are in a excited rotational state and can radiate photons. A microwave spectroscopist on earth, or a spectrometer on a space station, can detect this radiation and determine from its frequency the type of molecule emitting it.

Angular momentum also plays a role in explaining the properties of the electron in the hydrogen atom (the other two-body system studied here). The s, p, and d orbitals that you learned about in general chemistry describe states in which the electron has different energies and angular momenta.

It is very likely that you have learned that electrons and nuclei have a mysterious property called 'spin'. This is a misleading name, inherited from the time when it was thought that spin existed because the electron was a spinning sphere. It turns out that spin is an intrinsic property of a particle, like mass and charge. A variety of experiments have shown that the spin states of electrons and nuclei behave as if they are the eigenstates of an angular-momentum operator. Nowadays we say that the spin is an intrinsic angular momentum of the particle.

The existence of the nuclear spin made possible the development of nuclear magnetic resonance (NMR) spectroscopy. This technique has revolutionized chemistry and medicine, and one cannot imagine a decent chemistry laboratory or hospital without a good NMR facility. NMR is based on changing the spin state of the nuclei by absorption of very low frequency electromagnetic radiation. To understand the properties of these spin states it is necessary to understand angular momentum.

This chapter gives a brief introduction to orbital angular momentum. As with other dynamical quantities, we need to find the operators corresponding to the classical angular momentum and to calculate their eigenvalues and eigenstates. Knowing these will tell us what values the angular momentum can take and how to calculate the transitions produced by photon absorption.

Since the angular momentum $\mathbf{L}$ is a vector, it is fully described when we know its three components L_x, L_y, and L_z. We expect therefore that in quantum mechanics the angular momentum is a vector operator $\hat{\mathbf{L}}$, and its three components $\hat{L}_x$, $\hat{L}_y$, and $\hat{L}_z$ are all operators. Of these operators, two play an important role: $\hat{L}^2 \equiv \hat{L}_x^2 + \hat{L}_y^2 + \hat{L}_z^2$ and $\hat{L}_z$. The reason for this is peculiar to quantum mechanics. It turns out that we can only measure simultaneously the values of $\hat{L}^2$ and *one* of its components. It is conventional to work with $\hat{L}_z$. Since $\hat{L}_z$ does not commute with either $\hat{L}_x$ or $\hat{L}_y$, these components of $\hat{\mathbf{L}}$ cannot be measured simultaneously with $\hat{L}_z$. I will tell you later what that means to a person performing experiments. For now, just be prepared to hear some weird, but perfectly true and exhaustively verified, statements.

There is a theorem in quantum mechanics, which I have mentioned in Chapter 3, that says that two operators that commute (such as $\hat{L}^2$ and $\hat{L}_z$) can have the same eigenfunctions but different eigenvalues. The joint eigenfunctions of $\hat{L}^2$ and $\hat{L}_z$ are called spherical harmonics. They were discovered before the invention of quantum mechanics and appear frequently in other fields of physics.

The spin is a special kind of angular momentum, not connected to a rotational motion. The theory of spin is based on a generalization of the results obtained here for the orbital angular momentum. We postulate that the spin angular momentum is represented by three operators, analogous to $\hat{L}_x$, $\hat{L}_y$, and $\hat{L}_z$, which have the same commutation relations as $\hat{L}_x$, $\hat{L}_y$, and $\hat{L}_z$, but share no other property. Therefore, all orbital angular momentum properties that can be derived from the commutation relations are also valid for spin. Most of the concepts required for understanding spin are introduced in this chapter.

Finally, at the end of the chapter we show that the two-particle Hamiltonian derived in the previous chapter can be written in terms of the angular momentum squared, as suggested in Chapter 13, §15.

The Operators Representing the Angular Momentum in Quantum Mechanics

§2. *Angular Momentum in Classical Mechanics.* We have already encountered the angular momentum operator, very briefly, in Chapter 2, §15. Since we plan to study it extensively in this chapter, I will remind you a few important properties.

In classical mechanics the angular momentum of a particle is defined by

$$\mathbf{L} = \mathbf{r} \times \mathbf{p} \tag{14.1}$$

Here $\mathbf{r}$ is the position of the particle and $\mathbf{p}$ is its momentum. The symbol $\times$ appearing in Eq. 14.1 denotes the vector product (or cross-product) of $\mathbf{r}$ with $\mathbf{p}$. The components of the vector $\mathbf{L}$ are

$$L_x = y\,p_z - z\,p_y \tag{14.2}$$

$$L_y = z\,p_x - x\,p_z \tag{14.3}$$

$$L_z = x\,p_y - y\,p_x \tag{14.4}$$

where x, y, z are the components of the vector $\mathbf{r}$, and p_x, p_y, p_z are those of the vector $\mathbf{p}$.

Eq. 14.1 is the general definition of the angular momentum of a particle at position $\mathbf{r}$, having momentum $\mathbf{p}$. Keep in mind, however, that we are interested here in a two-particle system; in that case $\mathbf{r}$ is the vector pointing from particle 1 to particle 2, whose length is equal to the distance between the particles, and $\mathbf{p}$ is the vector with components $\mu\frac{dx}{dt}$, $\mu\frac{dy}{dt}$, $\mu\frac{dz}{dt}$, where μ is the reduced mass (see Chapter 13).

§3. *The Angular Momentum Operator in Quantum Mechanics.* As you have learned in Chapter 2, all classical dynamical quantities are represented in quantum mechanics by operators. The operators $\hat{x}$, $\hat{y}$, and $\hat{z}$, corresponding to the coordinates x, y, and z are defined by:

$$\hat{x}\,\psi(x,y,z) \equiv x\,\psi(x,y,z) \tag{14.5}$$

$$\hat{y}\,\psi(x,y,z) \equiv y\,\psi(x,y,z) \tag{14.6}$$

$$\hat{z}\,\psi(x,y,z) \equiv z\,\psi(x,y,z) \tag{14.7}$$

When applied to a function, these operators multiply the function by the corresponding coordinate.

The operators corresponding to the momentum components are:

$$\hat{p}_x\,\psi(x,y,z) \equiv \frac{\hbar}{i}\frac{\partial}{\partial x}\psi(x,y,z) \tag{14.8}$$

$$\hat{p}_y\,\psi(x,y,z) \equiv \frac{\hbar}{i}\frac{\partial}{\partial y}\psi(x,y,z) \tag{14.9}$$

$$\hat{p}_z\,\psi(x,y,z) \equiv \frac{\hbar}{i}\frac{\partial}{\partial z}\psi(x,y,z) \tag{14.10}$$

with $i = \sqrt{-1}$. Here we show how the operators act on an arbitrary function $\psi(x,y,z)$.

Replacing x, y, z, p_x, p_y, p_z in Eqs 14.2–14.4 with the operators defined by Eqs 14.5–14.7 and 14.8–14.10 gives us the operators for the components of the angular momentum:

$$\hat{L}_x \psi = \frac{\hbar}{i}\left(y\frac{\partial \psi}{\partial z} - z\frac{\partial \psi}{\partial y} \right) \tag{14.11}$$

$$\hat{L}_y \psi = \frac{\hbar}{i}\left(z\frac{\partial \psi}{\partial x} - x\frac{\partial \psi}{\partial z} \right) \tag{14.12}$$

$$\hat{L}_z \psi = \frac{\hbar}{i}\left(x\frac{\partial \psi}{\partial y} - y\frac{\partial \psi}{\partial x} \right) \tag{14.13}$$

§4. *Angular Momentum in Spherical Coordinates.* Angular momentum is connected to the rotational motion of a particle. As you have seen in the previous chapter, this motion is best described by using the spherical coordinates $\{r, \theta, \phi\}$ (see Chapter 13, Fig. 13.2). This means that in Eqs 14.11–14.13 we must replace x, y, and z with expressions containing r, θ, and ϕ and also replace $\frac{\partial}{\partial x}, \frac{\partial}{\partial y}, \frac{\partial}{\partial z}$ with expressions containing only functions of r, θ, and ϕ and the partial derivatives $\frac{\partial}{\partial r}, \frac{\partial}{\partial \theta}, \frac{\partial}{\partial \phi}$.

The result is:

$$\hat{L}_x = i\hbar\left(\sin\phi\,\frac{\partial}{\partial \theta} + \cot\theta\cos\phi\,\frac{\partial}{\partial \phi} \right) \tag{14.14}$$

$$\hat{L}_y = i\hbar\left(-\cos\phi\,\frac{\partial}{\partial \theta} + \cot\theta\sin\phi\,\frac{\partial}{\partial \phi} \right) \tag{14.15}$$

$$\hat{L}_z = -i\hbar\,\frac{\partial}{\partial \phi} \tag{14.16}$$

The algebra needed for performing this change of variables is straightforward but rather tedious. Supplement 14.1 shows, briefly, how such a change of coordinates is performed. The procedure uses the equations connecting the spherical coordinates to the Cartesian ones (Chapter 13, §11, Eqs 10.29–10.31) and the chain rule of calculus.

If you like performing lengthy and tedious calculations, the derivation of Eqs 14.14–14.16 from Eqs 14.11–14.13 is fun. If your definition of fun differs from mine, do not panic. Going back and forth between various coordinate systems is so common in all branches of physics (electrodynamics, quantum mechanics, fluid mechanics, theory of elasticity) that it is well covered in many books. A classic, by Morse and Feshbach, is mentioned on p. 200, where the change of coordinates is explained in Chapter I. A practicing physical chemist is most likely to look up in that book the necessary equations, rather than spending time doing the transformation.

§5. *The Operator* $\hat{L}^2$. In the classical analysis of rotational motion of a two-particle system (see Chapter 13), the rotational energy is proportional to the angular momentum squared. It is reasonable to expect that this quantity is an important player in the quantum theory of this system. In preparation for this, we determine here the expression for the operator $\hat{L}^2$.

The angular momentum is a vector and its length is given by

$$\hat{L}^2 = \hat{L}_x^2 + \hat{L}_y^2 + \hat{L}_z^2 \tag{14.17}$$

To obtain $\hat{L}^2$ in spherical coordinates, insert in Eq. 14.17 the expressions from Eqs 14.14–14.16 and simplify the result. This is another tedious (but straightforward) calculation that I will not do in detail. You can test that the result is:

$$\hat{L}^2 \psi = -\hbar^2 \left[\frac{1}{\sin\theta} \frac{\partial}{\partial\theta} \sin\theta \frac{\partial\psi}{\partial\theta} + \frac{1}{\sin^2\theta} \frac{\partial^2\psi}{\partial\phi^2} \right] \tag{14.18}$$

We will have an opportunity to see this operator at work shortly.

Exercise 14.1

Convince yourself that the procedure described above leads to Eq. 14.18.

The Commutation Relations between $\hat{L}^2$ and $\hat{L}_x$, $\hat{L}_y$, $\hat{L}_z$

§6. *Introduction.* When we studied the eigenvalue problem in Chapter 3, two general theorems were mentioned. Two operators commute if and only if they have essentially the same eigenfunctions. Moreover, if two operators commute then the magnitude of the quantities represented by these operators can both be measured without one measurement interfering with the other. I can measure L^2 alone or L_z alone or first L^2 and then L_z, and will get the same results in any of these measurements. We say that the two quantities can be measured simultaneously. If two operators do not commute, then measuring the magnitude of one quantity alters the value of the other one. For example, $\hat{L}_z$ and $\hat{L}_x$ do not commute. This means that if two identical systems are prepared and on one. I measuere L_z and then L_x, and on the other I measure L_x directly, I can get different results for L_x. We say that L_z and L_x cannot be measured simultaneously.

These theorems make clear that it is important to know whether the operators $\hat{L}^2$, $\hat{L}_x, \hat{L}_y, \hat{L}_z$ commute with each other.

There is, however, a deeper reason to study the commutation relations. We have introduced the angular momentum by replacing $\mathbf{r}$ and $\mathbf{p}$, in the classical definition $\mathbf{L} = \mathbf{r} \times \mathbf{p}$ of the orbital angular momentum, with the appropriate operators. This gives us the operators $\hat{L}^2$, $\hat{L}_x$, $\hat{L}_y$, and $\hat{L}_z$. Using these expressions we can calculate the commutators of each pair of operators (e.g. $[\hat{L}^2, \hat{L}_x]$). It turns out that these commutation relations provide a more general definition of the angular momentum than $\mathbf{L} = \mathbf{r} \times \mathbf{p}$. This means that we can derive, from the commutation relations, the same expressions for all physically relevant quantities that we can derive from $\mathbf{L} = \mathbf{r} \times \mathbf{p}$. In addition, the definition based on the commutation relations can be applied equally well for the spin, for which $\mathbf{L} = \mathbf{r} \times \mathbf{p}$ makes no sense.

§7. *The Commutation Relations.* Remember that the commutator $[\hat{A}, \hat{B}]$ of two operators $\hat{A}$ and $\hat{B}$ is

$$[\hat{A}, \hat{B}]\psi \equiv \hat{A}\hat{B}\psi - \hat{B}\hat{A}\psi \tag{14.19}$$

The commutator is an operator and I show it acting on an arbitrary function ψ. Two operators commute when their commutator satisfies

$$[\hat{A}, \hat{B}]\psi = 0$$

when applied to any function ψ.

It is tedious, but not difficult, to show that (use the definitions Eqs 14.14–14.16 and the definition of the commutator, Eq. 14.19)

$$\left[\hat{L}_x, \hat{L}_y\right] = i\hbar\hat{L}_z \tag{14.20}$$

$$\left[\hat{L}_y, \hat{L}_z\right] = i\hbar\hat{L}_x \tag{14.21}$$

$$\left[\hat{L}_z, \hat{L}_x\right] = i\hbar\hat{L}_y \tag{14.22}$$

The components of the angular momentum do not commute with each other. Therefore, we cannot measure them simultaneously.

It is again tedious but not difficult, to use the definition (Eq. 14.18) of $\hat{L}^2$ and the definition (Eqs 14.14–14.16) of $\hat{L}_x$, $\hat{L}_y$, $\hat{L}_z$, to show that

$$[\hat{L}^2, \hat{L}_x]\psi(r, \theta, \phi) = 0 \tag{14.23}$$

$$[\hat{L}^2, \hat{L}_y]\psi(r, \theta, \phi) = 0 \tag{14.24}$$

$$[\hat{L}^2, \hat{L}_z]\psi(r, \theta, \phi) = 0 \tag{14.25}$$

These equations show that the angular momentum squared (which is proportional to the rotational energy) commutes with each component of the angular momentum. This means that we can measure simultaneously the value of L^2 and that of one component. It is customary to choose this component to be L_z, although it makes no difference which component is used.

These results tell us that it is impossible to place the system in a state in which we can measure simultaneously the values of L_x, L_y, and L_z. We can only place the system in a state in which we know L^2 and one of the components (here we choose L_z). This means that it is impossible in quantum mechanics to create a state in which we can determine the direction of the angular momentum vector. We will discuss this in more detail when we examine the rotational motion of a diatomic molecule and the rotation of the electron in a hydrogen atom.

Exercise 14.2

Use a symbolic manipulation program to test that the commutation relations written above are correct, regardless of what function the commutator is applied to.

The Eigenvalue Equations for $\hat{L}^2$ and $\hat{L}_z$

§8. As you have learned in Chapter 4, if we want to know what values a certain dynamical variable can take in an experiment, we need to find the eigenvalues of the corresponding operator.

The eigenvalue equation for an arbitrary operator $\hat{O}$ (see Chapter 3) is

$$\hat{O}\psi(\theta, \phi) = \lambda\,\psi(\theta, \phi) \tag{14.26}$$

In writing this equation, I assumed that the operator acts on functions that depend on θ and ϕ, which is the case for the operators related to the angular momentum.

As explained in Chapter 3, there may be many pairs $\{\lambda_i, \psi_i(\theta, \phi)\}$, $i = 0, 1, 2, \ldots$ that satisfy this equation. The functions $\psi_i(\theta, \phi)$ are called the eigenfunctions of the operator $\hat{O}$ and the numbers λ_i are the eigenvalues of $\hat{O}$ corresponding to the eigenfunctions $\psi_i(\theta, \phi)$.

The eigenfunctions and the eigenvalues contain all the information needed for understanding the phenomena that involve the quantity O.

§9. *The Eigenvalue Problem for $\hat{L}^2$.* The general formula Eq. 14.26 applied to the operator eigenvalue problem for $\hat{L}^2$ is (replace $\hat{O}$ with $\hat{L}^2$ in Eq. 14.26 and then use Eq. 14.18 for $\hat{L}^2$)

$$\hat{L}^2 \psi(\theta, \phi) = -\hbar^2 \left[\frac{1}{\sin\theta} \frac{\partial}{\partial\theta} \sin\theta \frac{\partial \psi(\theta, \phi)}{\partial\theta} + \frac{1}{\sin^2\theta} \frac{\partial^2 \psi(\theta, \phi)}{\partial\phi^2} \right]$$
$$= \lambda\, \psi(\theta, \phi) \tag{14.27}$$

This equation appears in many branches of physics and was solved in the nineteenth century. The eigenfunctions are called *spherical harmonics* and are denoted by $Y_\ell^m(\theta, \phi)$. The eigenvalues are

$$\lambda_\ell = \hbar^2 \ell(\ell + 1), \quad \ell = 0, 1, 2, \ldots \tag{14.28}$$

This means that the spherical harmonics $Y_\ell^m(\theta, \phi)$ and the eigenvalues λ_ℓ satisfy the equation

$$\hat{L}^2\, Y_\ell^m(\theta, \phi) = \hbar^2 \ell(\ell + 1)\, Y_\ell^m(\theta, \phi) \tag{14.29}$$

$$\text{where} \quad \ell = 0, 1, 2, \ldots \tag{14.30}$$

$$\text{and} \quad m = -\ell, -\ell + 1, \ldots, \ell - 1, \ell \tag{14.31}$$

The eigenfunctions $Y_\ell^m(\theta, \phi)$ are labeled by *two indices*, ℓ and m. It is not unusual to have more than one index labeling an eigenfunction or even an eigenvalue. In Chapter 8 you have seen that the eigenfunctions and the eigenvalues of a particle in a box were labeled by three indices.

The values the indices ℓ and m can take are prescribed by Eqs 14.30 and 14.31. For every value of ℓ, m can take $2\ell + 1$ values. Since the eigenvalues are *independent of* m, there are $2\ell + 1$ eigenfunctions Y_ℓ^m (namely, $Y_\ell^{-\ell}$, $Y_\ell^{-(\ell-1)}$, ..., $Y_\ell^{\ell-1}$, Y_ℓ^ℓ) for each eigenvalue $\hbar^2\ell(\ell + 1)$. We say that the eigenvalue $\hbar^2\ell(\ell + 1)$ is *degenerate* and its *degeneracy* is equal to $2\ell + 1$.

We will discuss shortly the physical meaning of this result. But before doing this, we will look at the eigenvalue problem for the operator $\hat{L}_z$.

§10. *The Eigenvalue Problem for* $\hat{L}_z$. Since $\hat{L}_z$ and $\hat{L}^2$ commute, they must have the same eigenfunctions. Indeed, it can be shown that the equation

$$\hat{L}_z\, Y_\ell^m(\theta, \phi) = \hbar m\, Y_\ell^m(\theta, \phi) \tag{14.32}$$

is satisfied for any values of ℓ and m allowed by Eqs 14.30 and 14.31. Thus, the eigenvalues of $\hat{L}_z$ are

$$\hbar m \;\; \text{with } m = -\ell, -\ell - 1, \ldots, \ell - 1, \ell \tag{14.33}$$

For example, if $\ell = 1$, Y_1^{-1}, Y_1^0, Y_1^1 are eigenfunctions of L_z with the eigenvalues $-\hbar$, 0, and $\hbar$, respectively.

§11. *Spherical Harmonics.* The spherical harmonics $Y_\ell^m(\theta, \phi)$ have been extensively studied in many branches of physics. They have very interesting properties, but we do not have time to examine them here. The general formula is

$$Y_\ell^m(\theta, \phi) = \frac{(-1)^{\ell+m}}{2^\ell \ell!}\sqrt{\frac{2\ell + 1}{4\pi}\frac{(\ell - m)!}{(\ell + m)!}}e^{im\phi}(\sin\theta)^m$$

$$\times \frac{d^{\ell+m}}{d(\cos\theta)^{\ell+m}}\left(1 - \cos^2\theta\right)^\ell \tag{14.34}$$

The functional form of Y_ℓ^m for a few values of ℓ and m is given below.

$$Y_0^0(\theta, \phi) = \frac{1}{2\sqrt{\pi}} \tag{14.35}$$

$$Y_1^{-1}(\theta, \phi) = \frac{1}{2}e^{-i\phi}\sqrt{\frac{3}{2\pi}}\sin\theta \tag{14.36}$$

$$Y_1^0(\theta, \phi) = \frac{1}{2} \sqrt{\frac{3}{\pi}} \cos \theta \qquad (14.37)$$

$$Y_1^1(\theta, \phi) = -\frac{1}{2} e^{i\phi} \sqrt{\frac{3}{2\pi}} \sin \theta \qquad (14.38)$$

$$Y_2^{-2}(\theta, \phi) = \frac{1}{4} e^{-2i\phi} \sqrt{\frac{15}{2\pi}} \sin^2 \theta \qquad (14.39)$$

$$Y_2^{-1}(\theta, \phi) = \frac{1}{2} e^{-i\phi} \sqrt{\frac{15}{2\pi}} \cos \theta \, \sin 2\theta \qquad (14.40)$$

$$Y_2^0(\theta, \phi) = \frac{1}{4} \sqrt{\frac{5}{\pi}} (-1 + 3 \cos^2 \theta) \qquad (14.41)$$

$$Y_2^1(\theta, \phi) = -\frac{1}{2} e^{i\phi} \sqrt{\frac{15}{2\pi}} \cos \theta \, \sin \theta \qquad (14.42)$$

$$Y_2^2(\theta, \phi) = \frac{1}{4} e^{2i\phi} \sqrt{\frac{15}{2\pi}} \sin^2 \theta \qquad (14.43)$$

$$Y_3^{-3}(\theta, \phi) = \frac{1}{8} e^{-3i\phi} \sqrt{\frac{35}{\pi}} \sin^3 \theta \qquad (14.44)$$

$$Y_3^{-2}(\theta, \phi) = \frac{1}{4} e^{-2i\phi} \sqrt{\frac{105}{2\pi}} \cos \theta \, \sin^2 \theta \qquad (14.45)$$

$$Y_3^{-1}(\theta, \phi) = \frac{1}{8} e^{-i\phi} \sqrt{\frac{21}{\pi}} (-1 + 5 \cos^2 \theta) \sin \theta \qquad (14.46)$$

$$Y_3^0(\theta, \phi) = \frac{1}{4} \sqrt{\frac{7}{\pi}} (-3 \cos \theta + 5 \cos^3 \theta) \qquad (14.47)$$

$$Y_3^1(\theta, \phi) = -\frac{1}{8} e^{i\phi} \sqrt{\frac{21}{\pi}} (-1 + 5 \cos^2 \theta) \sin \theta \qquad (14.48)$$

$$Y_3^2(\theta, \phi) = \frac{1}{4} e^{2i\phi} \sqrt{\frac{105}{2\pi}} \cos \theta \, \sin^2 \theta \qquad (14.49)$$

$$Y_3^3(\theta,\phi) = -\frac{1}{8}\, e^{3i\phi}\, \sqrt{\frac{35}{\pi}}\, \sin^3\theta \qquad (14.50)$$

$$Y_4^{-4}(\theta,\phi) = \frac{3}{16}\, e^{-4i\phi}\, \sqrt{\frac{35}{2\pi}}\, \sin^4\theta \qquad (14.51)$$

$$Y_4^{-3}(\theta,\phi) = \frac{3}{8}\, e^{-3i\phi}\, \sqrt{\frac{35}{\pi}}\, \cos\theta\, \sin^3\theta \qquad (14.52)$$

$$Y_4^{-2}(\theta,\phi) = \frac{3}{8}\, e^{-2i\phi}\, \sqrt{\frac{5}{2\pi}}\, (-1 + 7\cos^2\theta)\sin^2\theta \qquad (14.53)$$

$$Y_4^{-1}(\theta,\phi) = \frac{3}{8}\, e^{-i\phi}\, \sqrt{\frac{5}{\pi}}\, (-3 + 7\cos^2\theta)\cos\theta\, \sin\theta \qquad (14.54)$$

$$Y_4^0(\theta,\phi) = \frac{3}{16\sqrt{\pi}}\, (3 - 30\cos^2\theta + 35\cos^4\theta) \qquad (14.55)$$

$$Y_4^1(\theta,\phi) = -\frac{3}{8}\, e^{-i\phi}\, \sqrt{\frac{5}{\pi}}\, (-3 + 7\cos^2\theta)\cos\theta\, \sin\theta \qquad (14.56)$$

$$Y_4^2(\theta,\phi) = \frac{3}{8}\, e^{2i\phi}\, \sqrt{\frac{5}{2\pi}}\, (-1 + 7\cos^2\theta)\sin^2\theta \qquad (14.57)$$

$$Y_4^3(\theta,\phi) = -\frac{3}{8}\, e^{3i\phi}\, \sqrt{\frac{35}{\pi}}\, \cos\theta\, \sin^3\theta \qquad (14.58)$$

$$Y_4^4(\theta,\phi) = \frac{3}{16}\, e^{4i\phi}\, \sqrt{\frac{35}{2\pi}}\, \sin^4\theta \qquad (14.59)$$

These expressions were generated in Workbook QM14.2 by using the function **SphericalHarmonicY**$[\ell, m, \theta, \phi]$ provided by **Mathematica**, which returns the expression for function $Y_\ell^m(\theta,\phi)$. In Workbook QM14.3, I tested that the general expression (Eq. 14.34) gives the same result.

If you use other sources to obtain expressions for the spherical harmonics, you should be aware that different authors sometimes use slightly different definitions. They are all eigenfunctions of $\hat{L}^2$ and $\hat{L}_z$, because they differ only by multiplicative factors.

§12. *The Physical Interpretation of these Eigenstates.* As seen in the previous chapters, the eigenstates of an operator can be used to study a variety of physical properties of the quantity that the operator represents. I postpone studying the physical properties of the angular momentum until we examine the properties of a diatomic molecule and those of a hydrogen atom. In those systems the angular momentum properties are connected to the rotation of the diatomic or the rotation of the electron around the nucleus. The angular momentum properties are entangled with the radial motion and are easier to understand in the context of specific systems.

Exercise 14.3

Show that if $\Psi_\ell^m(\theta, \phi)$ is an eigenfunction of an arbitrary operator, then $\alpha\, \Psi_\ell^m(\theta, \phi)$, where α is an arbitrary complex number, is also an eigenfunction, corresponding to the same eigenvalue as $\Psi_\ell^m(\theta, \phi)$.

Exercise 14.4

Take several values of ℓ and m and verify Eqs 14.29 and 14.32 for Y_ℓ^m given above. Some examples are given in Workbook QM14.2.

Workbook

Exercise 14.5

(a)   Show that the function

$$\sum_{m=-\ell}^{\ell} c_m(r) Y_\ell^m(\theta, \phi) \tag{14.60}$$

is an eigenfunction of $\hat{L}^2$ with the eigenvalue $\hbar^2 \ell(\ell + 1)$.

(b)   Show that the function defined by Eq. 14.60 is not an eigenfunction of L_z.

(c)   Choose the coefficients c_m in Eq. 14.60 to construct $2\ell + 1$ real (as opposed to imaginary or complex) eigenfunctions $\eta_\ell(r, \theta, \phi)$ of $\hat{L}^2$. For example, for $\ell = 1$ you should construct three real eigenfunctions η_1, η_2, η_3 corresponding to three different choices of c_{-1}, c_0, and c_1. All three should be eigenfunctions of $\hat{L}^2$ with the eigenvalue $1(1 + 1)\hbar^2$. For $\ell = 2$ there are five such real eigenfunctions, etc. Going beyond $\ell = 2$ becomes complicated, but perhaps you see a pattern that will help you out. The real functions constructed in this way give the angular part of the famous s, p, d orbitals you have learned about in general chemistry. We will construct and use them when we study the hydrogen atom and the chemical bond.

Supplement 14.1 A Brief Explanation of the Procedure for Changing Coordinates

§13. The change of variables from $\{x, y, z\}$ to $\{r, \theta, \phi\}$ is motivated by physics. However, the calculation is a tedious exercise in calculus. If you are interested in such manipulations, you will find here a hint of how they are done. However, you can skip this material without irreparable damage to your education.

To go from $\hat{L}_x \psi$ given by Eq. 14.11 to $\hat{L}_x \psi$ given by Eq. 14.14, I express x, y, and z in terms of r, θ, and ϕ by using Eq. 13.29. This is the easy part. It is harder to express $(\partial \psi / \partial z)_{x,y}$ and $(\partial \psi / \partial y)_{x,z}$ in terms of $(\partial \psi / \partial \theta)_{r,\phi}$, $(\partial \psi / \partial \phi)_{r,\theta}$, and $(\partial \psi / \partial r)_{\theta,\phi}$. The subscripts in these formulae are added to remind me which variables are held constant when the derivative is taken. The chain rule gives

$$\left(\frac{\partial \psi}{\partial x}\right)_{y,z} = \left(\frac{\partial \psi}{\partial r}\right)_{\theta,\phi} \left(\frac{\partial r}{\partial x}\right)_{y,z} + \left(\frac{\partial \psi}{\partial \theta}\right)_{r,\phi} \left(\frac{\partial \theta}{\partial x}\right)_{y,z}$$

$$+ \left(\frac{\partial \psi}{\partial \phi}\right)_{r,\theta} \left(\frac{\partial \phi}{\partial x}\right)_{y,z} \tag{14.61}$$

I will calculate derivatives $(\partial r / \partial x)_{y,z}$, $(\partial \theta / \partial x)_{y,z}$, and $(\partial \phi / \partial x)_{y,z}$ from Eq. 13.31, which expresses r, θ, and ϕ as functions of x, y, and z. From $r = \sqrt{x^2 + y^2 + z^2}$, I get

$$\left(\frac{\partial r}{\partial x}\right)_{y,z} = \frac{\partial}{\partial x}\sqrt{x^2 + y^2 + z^2} = \frac{2x}{2\sqrt{x^2 + y^2 + z^2}}$$

$$= \frac{x}{r} = \frac{r \sin \theta \cos \phi}{r} = \sin \theta \cos \phi \tag{14.62}$$

In this calculation, I used $x = r \sin \theta \cos \phi$ (Eq. 13.29).

To calculate $(\partial \theta / \partial x)_{y,z}$, I use $\cos \theta = z / \sqrt{x^2 + y^2 + z^2}$ (Eq. 13.31). Taking the derivative gives

$$\left(\frac{\partial \cos \theta}{\partial x}\right)_{y,z} = \frac{\partial}{\partial x}\left(\frac{z}{\sqrt{x^2 + y^2 + z^2}}\right)_{y,z} \tag{14.63}$$

which leads to

$$-\sin\theta\left(\frac{\partial\theta}{\partial x}\right)_{y,z} = -z\,\frac{1}{(\sqrt{x^2+y^2+z^2})^2}\,\frac{2x}{2\sqrt{x^2+y^2+z^2}} \tag{14.64}$$

Using $\sqrt{x^2+y^2+z^2}=r$, $x=r\sin\theta\cos\phi$, and $z=r\cos\theta$ (Eq. 13.29), this becomes

$$\left(\frac{\partial\theta}{\partial x}\right)_{y,z} = \frac{(r\sin\theta\cos\phi)(r\cos\theta)}{r^3\sin\theta} = \frac{\cos\theta\cos\phi}{r} \tag{14.65}$$

To calculate $(\partial\phi/\partial x)_{y,z}$, I use Eq. 13.31:

$$\left(\frac{\partial\tan\phi}{\partial x}\right)_{y,z} = \frac{\partial}{\partial x}\left(\frac{y}{x}\right)_{y,z} \tag{14.66}$$

Using $(\partial\tan\phi/\partial x)_{y,z} = \sec^2\phi\,(\partial\phi/\partial x)_{y,z}$ and $\frac{\partial}{\partial x}(y/x)_{y,z} = -y/x^2$ to perform the derivatives gives

$$\sec^2\phi\left(\frac{\partial\phi}{\partial x}\right)_{y,z} = -\frac{y}{x^2} \tag{14.67}$$

Using $y=r\sin\theta\sin\phi$, $x=r\sin\theta\cos\phi$, and $\sec^2\phi=1/\cos^2\phi$ in this equation gives

$$\left(\frac{\partial\phi}{\partial x}\right)_{y,z} = -\frac{\sin\phi}{r\sin\theta} \tag{14.68}$$

Let us see what we have achieved. I want to calculate $(\partial\psi/\partial x)_{y,z}$ by using Eq. 14.61. The derivatives $(\partial r/\partial x)_{y,z}$, $(\partial\theta/\partial x)_{y,z}$, and $(\partial\phi/\partial x)_{y,z}$ that appear in Eq. 14.61 are given by Eqs 14.62, 14.65, and 14.68. Putting these in Eq. 14.61 gives

$$\left(\frac{\partial\psi}{\partial x}\right)_{y,z} = \sin\theta\,\cos\phi\left(\frac{\partial\psi}{\partial r}\right)_{\theta,\phi} + \frac{\cos\theta\,\cos\phi}{r}\left(\frac{\partial\psi}{\partial\theta}\right)_{r,\phi}$$

$$-\frac{\sin\phi}{r\sin\theta}\left(\frac{\partial\psi}{\partial\phi}\right)_{r,\theta} \tag{14.69}$$

You can perform the same kind of calculations to get $(\partial\psi/\partial y)_{x,z}$ and $(\partial\psi/\partial z)_{x,y}$. This is a tedious exercise in calculus. You should try to do it to see if you can obtain Eqs 14.14–14.16. The intermediate results are

$$\left(\frac{\partial r}{\partial y}\right)_{x,z} = \sin\theta\,\sin\phi \tag{14.70}$$

$$\left(\frac{\partial \phi}{\partial y}\right)_{x,z} = \frac{\cos\phi}{r\sin\theta} \tag{14.71}$$

$$\left(\frac{\partial \theta}{\partial y}\right)_{x,z} = \frac{\cos\theta\sin\phi}{r} \tag{14.72}$$

$$\left(\frac{\partial r}{\partial z}\right)_{x,y} = \cos\theta \tag{14.73}$$

$$\left(\frac{\partial \phi}{\partial z}\right)_{x,y} = 0 \tag{14.74}$$

$$\left(\frac{\partial \theta}{\partial z}\right)_{x,y} = -\frac{\sin\theta}{r} \tag{14.75}$$

Since

$$\frac{\partial\psi}{\partial y} = \left(\frac{\partial\psi}{\partial r}\right)\left(\frac{\partial r}{\partial y}\right)_{x,z} + \left(\frac{\partial\psi}{\partial\theta}\right)\left(\frac{\partial\theta}{\partial y}\right)_{x,z} + \left(\frac{\partial\psi}{\partial\phi}\right)\left(\frac{\partial\phi}{\partial y}\right)_{x,z}, \tag{14.76}$$

I use Eqs 14.70–14.72 to obtain

$$\left(\frac{\partial\psi}{\partial y}\right)_{x,z} = \sin\theta\,\sin\phi\,\frac{\partial\psi}{\partial r} + \frac{\cos\theta\,\sin\phi}{r}\,\frac{\partial\psi}{\partial\theta} + \frac{\cos\phi}{r\sin\theta}\,\frac{\partial\psi}{\partial\phi} \tag{14.77}$$

Similarly, I get

$$\left(\frac{\partial\psi}{\partial z}\right)_{x,y} = \cos\theta\,\frac{\partial\psi}{\partial r} - \frac{\sin\theta}{r}\,\frac{\partial\psi}{\partial\theta} \tag{14.78}$$

Next, use Eq. 13.29 for x, y, z and Eqs 14.69, 14.77, and 14.78 for the derivatives $(\partial\psi/\partial x)_{y,z}$, $(\partial\psi/\partial y)_{x,z}$, and $(\partial\psi/\partial z)_{x,y}$ in the definitions (Eqs 14.11–14.13) of $\hat{L}_x\psi$, $\hat{L}_y\psi$, and $\hat{L}_z\psi$, and you will obtain Eqs 14.14–14.16. This is what textbook writers describe as a trivial but tedious calculation. In most cases, they mean: we know that this is right, but we are unwilling to prove it. If you go on and learn more about physical chemistry, you will find that there are specialized books that give everything you need to go from one coordinate system to another. You do not need to do this sort of thing yourself. It is more important to know when it is advantageous to use a particular coordinate system. Here the force has spherical symmetry and this suggests using spherical coordinates.

15

TWO-PARTICLE SYSTEMS: THE RADIAL AND ANGULAR SCHRÖDINGER EQUATIONS

Introduction

§1. Our analysis of classical mechanics showed that the square of the angular momentum (denoted L^2) plays a prominent role in interpreting the motion of a two-particle system. In particular, the Hamiltonian (i.e. the total energy) contained the term

$$\frac{L^2}{2\mu r^2}$$

That led us to believe that the Hamiltonian operator of the two-particle system should contain a similar term, with L^2 replaced by the corresponding operator $\hat{L}^2$. In Chapter 14 we studied angular momentum to determine the form of the operator $\hat{L}^2$ in the hope of identifying its presence in the Hamiltonian operator. In this chapter we perform this identification, and use it to write the Hamiltonian in a simple and useful form.

We have managed to treat the hydrogen atom and the diatomic molecule jointly, so far, because the potential energy $V(r)$ of both systems depends *only on the distance between the particles*. In this chapter we use this common property to show that we can "split" the two-particle Schrödinger equation into two equations: one describing the radial motion (i.e. the change of the distance between the particles) and another describing the orientation of the line joining the two particles. This is achieved by using the method of *separation of variables*, which you have first encountered when studying the particle in a box (Chapter 8). However, unlike the case of the particle in the box, these two equations are coupled: the rotational motion described by one of them affects the radial motion described by the other, and vice versa.

To make further progress in our study of the two-particle system, the specific form of $V(r)$ can no longer be ignored. The Coulomb interaction between the electron and the nucleus (in the hydrogen atom) is very different from the interaction between the atoms of a diatomic molecule. Because of this, the properties of a diatomic molecule are vastly different from those of the hydrogen atom. These deep differences are caused by differences in the potential energy and, to a lesser extent, by mass differences. Several future chapters are dedicated to solving the radial equation for these two systems and to discussing the physical consequences of the solutions.

The Schrödinger Equation in Terms of $\hat{L}^2$

§2. *The Hamiltonian of a Two-particle System in Terms of* $\hat{L}^2$. In Chapter 13 we compared the classical Hamiltonian to the quantum one and suggested that the quantum Hamiltonian

$$\hat{H}\psi = -\frac{\hbar^2}{2\mu}\left[\frac{1}{r^2}\frac{\partial}{\partial r}\left(r^2\frac{\partial\psi}{\partial r}\right)\right] - \frac{\hbar^2}{2\mu r^2}\left[\frac{1}{\sin\theta}\frac{\partial}{\partial\theta}\sin\theta\frac{\partial\psi}{\partial\theta} + \frac{1}{\sin^2\theta}\frac{\partial^2\psi}{\partial\phi^2}\right] + V\psi$$
$$= E\psi \tag{15.1}$$

can be written in the form

$$-\frac{\hbar^2}{2\mu}\frac{1}{r^2}\frac{\partial}{\partial r}r^2\frac{\partial\psi}{\partial r} + \frac{\hat{L}^2\psi}{2\mu r^2} + V\psi = E\psi \tag{15.2}$$

We are now in position to show that this suggestion is correct. Indeed, all we have to do to confirm Eq. 15.2 is to take a look at the definition (Eq. 14.18) of $\hat{L}^2$ and use it to rewrite the second term in Eq. 15.1. This leads directly to Eq. 15.2, which will be used in the rest of our work with two-particle systems.

The Radial and the Angular Schrödinger Equations

§3. *The Separation of Variables in the Schrödinger Equation.* To make progress in our study of the two-particle system, we must solve the Schrödinger equation (Eq. 15.2). In keeping with the method of separation of variables, I seek a solution of the form

$$\psi(r,\theta,\phi) = \frac{F(r)}{r} f(\theta,\phi) \tag{15.3}$$

Here $\psi(r,\theta,\phi)$, $F(r)$, and $f(\theta,\phi)$ are *unknown functions*; moreover, the eigenvalue E is an unknown number. The function ψ is an energy eigenstate of the two-particle system; $F(r)/r$ is an auxiliary wave function describing the radial properties of the system; and $f(\theta,\phi)$ describes the orientation of the axis of the two-particle system. We have written the radial part as $F(r)/r$ for later convenience. Taking this form (instead of writing $\psi = g(r)f(\theta,\phi)$, where $g(r)$ is an unknown function) is not the result of any deep insight; it is based on playing with the equations and seeing what leads to the simplest result. You may doubt, at this point, that replacing one unknown function $\psi(r,\theta,\phi)$ with a product of two unknown functions, F and f, is progress. Moreover, we are not yet sure that the equation has a solution of this form.

To see if this form is correct we introduce Eq. 15.3 into the Schrödinger equation Eq. 15.2, to verify that we do not run into a conflict. This replacement gives:

$$-f(\theta,\phi)\frac{\hbar^2}{2\mu r^2}\frac{\partial}{\partial r}\left[r^2\frac{\partial}{\partial r}\left(\frac{F(r)}{r}\right)\right] + \frac{F(r)}{r}\frac{\hat{L}^2 f(\theta,\phi)}{2\mu r^2}$$

$$+ f(\theta,\phi)V(r)\frac{F(r)}{r}$$

$$= E\frac{F(r)}{r}f(\theta,\phi) \tag{15.4}$$

A great simplification occurs if I choose

$$f(\theta,\phi) = Y_\ell^m(\theta,\phi) \tag{15.5}$$

where $Y_\ell^m(\theta,\phi)$ is a *spherical harmonic*, satisfying the equation (see Chapter 14)

$$\hat{L}^2 Y_\ell^m(\theta,\phi) = \hbar^2\ell(\ell+1)Y_\ell^m(\theta,\phi) \tag{15.6}$$

Using Eq. 15.6 in Eq. 15.4 and doing a bit of house-cleaning leads to:

$$-\frac{\hbar^2}{2\mu}\frac{\partial^2 F}{\partial r^2} + \left(\frac{\ell(\ell+1)\hbar^2}{2\mu r^2} + V(r)\right)F(r) = E\,F(r) \tag{15.7}$$

Let us take a deep breath and see what we have here. We assumed first that the eigenvalue $\psi(r,\theta,\phi)$ has the form $F(r)f(\theta,\phi)/r$. Then we made a further assumption that $f(\theta,\phi) = Y_\ell^m$. We conclude that if we can solve Eq. 15.7, to obtain $F(r)$, then

$$\psi(r,\theta,\phi) \equiv \frac{F(r)}{r}Y_\ell^m \tag{15.8}$$

is a solution of the Schrödinger equation for the two-particle system.

§4. *Additional Physical Conditions.* To make further progress we need to find the eigenfunction satisfying the radial Schrödinger equation (Eq. 15.7). In Chapter 3, I explained that not all eigenfunctions are physically acceptable. Besides being solutions of the eigenvalue equation they must also satisfy the normalization condition:

$$\int_{-\infty}^{+\infty} dx \int_{-\infty}^{+\infty} dy \int_{-\infty}^{+\infty} dz\ \psi^*(x,y,z)\psi(x,y,z) = 1 \tag{15.9}$$

If this condition is not satisfied, we cannot interpret $\psi^*\psi$ as a probability density.

Since we are working in spherical coordinates, this integral has to be transformed into one over r, θ, and ϕ. You have learned how to do this when you studied calculus: the volume element is

$$dx\,dy\,dz = r^2 \sin\theta\,dr\,d\theta\,d\phi \tag{15.10}$$

and Eq. 15.9 becomes

$$\int_0^\infty dr \int_0^\pi d\theta \int_0^{2\pi} d\phi\ \sin\theta\,r^2\psi^*(r,\theta,\phi)\psi(r,\theta,\phi) = 1 \tag{15.11}$$

If Eq. 15.8 is inserted in Eq. 15.11, I must have

$$\int_0^\infty dr\,r^2\frac{F(r)^*F(r)}{r^2} \int_0^\pi \sin\theta d\theta \int_0^{2\pi} d\phi\ Y_\ell^m(\theta,\phi)^*Y_\ell^m(\theta,\phi) = 1 \tag{15.12}$$

§5. *The Integral over the Angles.* It can be shown by direct integration that the equation

$$\int_0^{2\pi} d\phi \int_0^{\pi} \sin\theta \, d\theta \, Y_\ell^m(\theta,\phi)^* Y_\ell^m(\theta,\phi) = 1 \tag{15.13}$$

is valid regardless of the values of ℓ and m (as long as $\ell = 0,1,2,\ldots$ and $m = -\ell, -\ell+1, \ldots, \ell-1, \ell$).

Below I show that this equation is satisfied for $\ell = 3$ and $m = 2$. I picked this pair at random; you can provide a similar proof for other values.

The spherical harmonic $Y_3^2(\theta,\phi)$ is given by Eq. 14.49 and it is:

$$Y_3^2(\theta,\phi) = \frac{1}{4} e^{2i\phi} \sqrt{\frac{105}{2\pi}} \, \cos\theta \, \sin^2\theta \tag{15.14}$$

The integral we need to evaluate is:

$$\int_0^{2\pi} d\phi \int_0^{\pi} \sin\theta \, Y_3^2(\theta,\phi)^* Y_3^2(\theta,\phi)$$

$$= \int_0^{2\pi} d\phi \int_0^{\pi} d\theta \sin\theta \left\{ \frac{1}{4} \sqrt{\frac{105}{2\pi}} \, \cos\theta \, \sin^2\theta \right\}^2 \tag{15.15}$$

The integrand is no longer dependent on ϕ because $\exp(2i\phi)^* = \exp(-2i\phi)$ and therefore $\exp(2i\phi)^* \exp(2i\phi) = 1$. Since the integrand does not depend on ϕ, the integration over ϕ gives a factor equal to 2π.

The integration over θ is more complicated. If you want to do it by hand use $d\theta \sin\theta = d\cos\theta$ and change variable to $\eta = \cos\theta$. Then, perform the integration. It is of course easier to use a symbolic manipulation program to do the integral. This is done in Workbook QM15.2 for several values of ℓ and m. The result is that the integral in Eq. 15.15 is equal to 1, as stated.

§6. *The Radial Normalization.* The fact that the integral over the angles is equal to 1 (see Eq. 15.13) greatly simplifies the normalization condition Eq. 15.12, which becomes

$$\int_0^{\infty} F(r)^* F(r) dr = 1 \tag{15.16}$$

To find out the eigenstates and the eigenvalues of the radial Schrödinger equation we need to solve Eq. 15.7 subject to the condition imposed by Eq. 15.16. This is what we do in the next chapters, for the diatomic molecule and for the hydrogen atom.

§7. *Summary.* The Schrödinger equation for the wave function of a two-particle system, in the center-of-mass coordinates, is

$$
\hat{H}\psi = -\frac{\hbar^2}{2\mu}\left[\frac{1}{r^2}\frac{\partial}{\partial r}\left(r^2\frac{\partial\psi(r,\theta,\phi)}{\partial r}\right)\right]
$$
$$
-\frac{\hbar^2}{2\mu r^2}\left[\frac{1}{\sin\theta}\frac{\partial}{\partial\theta}\sin\theta\,\frac{\partial\psi(r,\theta,\phi)}{\partial\theta} + \frac{1}{\sin^2\theta}\frac{\partial^2\psi(r,\theta,\phi)}{\partial\phi^2}\right]
$$
$$
+V\psi(r,\theta,\phi)
$$
$$
= E\psi(r,\theta,\phi) \tag{15.17}
$$

Here r is the distance between the particles, θ is the polar angle, and ϕ is the azimuthal angle. The last two variables describe the orientation of the line connecting the two particles.

The effective mass is

$$
\mu = \frac{m_1 m_2}{m_1 + m_2} \tag{15.18}
$$

where m_1 and m_2 are the masses of the two particles.

The solution of the equation is of the form:

$$
\psi(r,\theta,\phi) = \frac{F(r)}{r}Y_\ell^m(\theta,\phi), \tag{15.19}
$$

where the radial wave function $F(r)$ satisfies the equation

$$
-\frac{\hbar^2}{2\mu}\frac{\partial^2 F}{\partial r^2} + \left(\frac{\ell(\ell+1)\hbar^2}{2\mu r^2} + V(r)\right)F(r) = E\,F(r) \tag{15.20}
$$

and the normalization condition

$$
\int_0^\infty F(r)^*F(r)dr = 1 \tag{15.21}
$$

This is how far we can go in the analysis of a two-particle system by making use of the spherical symmetry. Further progress can be made only by solving the radial equation Eq. 15.20 for a specific form of $V(r)$. We know analytical solutions for this equation for the hydrogen atom. We can obtain one for a diatomic molecule if we make certain approximations which essentially assume that the molecule does not have a large internal energy (internal means vibrational and rotational). These solutions will be presented in the next chapters.

16

THE ENERGY EIGENSTATES OF A DIATOMIC MOLECULE

Introduction

§1. In this chapter we solve the Schrödinger equation for a diatomic molecule. Our starting point is the material summarized in §7 of the previous chapter, which is reviewed here.

The wave function of a two-particle system has the form

$$\psi(r,\theta,\phi) \equiv \frac{F(r)}{r} Y_\ell^m(\theta,\phi) \tag{16.1}$$

where $Y_\ell^m(\theta,\phi)$ is a spherical harmonic function (defined in Chapter 14) and $F(r)$ is a radial wave function, determined by

$$-\frac{\hbar^2}{2\mu}\frac{\partial^2 F}{\partial r^2} + \left(\frac{\ell(\ell+1)\hbar^2}{2\mu r^2} + V(r)\right)F(r) = E\,F(r), \tag{16.2}$$

Here μ is the reduced mass:

$$\mu = \frac{m(1)m(2)}{m(1)+m(2)} \tag{16.3}$$

$m(1)$ and $m(2)$ are the masses of the atoms, E is the as-yet-unknown energy of the diatomic, r is the distance between the atoms, and θ and ϕ are the polar and the azimuthal angles between the axis of the molecule and the axes of the coordinate system.

The radial wave function $F(r)$ must satisfy the normalization condition

$$\int_0^\infty F(r)^* F(r)\,dr = 1 \qquad (16.4)$$

As with all eigenvalue problems, Eq. 16.2 has multiple solutions for E and $F(r)$; the eigenvalues E are the energies that the diatomic molecule can have.

In this chapter, I solve the radial Schrödinger equation approximately and examine some of the physical predictions implied by the solutions. The next chapter shows how the results obtained here can be used to explain (or predict) the absorption and the emission spectra of a diatomic molecule.

To solve the Schrödinger equation of a diatomic molecule analytically, we make the *rigid-rotor* and the *harmonic* approximations. Both are based on the observation that diatomic molecules are stiff and the distance between the atoms is unlikely to deviate much from a fixed value called the bond length. If the internal energy (i.e. vibrational and rotational) of the molecule is small, these approximations introduce small (but experimentally detectable) errors. When the internal energy of the molecule is large, the approximations fail and we must solve the Schrödinger equation numerically.

The Harmonic Approximation

§2. *The Form of the Function $V(r)$.* The radial Schrödinger equation is difficult to solve because the potential energy $V(r)$ is a fairly complicated function. Here we study its form and physical meaning with the aim of approximating it with an expression that allows us to solve the equation.

A reasonable approximation to $V(r)$ is provided by the *Morse potential*

$$V(r) = D_0 \left\{ 1 - \exp[\alpha(r - r_1)] \right\}^2 \qquad (16.5)$$

Here r is the interatomic distance, and D_0, α, and r_1 are parameters whose meaning will become clear as we proceed. The values of these parameters change from molecule to molecule.

In what follows I will use this formula, specialized for the CO molecule, to explain the general properties of $V(r)$. The parameters in the Morse potential for the CO are

$$\left.\begin{array}{l} D_0 = 11.2254 \text{ eV} \\ \alpha = 3.25145 \text{ Å}^{-1} \\ r_1 = 1.1283 \text{ Å} \end{array}\right\} \tag{16.6}$$

These values have been determined by adjusting D_0, α, and r_0 to give the best fit to spectroscopic data. A plot of $V(r)$ for CO is shown in Fig. 16.1 as the solid line.

§3. *The Physics Described by* $V(r)$. According to classical mechanics the force between the atoms is

$$f(r) = -\frac{\partial V(r)}{\partial r} \tag{16.7}$$

This is plotted, as a dashed line, in Fig. 16.1.

For all diatomic molecules, $V(r)$ has a minimum at a distance denoted r_0 and called the *bond length*. The bond length is calculated by solving, for r, the equation

$$\frac{dV(r)}{dr} = 0 \tag{16.8}$$

Combining Eqs 16.7 and 16.8 tells us that when $r = r_0$ no force acts between the atoms.

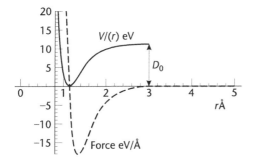

Figure 16.1 Solid line: The potential energy $V(r)$ in eV as a function of r in Å; dashed line: the force $-\partial V / \partial r$ in eV Å as a function of r in Å. The calculations were performed in Workbook QM16.1.

Workbook

Exercise 16.1

Show that for a Morse potential, $r_0 = r_1$.

We can understand how the atoms interact by examining how the force depends on r. When $r < r_0$, the force is positive (see Fig. 16.1): the atoms repel each other. When $r > r_0$, the force is negative: the atoms attract each other. This is a general feature of a pair of chemically bonded atoms: a love–hate relationship where the atoms seem to be saying "don't get too close, but don't get too far".

If the bond is stretched too much, the force of attraction between atoms becomes zero and the bond between them is broken; for CO this happens when r is approximately 3 Å. When the bond is broken, the energy of the two atoms is equal to D_0. It is for this reason that D_0 is called the dissociation energy.

The same story is told by the potential energy curve. If $r \leq r_0$ and I want to shorten the bond, the potential energy will have to be increased. This means that I must use force to push the atoms closer. If $r \geq r_0$ and I want to increase the bond length, the energy will have to be increased. This means that I must use force to pull the atoms apart. If $r \geq 3$ (for CO), it takes no energy at all to increase the bond length: the bond is broken.

Table 16.1 gives the dissociation energies D_0 for several diatomic molecules. The value $D_0 = 11.2254$ eV for CO is at the high end. Most diatomics require less energy to dissociate.

§4. *The Harmonic Approximation.* In §3 we concluded that it takes energy to make the distance between the atoms deviate from r_0. The larger the deviation, the larger the energy required to make it happen. It is therefore reasonable to assume that if the molecule has a small amount of radial energy, the interatomic distance r is close to the bond length r_0 and the deviation

$$q = r - r_0 \tag{16.9}$$

is small. This means that in the radial Schrödinger equation we can replace $V(r)$ with an expression that represents it well when $q = r - r_0$ is small. One such expression is provided by the power series expansion:

$$V(r) \approx V(r_0) + \frac{dV(r)}{dr}\bigg|_{r=r_0} (r - r_0) + \frac{1}{2} \frac{d^2V(r)}{dr^2}\bigg|_{r=r_0} (r - r_0)^2 \tag{16.10}$$

Table 16.1 Values of the bond length r_0, the force constant k (see Eq. 16.12), the energy $\hbar\omega$ (where ω is the frequency), the dissociation energy D_0 and the constants in Eq. 16.25 for some diatomic molecules. The isotopic species are ^1H, ^{19}F, ^{14}N, ^{12}C, ^{127}I, ^{16}O. CO* is an electronically excited CO molecule.

Molecule	r_0(Å)	k(mdyn/Å)	$\hbar\omega$ (cm^{-1})	D_0(cm^{-1})	B(cm^{-1})	A(cm^{-1})	X(cm^{-1})	C(cm^{-1})
H_2	0.7416	5.751	4401.20	38,292	60.85000	3.06000	121.300	4.7×10^{-2}
HF	0.9172	9.649	4137.20	49,314	20.94600	0.78900	88.700	2.2×10^{-3}
N_2	1.0976	22.940	2358.10	79,868	1.99870	0.01700	14.190	5.7×10^{-6}
CO	1.1283	19.018	2169.80	90,542	1.93127	0.01750	13.294	6.20×10^{-6}
CO*	1.3700	5.366	1152.60	29,522	1.30990	0.01680	7.281	5.8×10^{-6}
O_2	1.2074	11.768	1580.40	42,046	1.44570	0.01580	12.070	4.8×10^{-6}
I_2	2.6680	1.720	214.52	12,560	0.03734	0.00012	0.613	4.5×10^{-9}

Adapted from Table 3.2 (page 153) of I.N. Levine, *Quantum Chemistry, Volume II: Molecular Spectroscopy*, Allyn and Bacon, Boston, 1970, which cites: D. Steele, E. R. Lippincott, and J. T. Vanderslice, *Rev. Mod. Phys.* **34**, 239 (1962); P. H. Krupenie, *The Band Spectrum of Carbon Monoxide*, NBS publication NSRDS-NBS 5 (1966); J. V. Foltz, D. H. Rank, and T. A. Wiggins, *J. Mol. Spectr.* **21**, 203 (1966).

Since $q = r - r_0$ is small we have dropped all terms in which the power of q is higher than 2. Moreover, the definition of r_0 is such that $\left.\frac{dV}{dr}\right|_{r=r_0} = 0$ and we can drop the second term in Eq. 16.10, to obtain

$$V(r) \approx V(r_0) + \frac{k(r-r_0)^2}{2} \tag{16.11}$$

The quantity

$$k \equiv \left.\frac{d^2V(r)}{dr^2}\right|_{r=r_0} \tag{16.12}$$

is called the *force constant* of the diatomic molecule, or the force constant of the harmonic oscillator.

By now you have heard many times that adding or subtracting a constant from the energy causes no changes in the behavior of the system (the force corresponding to $V(r)$ is the same as that corresponding to $V(r)+$ constant). Therefore, we can drop the constant $V(r_0)$ from Eq. 16.11 and arrive at our final result:

$$V(r) \approx \frac{k(r - r_0)^2}{2} \tag{16.13}$$

This expression is the *harmonic approximation* to the potential $V(r)$.

One more piece of notation is customary: the *vibrational* frequency is

$$\omega \equiv \sqrt{\frac{k}{\mu}} \tag{16.14}$$

where μ is the reduced mass of the diatomic. Some authors call ω the *harmonic frequency* of the diatomic.

§5. *When is this Approximation Correct?* The harmonic approximation is based on the stiffness of the chemical bond. The approximation is valid if the terms that we neglected (in Eq. 16.10) are smaller than the terms that we kept:

$$\left| \frac{1}{3!} \frac{d^3V}{dr^3} \right|_{r=r_0} (r - r_0)^3 \right| \ll \left| \frac{1}{2!} \frac{d^2V}{dr^2} \right|_{r=r_0} (r - r_0)^2 \right|$$

This is equivalent to

$$|r - r_0| \ll \frac{3 \left| \frac{d^2V(r)}{dr^2} \right|_{r=r_0}}{\left| \frac{d^3V(r)}{dr^3} \right|_{r=r_0}}$$

Exercise 16.2

Use the Morse potential to derive algebraic expressions for the force constant k, the bond length r_0, and the vibrational frequency ω. After you have done this, take a look at Workbook QM16.2.

Exercise 16.3

Calculate, by using the Morse potential for CO, the numerical values of the force constant k, the bond length r_0, and the vibrational frequency ω. Use SI units, and also a system in which the energy is in eV and the distance is in Å. Compare your results with those in Workbook QM16.3.

---

§6. *The Harmonic Approximation to $V(r)$.* To understand how the harmonic approximation represents $V(r)$, I pretend that the Morse potential for CO is the exact potential and calculate the harmonic approximation to it. Then I plot $V(r)$ and the harmonic approximation on the same graph. The calculations are performed in Workbook QM16.4. The harmonic potential approximating the Morse potential for CO is

$$V_h(r) = 118.674(r - 1.1283)^2 \text{ eV} \qquad (16.15)$$

In this equation r is in Å and V_h is in eV.

In Fig. 16.2, I show $V(r)$ and the harmonic approximation $V_h(r)$ for CO, in the low energy range. As you can see, this is not a bad approximation as long as the energy is low, but it becomes very bad when the energy increases (see Fig. 16.3). In particular, the harmonic potential does not allow the dissociation of the molecule, no matter how much energy we give it. This is a disastrous failure if you want to study molecular dissociation.

§7. *Why is this a Harmonic Approximation?* The potential energy $V(r)$ controls the manner in which the distance between the atoms changes in time. In the

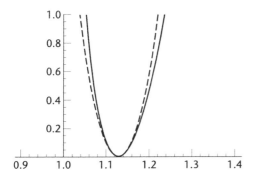

Figure 16.2 The solid line shows $V(r)$, the dashed line shows the harmonic approximation $V_h(r)$. The plot, made in Workbook QM16.4, is confined to low energy values.

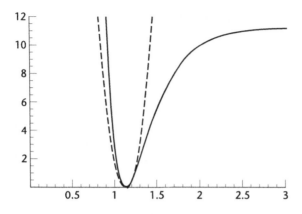

Figure 16.3 The solid line shows $V(r)$, the dashed line shows the harmonic approximation $V_h(r)$. In this plot, made in Workbook QM16.4, r varies over a wider range than in Fig. 16.2.

harmonic approximation the force of the interaction between the atoms is

$$f_h(r) = -\frac{dV_h(r)}{dr} = -k(r - r_0) \tag{16.16}$$

In mechanics this is called Hooke's law. It is the force that would exist if the two atoms were connected by an elastic spring whose rest length is r_0. The name of *force constant* for k comes from the mechanics of springs.

If we use Hooke's force in Newton's equation, we obtain (recall that $q = r - r_0$)

$$\mu \frac{d^2q}{dt^2} = f(r) = -kq$$

This equation is very easy to integrate. The result is

$$q(t) = A \cos(\omega t + \phi) \tag{16.17}$$

where ω is given by Eq. 16.14. We have found that the distance between the atoms oscillates with the frequency ω; now you see where ω got its name. This kind of motion is called a harmonic oscillation, hence the name "harmonic approximation" for the quadratic potential leading to it. The amplitude A and the phase ϕ of the oscillation are determined by the initial conditions.

The Rigid-Rotor Approximation

§8. After the harmonic approximation has been made, the radial Schrödinger equation (Eq. 16.2) becomes (use Eq. 16.14 to write $k = \mu\omega^2$)

$$-\frac{\hbar^2}{2\mu}\frac{d^2F(r)}{dr^2} + \frac{\mu\omega^2}{2}(r - r_0)^2 F(r) = \left\{E - \frac{\hbar^2\ell(\ell + 1)}{2\mu r^2}\right\} F(r) \tag{16.18}$$

To solve this equation analytically we need to make a further approximation, which sets

$$\frac{\hbar^2\ell(\ell + 1)}{2\mu r^2} \approx \frac{\hbar^2\ell(\ell + 1)}{2\mu r_0^2} \tag{16.19}$$

The reason for doing this is simple. We have established that when the molecule has a small internal energy, the interatomic distance r does not deviate much from the bond distance r_0. Therefore we can replace r in the left-hand side of Eq. 16.19 with r_0. This leads to a tremendous simplification since it "downgrades" the term in question from being an operator to being a mere constant.

With this *rigid-rotor* approximation, the radial Schrödinger equation (Eq. 16.18) becomes

$$-\frac{\hbar^2}{2\mu}\frac{\partial^2 F}{\partial r^2} + \frac{\mu\omega^2}{2}(r - r_0)^2 F(r) = E_v\, F(r) \tag{16.20}$$

with E_v defined by

$$E_v \equiv E - \frac{\hbar^2\ell(\ell + 1)}{2\mu r_0^2} \tag{16.21}$$

§9. *Physical Interpretation.* To understand the rigid-rotor approximation, let us take seriously the idea that r stays constant and equal to r_0. If r does not change then the radial kinetic energy must be zero. This means that the derivatives with respect to r in Eq. 16.18 should be zero. Furthermore, $r = r_0$ cancels the harmonic potential and Eq. 16.18 becomes

$$\frac{\hbar^2 \ell(\ell + 1)}{2\mu r_0^2} = E$$

This gives an expression for the total energy E of a rigid diatomic (i.e. one whose interatomic distance r is forced to stay equal to r_0).

What does this correspond to? Our rigid molecule can no longer vibrate, but it rotates. Its total energy E has to be equal to the rotational energy. We therefore identify the quantity

$$E_\ell \equiv \frac{\hbar^2 \ell(\ell + 1)}{2\mu r_0^2} \tag{16.22}$$

with the rotational energy of the rigid rotor. This energy depends on the quantum number ℓ; $\hbar^2 \ell(\ell + 1)$ are the allowed values of the angular momentum squared. This reminds us of the classical results giving rotational energy as $L^2/2\mu r_0^2$. The only difference is that the values of L^2 are now "quantized".

The symbol E_v appearing in Eq. 16.21 is the total energy E minus the rotational energy E_ℓ; this means that E_v is the energy of the vibrational motion.

§10. *The Rigid-Rotor Approximation Decouples Vibrational and Rotational Motion.* When we analyzed the classical theory of two-particle systems, in Chapter 13, §14, I explained that the term $L^2/2\mu r^2$ couples the rotational and the vibrational motions. I remind you very briefly what that means. The radial (vibrational) motion changes the magnitude of r. If r increases, the classical rotational energy $L^2/2\mu r^2$ becomes smaller and the rate of rotation decreases. A reduction in r has the opposite effect. One can also understand this effect by considering the total energy. If r is increased, the harmonic energy $\frac{\mu\omega^2}{2}(r - r_0)^2$ increases. Since no external force acts on the molecule, its total energy must stay constant. This means that an increase in the vibrational energy must cause a decrease in the rotational energy.

No matter how we look at it, the vibrational motion affects the rotation, through the term $L^2/2\mu r^2$. When this term is replaced with $L^2/2\mu r_0^2$, the rotational energy becomes independent of r: stretching the molecule no longer affects the rotation energy. The rotation and the vibration have become independent or decoupled.

The Eigenstates and Eigenvalues of the Radial Schrödinger Equation in the Harmonic and Rigid-Rotor Approximation

§11. *The Eigenvalues.* The harmonic and the rigid-rotor approximations have reduced the radial Schrödinger equation to Eq. 16.20. This eigenvalue problem was solved by Charles Hermite, a French mathematician (1822–1901). He found that the eigenvalues are

$$E_v = \hbar\omega\left(v + \frac{1}{2}\right), \quad v = 0, 1, 2, \dots \tag{16.23}$$

We can use these values for E_v in Eq. 16.21 to calculate the total energy E:

$$E_{v,\ell} = \hbar\omega\left(v + \frac{1}{2}\right) + \frac{\hbar^2\ell(\ell+1)}{2\mu r_0^2} \tag{16.24}$$

E is labeled now with the subscripts v and ℓ, to indicate its dependence on these two *quantum numbers*. The first term in this equation is the vibrational energy, the second term is the rotational energy.

Remember that the quantum numbers can take the values

$$v = 0, 1, 2, \dots$$

$$\ell = 0, 1, 2, \dots$$

Exercise 16.4

I found the following data for the H_2 molecule in a book by Herzberg[a]:

$$\omega = 4395 \text{ cm}^{-1}$$

and

$$B = 60.80 \text{ cm}^{-1}$$

---

[a] G. Herzberg *Spectra of Diatomic Molecules*, Van Nostrand Reinhold Co., New York, 1950, P. 532.

where

$$B \equiv \frac{\hbar^2}{2\mu r_0}$$

- Use this information to calculate the energy of the H_2 molecule for $v = 0, 1, 2$ and $\ell = 0, 1, 2, 3$. Make an energy-level diagram. Perform the calculations with energy units of cm^{-1}, eV, and joule. See Appendix 2 for the conversion factors.

- Use the data above to calculate the force constant of the H–H bond (in eV per Å^2) and the bond length in Å.

§12. *An Improved Formula for the Energy.* We have solved the Schrödinger equation in the harmonic and the rigid-rotor approximations. This works if v is small and ℓ is not exceedingly large. However, if the accuracy of the calculations is to match that of the measurements, we need to correct for the errors made by these approximations. A better equation for the eigenvalues of the diatomic molecule is:

$$E_{v,\ell} = \hbar\omega \left(v + \frac{1}{2} \right) + B\, \ell(\ell+1) - X \left(v + \frac{1}{2} \right)^2$$

$$- A \left(v + \frac{1}{2} \right) \ell(\ell+1) - C[\ell(\ell+1)]^2 \qquad (16.25)$$

This equation is not exact: it also breaks down when v or ℓ is large. However, it gives more accurate energies than the harmonic approximation when v and ℓ are small.

The quantum numbers in Eq. 16.25 take the values $v = 0, 1, 2, \ldots$ and $\ell = 0, 1, 2, \ldots$. The constants X, A, and C are given in Table 16.1, in units of *reciprocal centimeters* (i.e. cm^{-1}). This is the energy unit preferred by chemists interested in vibrational and rotational spectra of molecules.

While reading materials on spectroscopy one starts to suspect that, at sometime in the past, spectroscopists had a secret meeting convened to create energy units whose purpose is to exasperate chemists! You can convert these units to the ones you prefer, by using the table given in Appendix 2.

Exercise 16.5

Use the data given in Table 16.1 and Eq. 16.25 to calculate the total energy (vibrational and rotational) of H_2 for $v = 0, 1, 2, 3, 4$ and $\ell = 0, 1, 2, 3, 4$. Make an energy-level diagram. Compare the results with those obtained when you make the harmonic and the rigid-rotor approximations.

Each term in Eq. 16.25 has a physical meaning. $X(v + \frac{1}{2})^2$ takes into account the effect of the third- and fourth-order terms in the expansion of the potential energy in power series.

In the harmonic approximation the energy difference between two contiguous states, $E_{v+1} - E_v = \hbar\omega$, is independent of the quantum number v. This means that in an energy-level diagram the spacing between any two successive vibrational levels is the same. This is no longer true when the correction terms are taken into account. In this case the levels in the diagram get closer to each other as the vibrational energy increases.

The term $A\ell(\ell + 1)(v + \frac{1}{2})$ depends on both the vibrational and the rotational quantum numbers, and accounts for the interaction between rotation and vibration.

The value of $C[\ell(\ell + 1)]^2$ depends only on the rotational quantum number ℓ and accounts for the centrifugal distortion. If the molecule rotates rapidly, the centrifugal force will lengthen the molecule, increasing the moment of inertia and decreasing the rotational energy.

You should keep in mind that Eq. 16.25 makes *small* corrections to the harmonic approximation. If the molecule has high internal energy this equation will fail. If you are interested in the behavior of a diatomic molecule that has very high internal energy, you will have to leave $V(r)$ and $\hbar^2\ell(\ell + 1)/2\mu r^2$ intact (no power series expansion, no rigid-rotor approximation) and solve the radial Schrödinger equation numerically. This is quite possible nowadays.

But even this is not enough. A diatomic molecule is not a stick that rotates and vibrates. It contains electrons that have their own motion. In some cases, the only effect of the electrons is to generate the potential energy $V(r)$. But, if the electrons have a net angular momentum, life becomes complicated and we cannot ignore their effect on the rotational motion.

Finally, trouble arises for homonuclear molecules (molecules having identical nuclei, e.g. H_2 and O_2). Quantum mechanics imposes some restrictions on the wave functions of such molecules and these affect the rotational energies. The changes brought about by this effect depend on nuclear spin.

If you try to use what you have learned so far to understand the behavior of a diatomic molecule, you will sometimes fail. This happens because you do not know enough. At the current stage of development of quantum mechanics, we can calculate the properties of diatomic molecules made of light atoms by solving the Schrödinger equation for the electrons and the nuclei. Such calculations avoid all the approximations mentioned above and obtain excellent agreement with experiment. This can be done without using any information from experiment, other than the charge, mass, and spin of the nuclei and the electrons.

§13. *The Radial Eigenfunctions.* The eigenfunction $F_v(x)$ corresponding to the eigenvalue E_v is

$$F_v(r) = \sqrt{\beta}\; \Phi_v(y) \tag{16.26}$$

where

$$\Phi_v(q) = \frac{1}{\pi^{\frac{1}{4}}\sqrt{2^v v!}} \exp\left[-\frac{y^2}{2}\right] \mathrm{H}_v(y) \tag{16.27}$$

with

$$y \equiv \beta(r - r_0) \tag{16.28}$$

$\mathrm{H}_v(z)$ is a Hermite polynomial; both **Mathematica** and **Mathcad** provide functions that calculate it.

The quantity

$$\beta \equiv \sqrt{\frac{\mu\omega}{\hbar}} \tag{16.29}$$

has units of length^{-1} and therefore y is a dimensionless variable.

Since the exponential becomes very small when

$$\frac{y^2}{2} \gg 1,$$

the wave function is practically zero if

$$|r - r_0| \gg \frac{\sqrt{2}}{\beta}$$

Since the wave function $F_v(r)$ gives the probability that r can take a certain value, we conclude that it is unlikely that the interatomic distance r will reach values larger than

$$r_0 + \frac{\sqrt{2}}{\beta}$$

or smaller than

$$r_0 - \frac{\sqrt{2}}{\beta}$$

The functions $F_v(x)$ for a few values of v are given below. They were generated in Workbook QM16.5.

Workbook

$$F_0(y) = \sqrt{\beta}\, \frac{\exp[-y^2/2]}{\pi^{1/4}} \tag{16.30}$$

$$F_1(y) = \sqrt{\beta}\, \frac{\sqrt{2}\exp[-y^2/2]y}{\pi^{1/4}} \tag{16.31}$$

$$F_2(y) = \sqrt{\beta}\, \frac{\exp[-y^2/2](4y^2 - 2)}{2\sqrt{2}\pi^{1/4}} \tag{16.32}$$

$$F_3(y) = \sqrt{\beta}\, \frac{\exp[-y^2/2](8y^3 - 12y)}{4\sqrt{3}\pi^{1/4}} \tag{16.33}$$

$$F_4(y) = \sqrt{\beta}\, \frac{\exp[-y^2/2](16y^4 - 48y^2 + 12)}{8\sqrt{6}\pi^{1/4}} \tag{16.34}$$

$$F_5(y) = \sqrt{\beta}\, \frac{\exp[-y^2/2](32y^5 - 160y^3 + 120y)}{16\sqrt{15}\pi^{1/4}} \tag{16.35}$$

Exercise 16.6

1.   To have a physical meaning these eigenfunctions must be normalized (i.e. they must satisfy Eq. 16.4). Test that this equation is satisfied by the first four

wave functions F_v, $v = 0, 1, 2, 3$. Compare your results with those in Cell 2 of Workbook QM16.5.

2.   The eigenfunctions must be orthogonal to each other, which means that

$$\int_0^\infty F_n(r)F_m(r)dr = \delta_{m,n}$$

where $\delta_{m,n} = 1$ if $m = n$ and $\delta_{m,n} = 0$ otherwise. Verify this relation for several values of n and m. Compare your work with that in Cell 3 of Workbook QM16.5. *Hint*. These integrals cannot be evaluated analytically, but you can replace the lower limit of integration with $-\infty$ since the integrand is zero when $r \leq 0$. After that replacement, you can perform the integrals.

The Physical Interpretation of the Eigenfunctions

§14.   *Introduction*. We know now that

$$\Psi_{v,\ell,m}(r,\theta,\phi) = \frac{F_v(r)}{r} Y_\ell^m(\theta,\phi) \tag{16.36}$$

is an energy eigenfunction of a diatomic molecule. The vibrational wave functions $F_v(r)$ are given by Eqs 16.30–16.35. Expressions for the spherical harmonics $Y_\ell^m(\theta,\phi)$ can be found in Chapter 14, Eqs 14.35–14.59.

This section explains how to use these functions to extract information about the physical properties of the molecule. I assume that an experimentalist has prepared an ensemble of diatomic molecules in the state $\Psi_{v,\ell,m}(r,\theta,\phi)$. This is not easy to do, and I will discuss later how it might be done. Here we are concerned with the following question: if such a state has been prepared, what are the properties of the molecule?

Remember that the eigenstate in Eq. 16.36 describes the wave function of a molecule seen by an observer who is pinned to the center of mass and translates with it, but does not rotate with the molecule. The observer does not "see" the translational motion of the molecule, but can detect its vibration and rotation. His observations can be described with the help of the vector $\mathbf{r}$, whose length is equal to the distance between atoms and which lies along the axis joining the atoms. In Cartesian coordinates this vector has the components $\mathbf{r} = \{x, y, z\}$. In spherical coordinates the same information is conveyed by the variables r, θ and

ϕ; these quantities are connected to $\{x, y, z\}$ through the equations Eq. 13.29 (see Chapter 13, §11).

Exercise 16.7

Use **Mathematica** or **Mathcad** to show that

$$\int_{-\infty}^{+\infty} dx \int_{-\infty}^{+\infty} dy \int_{-\infty}^{+\infty} dz \, \Psi_{v,\ell,m}(x,y,z)^* \Psi_{s,p,k}(x,y,z)$$

$$= \int_{0}^{\infty} dr \int_{0}^{\pi} \sin(\theta) d\theta \int_{0}^{2\pi} F_n(r) F_s(r) Y_\ell^m(\theta,\phi)^* \, Y_p^k(\theta,\phi)$$

$$= \delta_{n,s} \, \delta_{\ell,p} \, \delta_{m,k}$$

---

§15.   *The Probability of Having a Given Interatomic Distance and Orientation.* The expression

$$p_{v,\ell,m}(x,y,z) \, dxdydz \equiv \Psi(x,y,z)^* \, \Psi(x,y,z) \, dxdydz \qquad (16.37)$$

is the probability that the components $\{x, y, z\}$ of $\mathbf{r}$ have values between x and $x + dx$, y and $y + dy$, and z and $z + dz$.

Because of the spherical symmetry of the system, it is more convenient to work in spherical coordinates. This change of variables turns Eq. 16.37 into:

$$p_{v,\ell,m}(r,\theta,\phi) \sin(\theta) \, drd\theta d\phi = \Psi(r,\theta,\phi)^* \Psi(r,\theta,\phi) r^2 \, dr \sin\theta \, d\theta d\phi \qquad (16.38)$$

Here I used the expression

$$dx \, dy \, dz = r^2 \sin\theta \, dr \, d\theta \, d\phi$$

for the volume element.

$p_{v,\ell,m}(r,\theta,\phi) \sin\theta \, drd\theta d\phi$ is the probability that the length of $\mathbf{r}$ takes values between r and $r + dr$ and the molecular axis is contained in the solid angle

$$d\Omega = \sin\theta \, d\theta \, d\phi \qquad (16.39)$$

centered around the angles θ and ϕ.

Introducing the expression for $\Psi(r,\theta,\phi)$ given by Eq. 16.36 into Eq. 16.38 leads to

$$p_{v,\ell,m}(r,\theta,\phi)\sin\theta\,dr\,d\theta d\phi = F_v(r)^2 Y_\ell^m(\theta,\phi)^* Y_\ell^m(\theta,\phi)dr\sin\theta\,d\theta\,d\phi \quad (16.40)$$

The expression on the left-hand side is a product of the probability

$$p_v(r)dr = F_v(r)^2 dr \quad (16.41)$$

that the distance between atoms has a value between r and $r + dr$, and the probability

$$p_{\ell,m}(\theta,\phi)\sin\theta\,d\theta\,d\phi = Y_\ell^m(\theta,\phi)^*\,Y_\ell^m(\theta,\phi)\sin\theta\,d\theta\,d\phi \quad (16.42)$$

that the molecular axis is in the solid angle $d\Omega$ centered around the angles θ and ϕ.

The probability that the molecular axis has a given orientation is independent of the probability that it has a given interatomic distance. This happens because the rigid-rotor approximation decoupled the rotation from the vibration.

Knowing this expression for the probability allows us to calculate how the molecules in the state $\Psi_{v,\ell,m}(r,\theta,\phi)$ are stretched and oriented. But this is predicated on preparing a system whose molecules are in a state with definite values of v, ℓ, and m. This is very hard to achieve even in a very well-equipped chemical physics laboratory. For this reason, I will not discuss here the orientation of the diatomic molecules and will only touch briefly on the information provided by $p_v(r)$. Such information is of little practical interest but it provides practice in the use of quantum mechanics.

§16. *The Probability of Having a Given Bond Length.* This probability is given by Eq. 16.41 and can be evaluated by making use of Eqs 16.26–16.29. In Workbook QM16.6, I calculated $p_v(r)$ for $v = 0,1,2,3$. The results are plotted in Fig. 16.4.

The calculation was performed as follows. I used the Morse potential (Eq. 16.5), with the parameters for CO given by Eq. 16.6, to calculate the bond length r_0 (use Eq. 16.8), the force constant k (use Eq. 16.12), the frequency ω (use Eq. 16.14), and the parameter β (use Eq. 16.29). After obtaining these quantities, I calculated $F_v(r)^2$ from Eqs 16.26 and 16.27, and plotted the result.

The graph of $p_0(r)$ is dull: it tells us that in the ground state, the most likely interatomic distance is $r = r_0$ (the probability peaks at $r = r_0 = 1.1283$ Å). It is

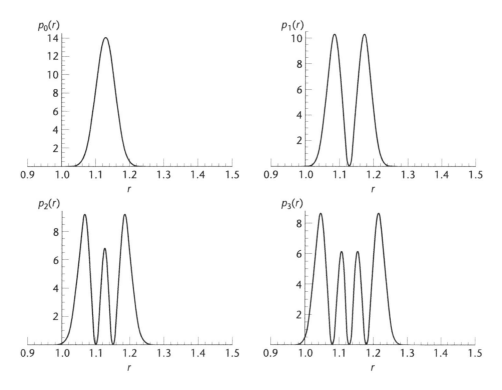

Figure 16.4 The probabilities $p_v(r)$ for $v = 0, 1, 2, 3$, for CO. The interatomic distance r is in Å. The plots were made in Cell 4 of Workbook QM16.6.

Workbook

also unlikely to find a CO molecule whose bond length falls outside the range $r \in$ [1.03, 1.21]. We expect this result as the whole theory was based on the assumption that the bond is stiff.

Something interesting happens when $v = 1$: the probability that $r = r_0$ is zero! The molecule has two most likely bond lengths, one smaller than r_0 and the other larger than r_0. As v increases, the number of peaks in $p_v(r)$ increases. This means that if we were to photograph all the molecules in a gas that have $v = 3$, we would find that most have one of four different lengths, for which $p_3(r)$ peaks. The largest number of molecules will have the shortest or the longest bond.

How do we understand these observations? It all depends on what you think 'understanding' means. We have observed the same behavior for a particle in a box: an excited particle prefers to be located in certain regions of the box, even though there is no force to guide it there. We noticed then that the intensity of a wave (acoustic or electromagnetic) created inside a box shows a similar behavior: the intensity peaks in various places in the box and there are "nodes" where the intensity is zero. A mathematically minded physicist would point out that the Schrödinger equation is very similar to the wave equation and the wave-like behavior of a particle should not surprise us.

There is another way of understanding the look of $p_v(r)$ at high v. The probability that we can find the bond length of a classical harmonic oscillator between r and $r + dr$ is the time that the bond length spends in this region divided by the total observation time (which is long). As the bond length increases, the potential energy $\frac{1}{2}k(r - r_0)^2$ increases and the kinetic energy $\frac{1}{2}mv^2$ decreases (the total energy is constant). Low kinetic energy means a small velocity and therefore the oscillator spends a longer time in regions with large r. By calculating the probability defined in this way, we will obtain a smooth curve that peaks at small and large r and dips at $r = r_0$. If v is large this curve resembles the "envelope" of the quantum result for $p_v(r)$. So, there is some correspondence between the quantum and the classical behavior of the oscillator.

I leave it to you to decide whether such remarks mean that you better understand the behavior of the vibrating diatomic molecule.

§17. *Turning Point and Tunneling.* The quantum probability $p_v(r)$ has another remarkable behavior, which contradicts classical mechanics. The energy of a classical oscillator is

$$E = \frac{\mu v^2}{2} + \frac{k(r - r_0)^2}{2} \tag{16.43}$$

By solving this equation for the velocity v, I obtain

$$v = \pm\sqrt{\frac{2E - k(r - r_0)^2}{\mu}}$$

If

$$2E = k(r - r_0)^2 \tag{16.44}$$

then the velocity is zero.
If

$$2E < k(r - r_0)^2 \tag{16.45}$$

then the velocity is imaginary. This makes no sense in mechanics and we say that the particle is *classically forbidden* from reaching values of r that satisfy Eq. 16.45. We have no reason to reject positions r for which

$$2E > k(r - r_0)^2 \qquad (16.46)$$

and the values of r that satisfy Eq. 16.46 are said to be *classically allowed*.

By solving Eq. 16.44 for r we obtain

$$r = r_0 \pm \sqrt{\frac{2E}{k}} \qquad (16.47)$$

These are the points at the border between the classically allowed and classically forbidden regions. They are called the *classical turning points*.

This may sound subtle but it is not. If we solve Newton's equation for an oscillator of fixed energy E, we find that $r(t)$ oscillates between $r_- = r_0 - \sqrt{2E/k}$ and $r_+ = r_0 + \sqrt{2E/k}$. The only point that you should retain from this discussion is that, in the classical world, r *cannot be smaller than* r_- *or larger than* r_+.

Is this true in quantum mechanics? If it were true, then the probability that r is smaller than r_- or larger than r_+ should be zero. It is easy to calculate this probability and when we do so, we find that it is not zero: r penetrates into the classically forbidden region! This process, called *tunneling*, was discussed in Chapter 7.

Exercise 16.8

Use Eq. 16.5 and the data given in Table 16.1 for H_2.

1. Calculate the classical turning points of the H_2 molecule for $v = 4$.

2. Calculate the probability that the bond length penetrates into the classically forbidden region when $v = 4$. *Hint.* Use

$$\pi_v \equiv \int_0^{r_-(v)} p_v(r)dr + \int_{r_+(v)}^{\infty} p_v(r)dr$$

for $v = 4$. You can calculate the integral numerically. What do you think: does π_v increase or decrease with v?

3. Perform the same calculation under the assumption that the force constant k is that of H_2 but the reduced mass is that of the electron.

4.   Examine the formula giving $p_v(r)$ and show that if μ becomes smaller while all other parameters are held constant, the oscillator can reach larger values of r.

§18.   *The Average Values of $r - r_0$ and $(r - r_0)^2$.* Once we know the probability that r has a certain value, we can calculate the average value $\langle f(r)\rangle_v$ of any function of r, by using the formula

$$\langle f(r)\rangle_v = \int_0^\infty f(r)p_v(r)dr \tag{16.48}$$

We can evaluate such integrals numerically by using a computer. Note, however, that $p_v(r)$ is zero when $r \le 0$. This is physically reasonable since making $r = 0$ will place the atoms on top of each other, which is unlikely at any energy for which the molecule is stable (i.e. energies less than the dissociation energy D_0). Mathematically, the same property is ensured by the factor $\exp\left[-\beta^2(r - r_0)^2\right]$ that is present in $p_v(r)$: if $r \le 0$, the values of β and r_0 are such that this exponential is practically zero. This means that we can extend the lower integration limit to $-\infty$ without changing the value of the integral. Therefore

$$\langle f(r)\rangle_v \cong \int_{-\infty}^\infty f(r)p_v(r)dr \tag{16.49}$$

This integral can often be evaluated analytically. As two examples, let us calculate the mean values of $r - r_0$ and $(r - r_0)^2$.

The mean value of $r - r_0$ is

$$\langle r - r_0\rangle_v \cong \int_{-\infty}^\infty (r - r_0)p_v(r)dr \tag{16.50}$$

$$= \int_{-\infty}^\infty (r - r_0)\beta\Phi_v(\beta(r - r_0))^2 dr \tag{16.51}$$

$$= \int_{-\infty}^\infty y\,\Phi_v(y)^2\frac{dy}{\beta} \tag{16.52}$$

$$= \frac{1}{\beta}\int_{-\infty}^\infty y\Phi_v(y)^2 dy \tag{16.53}$$

In going from Eq. 16.51 to Eq. 16.52, we have changed the integration variable from r to $y = \beta(r - r_0)$. The integral in Eq. 16.53 is dimensionless and $\langle r - r_0 \rangle_v$ has the dimensions of $1/\beta$, which is a length, as it should be.

We have done all this mathematics for naught, since physics tells me that $\langle r - r_0 \rangle_v$ should be equal to zero. Indeed, there is no physical reason why $r - r_0 = \delta$ is prefered to $r - r_0 = -\delta$; the potential and kinetic energies at these two points are the same. There is no difference between compressing and stretching the spring by the same amount δ. Therefore, $r - r_0 = \delta$ and $r - r_0 = -\delta$ are equally probable values of $r - r_0$, and when we calculate the average value they cancel to give $\langle r - r_0 \rangle_v = 0$.

Mathematically, we note that $\Phi_v(y) = \Phi_v(-y)$ (see Eq. 16.27), which means that $g(y) \equiv y\Phi(y)$ has the property that $g(-y) = -g(y)$. Because of this, $\int_{-a}^{+a} g(y) dy$ is zero regardless of the value of a.

Workbook

In Workbook QM16.7, I calculated $\langle r - r_0 \rangle_v$ numerically and found it to be zero for all the values of v that I tested.

When you start a calculation, it sometimes pays to spend a little time examining its physical meaning and symmetry properties; you might be able to guess the result without laborious calculation.

Following the reasoning outlined in Eqs 16.50–16.53, we find that

$$\langle (r - r_0)^2 \rangle_v \cong \frac{1}{\beta^2} \int_{-\infty}^{\infty} y^2 \Phi_v(y)^2 dy \tag{16.54}$$

$$= \frac{1}{\beta^2} \frac{2v + 1}{2} \tag{16.55}$$

It would take too long to show how to go from Eq. 16.54 to Eq. 16.55. Numerical calculations (Cell 4 of Workbook QM16.7) show that this equation is correct for all values of v for which I performed calculations. The numerical values of $\langle (r - r_0)^2 \rangle_v$ for CO, for $v = 0, 1, 2, 3$, are 0.028, 0.049, 0.063, 0.075 Å^2, respectively.

Exercise 16.9

1. Use physics to argue that $\langle (r - r_0)^2 \rangle_v$ cannot be zero.

2. Use Eq. 16.54 to argue that $\langle (r - r_0)^2 \rangle_v$ cannot be zero.

3. Discuss how $\langle (r - r_0)^2 \rangle_v$ changes when k, μ, or ω changes.

Exercise 16.10

1. Guess the value of $\langle (r - r_0)^{2n+1} \rangle_v$ when n is a positive integer.

2. Prove that $\langle (r - r_0)^{2n} \rangle_v$ cannot be zero when n is a positive integer.

3. (Difficult) It is known that if n is a positive integer and $p > 0$ then

$$\int_{-\infty}^{+\infty} x^{2n} \exp\left[-px^2\right] dx = \frac{(2n-1)!}{(2p)^n} \sqrt{\frac{\pi}{p}}$$

Use this fact to show that for any positive integer m, $\langle (r - r_0)^{2m} \rangle_v$ is a function only of $\langle (r - r_0)^2 \rangle_v$. *Note.* These Gaussian integrals are very common in probability theory and statistical mechanics.

§19. *The Energy Eigenstates and Eigenvalues of a Diatomic Molecule: Summary.* To find the energy eigenfunction and eigenvalues we have implemented the following steps.

- We have assumed that the Schrödinger equation

$$-\frac{\hbar^2}{2\mu r^2} \frac{\partial}{\partial r}\left(r^2 \frac{\partial \psi}{\partial r}\right) + \frac{\hat{L}^2 \psi}{2\mu r^2} + V(r)\psi = E\psi \tag{16.56}$$

  has a solution of the form

$$\psi(r, \theta, \phi) = \frac{F(r)}{r} f(\theta, \phi) \tag{16.57}$$

- We found that

$$f(\theta, \phi) = Y_\ell^m(\theta, \phi) \tag{16.58}$$

  where $Y_\ell^m(\theta, \phi)$ is a spherical harmonic (see Chapter 14), and that the radial function $F(r)$ satisfies the equation

$$-\frac{\hbar^2}{2\mu} \frac{\partial^2 F}{\partial r^2} + \left(\frac{\ell(\ell+1)\hbar^2}{2\mu r^2} + V(r)\right) F(r) = E F(r) \tag{16.59}$$

- To solve Eq. 16.59 we made the harmonic approximation

$$V(r) \approx \frac{k(r - r_0)^2}{2} \equiv \frac{kq^2}{2}, \tag{16.60}$$

where k is the force constant and r_0 is the bond length, and the rigid-rotor approximation

$$\frac{\hbar^2 \ell(\ell + 1)}{2\mu r^2} \approx \frac{\hbar^2 \ell(\ell + 1)}{2\mu r_0^2} \tag{16.61}$$

These approximations decouple the rotation from the vibration and are valid only when the internal energy of the diatomic is small.

- These approximations reduced the radial Schrödinger equation to the Hermite equation, which can be solved analytically to obtain the total energy

$$E_{v,\ell} = \hbar\omega \left(v + \frac{1}{2} \right) + \frac{\hbar^2 \ell(\ell + 1)}{2\mu r_0}$$

$$v = 0, 1, 2, \ldots$$

$$\ell = 0, 1, 2, \ldots \tag{16.62}$$

The first term is the vibrational energy and the second is the rotational energy.

- The radial wave function is given by

$$F_v(q) = \left(\frac{\beta^2}{\pi} \right)^{1/4} \frac{1}{\sqrt{2^v v!}} \exp\left[-\frac{(\beta q)^2}{2} \right] \mathrm{H}_v(\beta q) \tag{16.63}$$

where $q = r - r_0$, $H_v(q)$ is a Hermite polynomial and $\beta = \sqrt{\mu\omega/\hbar}$.

- The final eigenfunction is therefore

$$\psi_{v,\ell,m}(r, \theta, \phi) = \frac{F_v(q)}{r} Y_\ell^m(\theta, \phi) \tag{16.64}$$

- In §12, I gave an improved formula for the total energy, in which some of the errors made by the harmonic and the rigid-rotor approximations have been corrected. This equation works well only when the internal energy of the molecule is small.

These results will be used in the next chapter to calculate the absorption and emission spectra of a diatomic molecule.

Exercise 16.11

You know that for CO (with ^{12}C and ^{16}O)

$$\hbar\omega = 2170.2 \text{ cm}^{-1}$$

and

$$r_0 = 1.1281 \text{ Å}$$

1. Calculate the energy eigenvalues of the molecule (for a few values of v and ℓ).

2. Calculate the force constant and the moment of inertia of the molecule.

3. Plot the probability distribution of the variable $q = r - r_0$ when the molecule is in state $v = 2$ on the same graph with the harmonic potential.

4. Consider now the molecule ^{12}C^{18}O. This differs from ^{12}C^{16}O because two neutrons have been added to the oxygen nucleus. Would this change the force constant of the molecule, its bond length, its reduced mass, or its moment of inertia?

5. Calculate the vibrational frequency of ^{12}C^{18}O and plot the probability distribution of the variable $q = r - r_0$ when the molecule is in state $v = 2$ on the same graph with the harmonic potential.

6. Explain what the differences are between ^{12}C^{16}O and ^{12}C^{18}O and where they come from.

7. Say as much as you can about the difference between H_2 and D_2, where D is deuterium.

8. Discuss semiquantitatively whether you expect large differences between the C–H vibrational frequency and the C–D vibrational frequency in the CH_4 and CD_4 molecules.

17

DIATOMIC MOLECULE: ITS SPECTROSCOPY

Introduction

§1. We have seen in the previous chapter that the internal energy of a diatomic molecule is the sum of a vibrational and a rotational energy. If a molecule absorbs a photon, it will either rotate faster, or vibrate faster, or both. For most molecules, an increase in the rotational energy takes place when the energy of the absorbed photon is on the order of 1 cm^{-1} (Appendix 2 shows how to convert this unit of energy to other units). This falls in the microwave range of the spectrum. To increase vibrational energy, the photon energy must be of order of a hundred or a thousand cm^{-1}; these are infrared photons. Microwave and infrared spectroscopy use different sources of radiation and different detectors and require a different expertise, from spectroscopy using visible or ultra-violet light.

What can we learn from such measurements? In the microwave range, we measure transitions in which the quantum number ℓ changes by 1 (you will see soon why it does not change by 2 or more). By determining the frequency of the absorbed photon, we can determine the bond length r_0, since the rotational energy is $\hbar^2 \ell(\ell + 1)/2\mu r_0^2$.

If we use infrared light, photon absorption causes transitions from the state $\{v, \ell\}$ to $\{v + 1, \ell + 1\}$ or to $\{v + 1, \ell - 1\}$. By measuring the frequency of the absorbed photons we obtain the vibrational frequency ω. Since $\omega = \sqrt{k/\mu}$, we can calculate the force constant k and determine how stiff the bond is. A high value of k means that it takes a lot of energy to stretch the bond.

Isn't this amazing? By detecting that radiation of certain frequency disappears when it passes through a gas of diatomic molecules, we can measure the length of an object (the molecule) that is ten thousand times smaller than the thickness of a human hair, and we can find how much energy it takes to stretch the chemical bond.

If we abandon the harmonic, rigid-rotor approximation, we can do even better: we can determine the potential energy $V(r)$. This means that we now know the dissociation energy of the molecule, a quantity dear to a chemist's heart.

Such measurements are useful for testing the theories of the chemical bond, which calculate $V(r)$. There are also practical applications. Each diatomic molecule has a specific rotational and vibrational spectrum. Occasionally you hear that astrophysicists have found some exotic molecule in a distant point of the galaxy or in the outermost layer of the atmosphere. They do this by measuring the emission from transitions $\{v = 0, \ell\} \to \{v = 0, \ell - 1\}$; the energy of the emitted photons is an excellent fingerprint of each molecule.

You also hear sometimes that the temperature in some remote corner of the universe is 1 K. Who put a thermometer there? No one. The intensity of the photon emitted by transitions $\ell \to \ell - 1$ depends on temperature. If a distant point in the universe has diatomic molecules, there is a thermometer there!

Perhaps you are not interested in the temperature of the universe, but you might want to know the temperature of a point in a flame, where it is too hot to put a thermometer. No problem. Squirt diatomic molecules into the flame and measure their emission spectrum.

We do not have time to go into the vibrational and the rotational motion of molecules with more than two atoms, but the same principles apply to them. The only difference is that complicated molecules have more than one kind of vibrational motion, and each type of motion involves various parts of the molecule. The presence of a CH_3 group causes absorption of light of certain frequency, a ketone group absorbs at another frequency, etc. By measuring the frequencies of the infrared radiation absorbed by a molecule, you can identify the parts that the molecule is made of. Infrared spectroscopy is a major tool in analytical and synthetic chemistry.

You can also use it to do kinetics. If you have a reaction in which a ketone group is destroyed by a reagent, you can monitor how the intensity of the ketone absorption peak decays in time. The peak intensity is proportional to the concentration of the ketone groups.

Diatomic molecules absorb microwave, infrared, visible, and ultraviolet photons. Absorption and emission of visible or ultraviolet light takes place because of changes in the state of the *electrons* in the molecule. These transitions are not covered by the theory described here.

§2. *The Experiments.* A typical absorption experiment shines light on a sample (a gas of diatomics in our case) and measures the intensity of the light passing through, as a function of light frequency. If the frequency is such that photons are absorbed, the measurement registers a dip in the intensity. Fig. 17.1 shows the spectrum obtained by passing infrared light through a gas of $^{12}C^{16}O$ molecules. The photon frequencies for which dips are present in the spectrum are given in Table 17.1, in the columns labeled "experimental." The figure shows only the spectrum for transitions in which the vibrational quantum number changes from $v = 0$ to $v = 1$ and the rotational quantum number changes from ℓ to either $\ell + 1$ or to $\ell - 1$. For example, the peak at $6 \to 5$ corresponds to a transition from $\{v = 0, \ell = 6\}$ to $\{v = 1, \ell = 5\}$ (the figure is copied from a book that used the letter J for the quantum number ℓ).

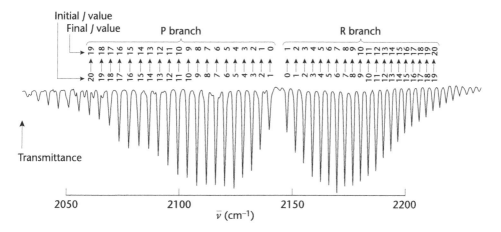

Figure 17.1 High-resolution infrared spectrum of gaseous carbon monoxide. D.C. Harris and M.D. Bertolucci, From *Symmetry and Spectroscopy: An Introduction to Vibrational and Electronic Spectroscopy*, Dover, New York, 1989, Figure 3–19.

Table 17.1 The energy of the absorbed photons, calculated for CO within harmonic approximation, are compared to the experimental values.

ℓ	ω_R (cm^{-1}), calculated	ω_R (cm^{-1}), experimental	ω_P (cm^{-1}), calculated	ω_P (cm^{-1}), experimental
0	2173.86	2147.08	0	0
1	2177.73	2150.86	2166.14	2139.43
2	2181.59	2154.60	2162.27	2135.55
3	2185.46	2158.30	2158.41	2131.63
4	2189.32	2161.97	2154.54	2127.68
5	2193.19	2165.60	2150.68	2123.70
6	2197.05	2169.20	2146.81	2119.68
7	2200.92	2172.76	2142.95	2115.63
8	2204.78	2176.29	2139.08	2111.54
9	2208.65	2179.77	2135.22	2107.43

The transitions from a state with $v = 0$ and a given ℓ to a state with $v = 1$ and $\ell - 1$ belong to the sfit P branch and the frequency of a photon absorbed in such a transition is labeled ω_P. The transitions from a state with $v = 0$ and a given ℓ to a state with $v = 1$ and $\ell + 1$ belong to the R branch and the frequency of a photon absorbed in such a transition is labeled ω_R. For example, the frequency of the transition from $v = 0$ and $\ell = 5$ to the state $v = 1$ and $\ell = 4$ is equal to $\omega_P = 2123.7$ cm^{-1} and that for the transition $v = 0$ and $\ell = 5$ to $v = 1$ and $\ell = 6$ is equal to $\omega_R = 2165.6$ cm^{-1}.

The family of dips at the left of Fig. 17.1 (lower frequency) is called the *P-branch* of the spectrum; the family at the right (high frequency) is the *R-branch*.

The spectrum has a fairly strong temperature dependence: as the temperature is lowered, many dips in the spectrum disappear.

These experimental results pose interesting questions.

1. Can theory predict, with a reasonable accuracy, the frequencies at which photon absorption takes place?

2. Can we understand why we only see transitions that change v or ℓ by one unit?

3. Will the theory explain the temperature dependence of the spectrum?

4. Can we calculate the relative magnitude of the dips?

This chapter shows how quantum mechanics answers these questions.

Collect the Necessary Equations

§3. *The Frequencies of the Absorbed Photons.* To calculate the frequencies

$$\Omega = \frac{E_f - E_i}{\hbar} \tag{17.1}$$

of the photons that a molecule can absorb, we need to know the internal energies E_f and E_i, which depend on the vibrational quantum number v, the rotational quantum number ℓ, and the parameters of the molecule. If the harmonic and the rigid-rotor approximations are made the energy is:

$$E(v, \ell) = \hbar\omega \left(v + \tfrac{1}{2}\right) + \frac{\ell(\ell + 1)\hbar^2}{2\mu r_0^2} \equiv \hbar\omega \left(v + \tfrac{1}{2}\right) + B\ell(\ell + 1) \tag{17.2}$$

The quantum numbers in these equations take the values $v = 0, 1, 2, \ldots$ and $\ell = 0, 1, 2, \ldots$. The value of the vibrational energy $\hbar\omega$ and that of the *rotational constant* B are given in Table 16.1, in units of cm^{-1}. For CO, Table 16.1 gives

$$\hbar\omega = 2169.8 \text{ cm}^{-1}, \quad B = 1.93127 \text{ cm}^{-1} \tag{17.3}$$

Exercise 17.1

Express these energies in units of eV, joule, and joule/mol.

We can calculate the frequencies of various transitions, for the CO molecule, by combining Eq. 17.1 for the absorbed photon frequency, with the energy formula Eq. 17.2 and the data given by Eq. 17.3. For example, the frequency Ω of a photon absorbed when the molecule undergoes the transition

$$\{v_i, \ell_i\} \rightarrow \{v_f, \ell_f\}$$

is

$$\hbar\Omega = E(v_f, \ell_f) - E(v_i, \ell_i) = \hbar\omega(v_f - v_i) + B(\ell_f - \ell_i)(1 + \ell_f + \ell_i)$$

As we discussed in Chapter 11, not all photons whose frequency satisfies this relationship will be absorbed. You will soon see that absorption takes place only if $v_f = v_i + 1$ and $\ell_f = \ell_i + 1$ or $\ell_f = \ell_i - 1$.

Exercise 17.2

Calculate the energy $\hbar\Omega$ of the photon absorbed in the transition $\{v_i = 0, \ell_i = 5\} \to \{v_i = 1, \ell_f = 10\}$ and in the transition $\{v_i = 0, \ell_i = 5\} \to \{v_f = 0, \ell_f = 10\}$. Give the result in cm^{-1} and meV.

In Workbook QM17.1, I calculated the frequencies of the photons absorbed by CO, for the transitions indicated in Fig. 17.1. Table 17.1 gives the calculated values. As you can see, Eq. 17.2 reproduces the order of magnitude and the trends observed in the experiments, but it does not give accurate frequencies. Modern spectrometers can resolve the peak frequencies in infrared spectroscopy with an accuracy of about 0.02 cm^{-1}; the error made by theory is larger than this value. The error is caused by the harmonic and the rigid-rotor approximation.

A more accurate equation for the energy of the diatomic is

$$E_{v,\ell} = \hbar\,\omega\,\left(v + \tfrac{1}{2}\right) + B\,\ell(\ell+1) - X\,\left(v + \tfrac{1}{2}\right)^2$$
$$- A\,\left(v + \tfrac{1}{2}\right)\ell(\ell+1) - C[\ell(\ell+1)]^2 \qquad (17.4)$$

This is obtained by correcting some of the errors made by the harmonic and the rigid-rotor approximations, and is accurate only when v and ℓ are small. The values of the constants $\hbar\omega$, B, X, A, and C are given in Table 16.1, in units of cm^{-1}. For CO that table gives:

$$\hbar\omega_e = 2169.8 \text{ cm}^{-1}$$

$$B = 1.93127 \text{ cm}^{-1}$$

$$A = 0.0175 \text{ cm}^{-1}$$

$$X = 13.294 \text{ cm}^{-1}$$

$$C = 6.2 \times 10^{-6} \text{ cm}^{-1}$$

The frequencies obtained by using Eq. 17.4 are compared to experiment in Table 17.2 (see Workbook QM17.3). The accuracy is better that when we use the approximate equation (Eq. 17.2), but it is not perfect. To do better, we would have to solve the Schrödinger equation by using accurate numerical calculations, which is quite possible.

Table 17.2 A comparison of the peak frequencies, calculated for CO with the corrected formula (Eq. 17.4), to the experimental values.

ℓ	ω_R (cm^{-1}), calculated	ω_R (cm^{-1}), experimental	ω_P (cm^{-1}), calculated	ω_P (cm^{-1}), experimental
0	2147.13	2147.08	0	0
1	2151.08	2150.86	2139.33	2139.43
2	2155.06	2154.60	2135.49	2135.55
3	2159.08	2158.30	2131.68	2131.63
4	2163.13	2161.97	2127.90	2127.68
5	2167.22	2165.60	2124.16	2123.70
6	2171.34	2169.20	2120.16	2119.68
7	2175.50	2172.76	2116.80	2115.63
8	2179.69	2176.29	2113.16	2111.54
9	2183.91	2179.77	2109.57	2107.43

§4. *Photon Absorption and Emission Probabilities.* To calculate the probability that a photon of a given frequency is absorbed, I use the theory developed in Chapter 11.

I explained there (see §3–§5 of Chapter 11) that continuous light consists of overlapping light pulses. The probability that the molecule absorbs a photon from such a pulse and undergoes a transition from an initial state ψ_i to a final state ψ_f is

$$P_{i \to f} = \frac{\pi}{\epsilon_0 c \hbar^2} \left| \langle \psi_f | \hat{\mathbf{m}} \cdot \mathbf{e} | \psi_i \rangle \right|^2 N(\omega_{if}) \tag{17.5}$$

(I have taken $L(\Omega) = 1$ if $\Omega = \omega_{if}$ and 0 otherwise.) Besides the initial and final wave functions, the molecule is represented in this formula by the *dipole operator*

$$\hat{\mathbf{m}} = \sum_{\alpha=1}^{n} q \hat{\mathbf{r}}(\alpha) - \sum_{\beta=1}^{2} q Z_\beta \hat{\mathbf{R}}(\beta) \tag{17.6}$$

Here $\hat{\mathbf{r}}(1), \ldots, \hat{\mathbf{r}}(n)$ are the operators representing the positions of the n electrons in the molecule, and $\hat{\mathbf{R}}(1)$ and $\hat{\mathbf{R}}(2)$ are the position operators for the two nuclei. q is the electron charge (a negative number) and Z_β is the number of protons in nucleus β.

The other quantities in Eq. 17.5 represent the properties of light. $\mathbf{e}$ is a unit vector that gives the orientation of the electric-field vector of the linearly polarized light. $N(\omega)d\omega$ is the total energy of the photons in the pulse having a frequency between ω and $\omega + d\omega$. If the pulse contains no photons of frequency ω' then $N(\omega')$ is zero. A plot of $N(\omega)$ vs frequency (see Fig. 11.1) has a sharp peak, and the experimentalist can tune the laser to change the peak frequency. A transition occurs only if $N(\omega)$ is such that

$$N(\omega_{\text{if}}) \neq 0$$

for one of the transition frequencies

$$\omega_{\text{if}} = \frac{E_{\text{f}} - E_{\text{i}}}{\hbar}$$

Another necessary condition for photon absorption, in a transition from ψ_{i} to ψ_{f}, is that the matrix element $\langle \psi_{\text{f}} \,|\, \hat{\mathbf{m}} \cdot \mathbf{e} \,|\, \psi_{\text{i}} \rangle$ is not zero. If the photon frequency matches a transition frequency ω_{if} but the matrix element $\langle \psi_{\text{f}} \,|\, \hat{\mathbf{m}} \cdot \mathbf{e} \,|\, \psi_{\text{i}} \rangle$ is zero, then the absorption probability $P_{i \to f}$ is zero; the photon passes through the sample without being absorbed. No dip is seen in the spectrum, and we say that the transition $\psi_{\text{i}} \to \psi_{\text{f}}$ is *forbidden*.

Eq. 17.5 is in MKSA units. In this system, the electron charge q is -1.60×10^{-19} C, the permittivity of vacuum is $\epsilon_0 = 10^7/c^2$ F/m, the speed of light is $c = 2.997 \times 10^8$ m/s, and Planck's constant is $\hbar = 1.054 \times 10^{-34}$ J s.

In what follows, I will use Eq. 17.5 to calculate the relative intensities in the absorption spectrum.

§5. *The Wave Function for the Nuclear Motion.* To calculate the magnitude of the probability $P_{i \to f}$ with Eq. 17.5, we need the wave function of the molecule. In Chapter 16, I have shown that this is (see Eq. 16.36)

$$\Psi_{v,\ell,m}(r,\theta,\phi) = Y_\ell^m(\theta,\phi)\frac{F_v(r)}{r} \tag{17.7}$$

The spherical harmonics $Y_\ell^m(\theta,\phi)$, given by Eqs 14.35–14.59, describe the orientation of the molecular axis. The vibrational wave function $F_v(r)$ is (see Eqs 16.26–16.35)

$$F_v(r) = \sqrt{\beta}\,\Phi_v(y) \tag{17.8}$$

with

$$\Phi_v(y) = \frac{1}{\pi^{1/4}\sqrt{2^v v!}} \exp\left[-\frac{y^2}{2}\right] H_v(y) \tag{17.9}$$

Here $H_v(y)$ is the Hermite polynomial of order v,

$$y \equiv \beta(r - r_0) \tag{17.10}$$

and

$$\beta \equiv \sqrt{\frac{\mu\omega}{\hbar}} \tag{17.11}$$

The molecular properties present in these equations are the bond length r_0, the reduced mass μ, and the frequency ω. These are defined and explained in Chapter 16.

Mathematica and **Mathcad** provide built-in functions that evaluate spherical harmonics $Y_\ell^m(\theta,\phi)$ and Hermite polynomials $H_v(y)$. The expressions for $F_v(y)$, for $v = 0, \ldots, 5$, are given in Eqs 16.30–16.35.

§6. *The Electronic Wave Function.* If you are one of those people who wants to connect the equations closely to reality, you ought to have wondered why we never mention the electrons when we discuss the diatomic molecule. One can hardly imagine that they are unimportant: without them, the nuclei would stay as far from each other as possible, because of Coulomb repulsion.

In the quantum mechanics of a molecule, electrons play several roles. One is to keep the nuclei together. In our treatment so far, the nuclei stay together because their potential energy $V(r)$ had a minimum (see Chapter 16, §2–§4). Where does this come from? Quantum mechanics shows that $V(r)$ is the energy of the electrons plus the repulsion between nuclei, when the internuclear distance is r.

When calculated for a fixed internuclear distance r_α, the energy of the electrons takes discrete values $V_1(r_\alpha)$, $V_2(r_\alpha)$, ... It so happens that for a diatomic molecule, $V_2 - V_1$ is on the order of several eV, for all distances between nuclei that the molecule is likely to have. Because of this, the electrons can be excited only by visible or UV photons. If we plan experiments in which we shine IR or microwave radiation on the molecule, the electrons will not be excited. We can therefore calculate the *frequency* of various excitations of the nuclear motion, by microwave and IR radiation, without bothering with the electrons.

The calculation of the probability of photon absorption is a different matter. The dipole operator $\hat{\mathbf{m}}$ depends on the positions $\mathbf{R}(\beta)$, $\beta = 1, 2$, of the nuclei *and* on the positions $\mathbf{r}(\alpha)$, $\alpha = 1, 2, \ldots, n$, of the electrons. This is physically reasonable. The operator $\hat{\mathbf{m}}$ appears in the theory through the energy of interaction between the electric field of the light and the electric charges (nuclear or electronic) in the molecule. The nuclei are excited by the infrared radiation partly because light acts on the electrons, which are, in turn, interacting strongly with the nuclei. The ability of light to excite the nuclei ought therefore to depend on the state of the electrons. It is for this reason that the initial and the final state of the molecule must include the wave function of the electrons. The wave functions of the diatomic that must be used in Eq. 17.5 are

$$\psi_i(\mathbf{r}(1), \ldots, \mathbf{r}(n), r, \theta, \phi) = \chi_0(\mathbf{r}(1), \ldots, \mathbf{r}(n); r) Y_{\ell_i}^{m_i}(\theta, \phi) \frac{F_{v_i}(r)}{r} \tag{17.12}$$

and

$$\psi_f(\mathbf{r}(1), \ldots, \mathbf{r}(n), r, \theta, \phi) = \chi_0(\mathbf{r}(1), \ldots, \mathbf{r}(n); r) Y_{\ell_f}^{m_f}(\theta, \phi) \frac{F_{v_f}}{r} \tag{17.13}$$

Here $\chi_0(\mathbf{r}(1), \ldots, \mathbf{r}(n); r)$ is the lowest-energy eigenstate of the electrons when the distance between the nuclei is equal to r. This ground state appears in both the initial and final state because the absorption of an IR or microwave photon does not excite the electrons and therefore their state is not changed in the transition.

§7. *The Dipole Moment of the Molecule.* As we have discussed several times in this book, the symbol $\langle \psi_f | \hat{O} | \psi_i \rangle$ represents an integral over all coordinates of all relevant particles in the system. In the case of Eq. 17.5, with the wave functions given by Eqs 17.12 and 17.13, this symbol becomes

$$\langle \psi_f | \hat{\mathbf{m}} | \psi_i \rangle = \int d\mathbf{r}(1) \cdots \int d\mathbf{r}(n) \int_0^\infty r^2 dr \int_0^\pi \sin(\theta) d\theta \int_0^{2\pi} d\phi$$
$$\times \chi_0(\mathbf{r}(1), \ldots, \mathbf{r}(n); r)^* Y_{\ell_f}^{m_f}(\theta, \phi)^* \frac{F_{v_f}}{r} \hat{\mathbf{m}}$$
$$\times \chi_0(\mathbf{r}(1), \ldots, \mathbf{r}(n); r) Y_{\ell_i}^{m_i}(\theta, \phi) \frac{F_{v_i}}{r} \tag{17.14}$$

The integral over r, ϕ, and θ is the one encountered in Chapter 16 and has to do with the states of the nuclei. The integral over the electronic coordinates is new to us.

In what follows, I will use the notation

$$\mathbf{m}(r) = \int d\mathbf{r}(1) \cdots \int d\mathbf{r}(n) \chi_0(\mathbf{r}(1), \ldots, \mathbf{r}(n); r)^* \, \hat{\mathbf{m}}$$
$$\times \chi_0(\mathbf{r}(1), \ldots, \mathbf{r}(n); r) \qquad (17.15)$$

which defines the *dipole moment* of the molecule. Because the electronic wave function depends on the distance r between the nuclei, $\mathbf{m}$ is also a function of r. If we stretch the molecule, we change its dipole moment.

I am sure that in one of your introductory chemistry courses, you learned that some neutral molecules create an electric field around them and also interact with an external electric field. Why would a neutral molecule create an electric field or interact with one? Even though the negative and the positive charges in the molecule are equal in magnitude, they do not overlap perfectly in space. As a consequence, the molecule interacts with an electric field as if it consists of two charges of equal magnitude α and of opposite sign, separated by a distance ρ; the dipole moment of such a molecule is equal to $\alpha\rho$. Such a crude model is meant to represent the quantity defined by Eq. 17.15.

The dipole moment is an important quantity in chemistry. Liquids made of molecules with a high dipole moment (such as water) are very good *polar* solvents. Life on Earth would be extremely different if the dipole moment of water were not as large as it is; electrolytes would not dissociate and there would be no ions in your body. In what follows we will see that the dipole moment is also an essential player in microwave and infrared spectroscopy.

Note that the dipole moment is a vector; for most (but not all) diatomic molecules, $\mathbf{m}$ points along the molecular axis and we will consider only this case.

§8. *The Separation of the Transition Dipole Matrix Element into a Rotational and a Vibrational Contribution.* If we use Eq. 17.15 in Eq. 17.14, we find that

$$\langle \psi_f \mid \hat{\mathbf{m}} \mid \psi_i \rangle \cdot \mathbf{e} = \int_0^\pi \sin(\theta) d\theta \int_0^{2\pi} d\phi \int_0^\infty dr$$
$$\times F_{v_f}(r) Y_{\ell_f}^{m_f}(\theta, \phi)^* \, \mathbf{m}(r) \cdot \mathbf{e} \, F_{v_i}(r) Y_{\ell_i}^{m_i}(\theta, \phi) \qquad (17.16)$$

I will now take the OZ axis of the coordinate system to coincide with the polarization vector $\mathbf{e}$ of the laser. The dot product $\mathbf{m} \cdot \mathbf{e}$ is therefore

$$\mathbf{m} \cdot \mathbf{e} = m(r) \cos(\theta) \qquad (17.17)$$

where θ is the polar angle (between the OZ axis and the axis of the molecule) and $m(r)$ is the magnitude (length) of the vector $\mathbf{m}$ (the length of $\mathbf{e}$ is 1).

When we insert Eq. 17.17 in Eq. 17.16, we obtain

$$\langle \psi_f \,|\, \hat{\mathbf{m}} \,|\, \psi_i \rangle \cdot \mathbf{e} = \int_0^\pi \sin(\theta)d\theta \int_0^{2\pi} d\phi \, Y_{\ell_f}^{m_f}(\theta, \phi)^* \cos(\theta) Y_{\ell_i}^{m_i}(\theta, \phi)$$

$$\times \int_0^\infty dr \, F_{v_f}(r) m(r) F_{v_i}(r) \tag{17.18}$$

We can rewrite this expression with the notation used previously:

$$\langle \psi_f \,|\, \hat{\mathbf{m}} \,|\, \psi_i \rangle \cdot \mathbf{e} = \langle \ell_f, m_f \,|\, \cos(\theta) \,|\, \ell_i, m_i \rangle \, \langle v_f \,|\, m(r) \,|\, v_i \rangle \tag{17.19}$$

The integral

$$\langle \ell_f, m_f \,|\, \cos(\theta) \,|\, \ell_i, m_i \rangle \equiv \int_0^\pi \sin(\theta)d\theta \int_0^{2\pi} d\phi \, Y_{\ell_f}^{m_f}(\theta, \phi)^*$$

$$\times \cos(\theta) Y_{\ell_i}^{m_i}(\theta, \phi) \tag{17.20}$$

contains information only about the rotational motion. The integral

$$\langle v_f \,|\, m(r) \,|\, v_i \rangle \equiv \int_0^\infty dr \, F_{v_f}(r) m(r) F_{v_i}(r) \tag{17.21}$$

depends only on the vibrational state of the molecule.

When we use Eq. 17.19 in Eq. 17.5, we obtain

$$P_{i \to f} = \frac{\pi}{\epsilon_0 c \hbar^2} \left| \langle \ell_f, m_f \,|\, \cos(\theta) \,|\, \ell_i, m_i \rangle \right|^2 \left| \langle v_f \,|\, m(r) \,|\, v_i \rangle \right|^2 N(\omega_{if}) \tag{17.22}$$

To determine the probability of photon absorption, we must calculate the integrals defined by Eqs 17.20 and 17.21.

Before we do that, I note that it is traditional to distinguish two kinds of transitions described by this formula. Transitions of the type

$$\{v_i, \ell_i, m_i\} \to \{v_i, \ell_f, m_f\}$$

excite only the rotational motion, and require photons whose frequency is in the microwave region. Transitions of the type

$$\{v_i, \ell_i, m_i\} \rightarrow \{v_f, \ell_f, m_f\}, \ v_f \neq v_i$$

excite both vibrations and rotations. Such transitions are caused by the absorption of infrared photons. It may seem silly to discuss separately phenomena described by the same formula. However, the light sources, the light detectors, and the optical equipment used in these two types of measurements are so different that the distinction is justified.

§9. *The Harmonic Approximation for the Dipole Moment.* To perform the integral $\langle v_f \,|\, m(r) \,|\, v_i \rangle$ appearing in $P_{i \to f}$ (Eq. 17.22), we must know how the dipole $m(r)$ depends on r. It is possible to calculate this dependence from Eq. 17.15 but doing so is rarely necessary. When we discussed the harmonic approximation in Chapter 16, we used the fact that the interatomic distance r does not deviate much from the bond length r_0. In other words, $r - r_0$ is small. This means that we can expand $m(r)$ in a power series in $r - r_0$ and keep only the largest terms:

$$m(r) \approx m(r_0) + (r - r_0) \left. \frac{\partial m(r)}{\partial r} \right|_{r=r_0} \tag{17.23}$$

Inserting this in $\langle v_f \,|\, m(r) \,|\, v_i \rangle$ leads to

$$\langle v_f \,|\, m(r) \,|\, v_i \rangle = m(r_0)\langle v_f \,|\, v_i \rangle + \left. \frac{\partial m(r)}{\partial r} \right|_{r=r_0} \langle v_f \,|\, (r - r_0) \,|\, v_i \rangle \tag{17.24}$$

But the vibrational wave functions are orthonormal, which means that the integral in the first term of Eq. 17.24 satisfies:

$$\langle v_f \,|\, v_i \rangle \equiv \int_0^\infty dr F_{v_f}(r) F_{v_i}(r) = \delta(v_f, v_i) \tag{17.25}$$

where

$$\delta(v_f, v_i) = \begin{cases} 0 & \text{if } v_f \neq v_i \\ 1 & \text{if } v_f = v_i \end{cases} \tag{17.26}$$

I have changed the notation for the Kronecker delta from $\delta_{a,b}$ to $\delta(a,b)$ to avoid subscripts on subscripts.

The result tells us that the term $\langle v_f | v_i \rangle$ plays a role in the absorption probability only when $v_f = v_i$. This means that it contributes only to the microwave spectrum.

The term $\langle v_f | (r - r_0) | v_i \rangle$ in Eq. 17.24 can be calculated as follows:

$$\langle v_f | (r - r_0) | v_i \rangle \equiv \int_0^\infty dr F_{v_f}(r)(r - r_0)F_{v_i}(r) \tag{17.27}$$

$$\approx \int_{-\infty}^\infty dr F_{v_f}(r)(r - r_0)F_{v_i}(r) \tag{17.28}$$

$$= \int_{-\infty}^\infty \frac{dy}{\beta} \left(\sqrt{\beta}\Phi_{v_f}(y) \right) \frac{y}{\beta} \left(\sqrt{\beta}\Phi_{v_i}(y) \right) \tag{17.29}$$

$$\approx \frac{1}{\beta} \int_{-\infty}^\infty dy\, \Phi_{v_f}(y)\, y\, \Phi_{v_i}(y) \tag{17.30}$$

To go from Eq. 17.27 to Eq. 17.28, I made use of the observation that the wave function $F_v(r)$ is zero when $r < 0$. This is easy to understand physically. It would take an enormous energy to make $r = 0$, since that would mean that the atoms were on top of each other. This will never happen at the energies of interest to us. This fact is reflected in the wave function, which will become zero when r approaches 0. This is mathematically achieved through the Gaussian function $\exp\left[-y^2\right] \equiv \exp\left[-\beta^2 (r - r_0)^2\right]$ that is present in Φ_v and the fact that β is fairly large (so that $\exp\left[-\beta^2 (r_0)^2\right]$ is practically zero). This exponential is zero also for all negative values of r. This means that I can extend the lower limit of integration in Eq. 17.27 from zero to $-\infty$, as is done in Eq. 17.28, without changing the value of the integral.

To go from Eq. 17.28 to Eq. 17.29, I changed variables from r to $y = \beta(r - r_0)$, and then I cleaned up the equation to reach Eq. 17.30.

The integral in Eq. 17.30 is easily performed with either **Mathematica** or **Mathcad**, when v_i and v_f have numerical values. Since Φ_v is a Gaussian multiplied by a polynomial, the integral can also be performed analytically. The result is

$$\langle v_f | r - r_0 | v_i \rangle = \frac{1}{\beta\sqrt{2}} \left[\delta(v_f, v_i - 1)\sqrt{v_i} + \delta(v_f, v_i + 1)\sqrt{v_i + 1} \right] \tag{17.31}$$

Using Eqs 17.31 and 17.25 in Eq. 17.24 gives

$$\langle v_f \mid m(r) \mid v_i \rangle = m(r_0)\delta(v_f, v_i)$$

$$+ \frac{1}{\beta\sqrt{2}} \left[\delta(v_f, v_i - 1)\sqrt{v_i} + \delta(v_f, v_i + 1)\sqrt{v_i + 1} \right] \qquad (17.32)$$

The three terms represent three different transitions. If $v_f = v_i$, only the first term in Eq. 17.32 survives; this represents a transition in which only the rotational energy changes. If $v_f = v_i - 1$, only the middle term survives; this term represents the emission of an infrared photon. If $v_f = v_i + 1$, only the third term survives; this represents the absorption of an infrared photon.

Inserting Eq. 17.32 in Eq. 17.22 gives

$$P_{i \to f} = P^r_{i \to f} + P^e_{i \to f} + P^a_{i \to f} \qquad (17.33)$$

with

$$P^r_{i \to f} = \frac{\pi}{\epsilon_0 c \hbar^2} |m(r_0)|^2 \left| \langle \ell_f, m_f \mid \cos(\theta) \mid \ell_i, m_i \rangle \right|^2 \delta(v_f, v_i) N(\omega_{if}) \qquad (17.34)$$

$$P^e_{i \to f} = \frac{\pi}{\epsilon_0 c \hbar^2} \left| \frac{dm(r)}{dr} \right|^2_{r=r_0} \left| \langle \ell_f, m_f \mid \cos(\theta) \mid \ell_i, m_i \rangle \right|^2 \frac{v_i}{2\beta^2}\delta(v_f, v_i - 1) N(\omega_{if}) \qquad (17.35)$$

$$P^a_{i \to f} = \frac{\pi}{\epsilon_0 c \hbar^2} \left| \frac{dm(r)}{dr} \right|^2_{r=r_0} \left| \langle \ell_f, m_f \mid \cos(\theta) \mid \ell_i, m_i \rangle \right|^2 \frac{v_i + 1}{2\beta^2}\delta(v_f, v_i + 1) N(\omega_{if}) \qquad (17.36)$$

In deriving Eqs 17.33–17.36, I used Eq. 17.31 for $\langle v_f \mid r - r_0 \mid v_i \rangle$. Due to the presence of the Kronecker deltas in the three terms giving $\langle v_f \mid r - r_0 \mid v_i \rangle$, at most one of these probabilities can differ from zero for any fixed values of v_f and v_i. For that reason, when we take the square of the sum of terms, we can neglect the cross terms (in $(a + b)^2 = a^2 + b^2 + 2ab$ the cross term is $2ab$). Furthermore, you can easily verify that $\delta(a, b)^2 = \delta(a, b)$.

§10. *Physical Interpretation.* Each term in Eq. 17.33 corresponds to the probability of a separate physical process.

$P^{r}_{i \to f}$ is the probability that a photon is absorbed in a transition in which the vibrational state does not change; these are transitions in which only the rotational state changes.

$P^{e}_{i \to f}$ is the probability of a transition in which $v_f = v_i - 1$; this means that the vibrational quantum number decreases by 1, the energy of the molecule is lowered, and an infrared photon is emitted.

Finally, $P^{a}_{i \to f}$ is the probability of a transition in which $v_f = v_i + 1$; the vibrational quantum number increases by one, the energy of the molecule increases, and an infrared photon is absorbed.

§11. *The integral* $\langle \ell_f, m_f | \cos(\theta) | \ell_i, m_i \rangle$. To evaluate the magnitude of these probabilities, we must calculate the integral $\langle \ell_f, m_f | \cos(\theta) | \ell_i, m_i \rangle$, defined by Eq. 17.20. This is a fairly tedious calculation, which would take me too much time to explain. (It does not seem profitable to spend a lot of time learning how to perform integrals, now that computers can do them in a fraction of a second.)

The result is

$$\langle \ell_f, m_f | \cos(\theta) | \ell_i, m_i \rangle = \delta(m_f, m_i) [\delta(\ell_f, \ell_i + 1) A_+(\ell_i, m_i)$$

$$+ \delta(\ell_f, \ell_i - 1) A_-(\ell_i, m_i)] \tag{17.37}$$

with

$$A_+(\ell_i, m_i) = \sqrt{\frac{(\ell_i + m_i + 1)(\ell_i - m_i + 1)}{(2\ell_i + 1)(2\ell_i + 3)}} \tag{17.38}$$

and

$$A_-(\ell_i, m_i) = \sqrt{\frac{(\ell_i + m_i)(\ell_i - m_i)}{(2\ell_i + 1)(2\ell_i - 1)}} \tag{17.39}$$

The integral $\langle \ell_f, m_f | \cos(\theta) | \ell_i, m_i \rangle$ is easily calculated by **Mathematica** or **Mathcad** when the numerical values of the initial and final quantum numbers are specified.

Exercise 17.3

Use **Mathematica** or **Mathcad** to perform the integrals $\langle \ell_f, m_f | \cos(\theta) | \ell_i, m_i \rangle$ for $\ell_i = 0, 1, 2, 3$ and $\ell_f = 0, 1, 2, 3$ and all values of m_i and m_f (remember that m takes one of the values $-\ell$, $-\ell + 1$, ..., $\ell - 1$, ℓ). Test whether the results agree with Eqs 17.37–17.39.

Exercise 17.4

Use Eq. 14.35 to show that

$$\int_0^{2\pi} d\phi\, Y_\ell^m(\theta, \phi)^* f(\theta)\, Y_{\ell'}^{m'}(\theta, \phi)$$

is proportional to $\delta(m, m')$ regardless of the values of ℓ and ℓ' and of the form of the function $f(\theta)$.

Eq. 17.37 tells us that the absorption or emission of a *linearly polarized* photon cannot change the quantum number m. (In contrast, if the photon is circularly polarized, its absorption or emission can change m.) The term in Eq. 17.37 containing $\delta(\ell_f, \ell_i + 1)$ differs from zero only in a transition in which the rotational quantum number ℓ increases by one; since the energy of the molecule increases, this transition corresponds to photon absorption. The term containing $\delta(\ell_f, \ell_i - 1)$ corresponds to photon emission, in a transition in which the rotational quantum number ℓ decreases by one.

In the remainder of this chapter, I use these equations to discuss the excitation of vibrational and rotational motion by absorption of an infrared photon.

Vibrational and Rotational Excitation by Absorption of an Infrared Photon

§12. *Collect the Equations Needed.* The probability $P_{i \to f}^a$ that the molecule absorbs an infrared photon and changes its vibrational and rotational state is given by Eq. 17.36. The matrix element $\langle \ell_f, m_f \mid \cos(\theta) \mid \ell_i, m_i \rangle$, which appears in this expression, is given by Eqs 17.37–17.39. Combining these results gives

$$P_{i \to f}^a = \frac{\pi}{\epsilon_0 c \hbar^2} \left| \frac{dm(r)}{dr} \right|_{r=r_0}^2 \frac{v_i + 1}{2\beta^2} \delta(v_f, v_i + 1) N(\omega_{if})$$

$$\times \left\{ \delta(m_f, m_i) \left| [\delta(\ell_f, \ell_i + 1)A_+(\ell_i, m_i) \right. \right.$$

$$\left. \left. + \delta(\ell_f, \ell_i - 1)A_-(\ell_i, m_i)] \right|^2 \right\} \tag{17.40}$$

Note that the product $\delta(\ell_f, \ell_i + 1)\delta(\ell_f, \ell_i - 1)$ is always zero since ℓ_f can not be simultaneously equal to $\ell_i + 1$ *and* $\ell_i - 1$. For this reason the cross term disappears when we calculate the square of the expression containing A_+ and A_- (i.e. $(a + b)^2 = a^2 + b^2$ in this case). Recall that $\delta(a, b)^2 = \delta(a, b)$. With these facts in mind, I rewrite Eq. 17.40 as

$$P^a_{i \to f} = P^R_{i \to f} + P^P_{i \to f} \tag{17.41}$$

with

$$P^R_{i \to f} \equiv \frac{\pi}{\epsilon_0 c \hbar^2} \left| \frac{dm(r)}{dr} \right|^2_{r=r_0} \frac{v_i + 1}{2\beta^2} \delta(v_f, v_i + 1)\delta(\ell_f, \ell_i - 1)$$

$$\times N(\omega_{if})A_+(\ell_i, m_i)^2 \tag{17.42}$$

$$P^P_{i \to f} \equiv \frac{\pi}{\epsilon_0 c \hbar^2} \left| \frac{dm(r)}{dr} \right|^2_{r=r_0} \frac{v_i + 1}{2\beta^2} \delta(v_f, v_i + 1)\delta(\ell_f, \ell_i + 1)$$

$$\times N(\omega_{if})A_-(\ell_i, m_i)^2 \tag{17.43}$$

The transitions described by $P^R_{i \to f}$ constitute the R-branch of the absorption spectrum, while $P^P_{i \to f}$ describes the P-branch. The R- and P-branches of the vibrational spectrum of CO are shown in Fig. 17.1.

These formulae tell us that the ability of a molecule to absorb a photon and change its *vibrational* state is proportional to the square of the derivative of its dipole moment with respect to the interatomic distance r, calculated at the bond-length value $r = r_0$. This derivative tells us how much the dipole moment changes when the interatomic distance changes. This quantity is very different from the dipole moment. For example, CO has a very small dipole moment but it has an unusually large value of $(dm(r)/dr)_{r=r_0}$.

If it happens that $(dm(r)/dr)_{r=r_0}$ is zero, we say that the molecule is not infrared-active; if the approximations we have made are correct, there is nothing we can do to force the molecule to absorb an IR photon and change its vibrational state.

Exercise 17.5

Show that a molecule's ability to absorb microwave photons in a transition in which the molecule changes only its rotational state is proportional to $m(r_0)^2$. Thus, the intensity of microwave absorption is proportional to the dipole moment squared: no dipole, no rotational transition.

Exercise 17.6

Analyze the factor $1/\beta^2$. Use Eq. 17.11 for β, Eq. 16.14 for ω, and Eq. 16.16 to show that, all other things being equal, the stiffer the bond of the diatomic, the larger the ability of the molecule to absorb an IR photon and change its vibrational state.

Exercise 17.7

Use the equations mentioned in the previous exercise and the definition of the reduced mass μ to decide which of HCl and DCl has the higher photon-absorption probability.

The Kronecker deltas in Eqs 17.42 and 17.43 impose severe restrictions on the possible transitions caused by IR-photon absorption. The allowed transitions are shown in Fig. 17.2.

The absorption of linearly polarized photons can only proceed by transitions in which the vibrational quantum number is increased by one ($v_f = v_i + 1$), the rotational quantum number changes by one (either $\ell_f = \ell_i - 1$ or $\ell_f = \ell_i + 1$), and the quantum number m is unchanged. In addition, if $(dm(r)/dr)_{r=r_0} = 0$, it is not possible to absorb an IR photon to change the vibrational state.

One must keep in mind that these conclusions are valid only if the approximations used to reach them are valid. The matrix element $\langle v_f \,|\, r - r_0 \,|\, v_i \rangle$ is proportional to $\delta(v_f, v_i + 1)$ for absorption (or to $\delta(v_f, v_i - 1)$ for emission) because we have made the harmonic approximation (we replaced $V(r)$ with $k(r - r_0)^2/2$). When this

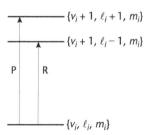

Figure 17.2 The transitions allowed by Eqs 17.41–17.43. The label 'R' indicates a transition contributing to the R-branch of the spectrum; 'P' is a P-branch transition.

approximation is removed, $\langle v_i + 2 \mid r - r_0 \mid v_i \rangle$, $\langle v_i + 3 \mid r - r_0 \mid v_i \rangle$, etc., are not zero, but are much smaller than $\langle v_i + 1 \mid r - r_0 \mid v_i \rangle$. This means that we can observe absorption caused by the transitions $v_i \rightarrow v_i + 2$, $v_i \rightarrow v_i + 3$, etc., if the sample is very thick, so that the light passing through it interacts with many molecules. In the 1930s, such absorption was detected by using a benzene sample 30 m in length.

A second approximation assumed that $m(r) = m(r_0) + (r - r_0)(dm(r)/dr)_{r=r_0}$. We had ignored the term $\frac{(r - r_0)^2}{2}(d^2 m(r)/dr^2)_{r=r_0}$ in the Taylor expansion. Had we kept it, the probability of the transition $v_i \rightarrow v_i + 2$ would have been different from zero (but small).

Exercise 17.8

Calculate $\langle v_i + 2 \mid (r - r_0)^2 \mid v_i \rangle$ for $v_i = 0, 1, 2$. Compare it to $\langle v_i + 1 \mid (r - r_0) \mid v_i \rangle$.

§13. *The Theory Seems to Disagree with the Experiments.* We conclude that it is possible to see forbidden transitions in the spectrum but that their probability is much smaller than that of the allowed transitions.

This conclusion seems to contradict experiment. The theory says that if the molecules in the gas are in the ground state (i.e. $v_i = 0$, $\ell_i = 0$) then only the transition

$$\{v_i = 0, \ell_i = 0, m_i\} \rightarrow \{v_f = 1, \ell_f = 1, m_i\}$$

will be observed; the spectrum must have one line. Fig. 17.1 shows that this is not the case.

In addition, the experiments show that the depth of the dips in the spectrum depend on temperature. This dependence is rather strong, and as the temperature is lowered, some peaks disappear because the number of photons absorbed at those frequencies is below the detection limit. Our formulae do not contain the temperature and therefore are unable to explain this behavior.

I will show next that the presence of a large number of dips in the spectrum and its temperature dependence are related. Both exist because the molecules in the gas have high kinetic energy, and collide harshly with each other. As a result many of them have high rotational and vibrational energy. These *thermally excited* molecules have a variety of initial states and absorb photons of difference frequencies. This is why there are so many lines in the absorption spectrum. The probability that a molecule is excited depends on temperature. The higher the temperatures, the greater the variety of excited states in the sample and the larger the number of dips in the spectrum.

In what follows, I will develop a quantitative theory of these qualitative ideas.

§14. *The Molecules in a Gas have a Variety of States.* Let us posit that the molecules in a gas have the state $\{v_i, \ell_i, m_i\}$ with the probability $\wp(v_i, \ell_i, m_i)$. If the number of possible initial states increases, so will the number of transitions allowed by Eqs 17.41–17.43. This is illustrated by Fig. 17.3, which shows all the transitions possible if some molecules in the gas are in the state $\{v = 0, \ell = 0\}$, some are in $\{v = 0, \ell = 1\}$, and some are in $\{v = 0, \ell = 2\}$.

The molecules in the initial state $\{v = 0, \ell = 0\}$ can undergo only the transition

$$\{v_i = 0, \ell_i = 0\} \rightarrow \{v_f = 1, \ell_f = 1\}$$

Since the quantum number ℓ in the final state is higher than that of the initial state, this transition belongs to the P-branch of the spectrum (see Eq. 17.43). I ignore the quantum number m in this discussion, because no transition in which m changes is allowed, if the light is linearly polarized.

If the initial state is $\{v_i = 0, \ell_i = 1\}$, the molecule is allowed to make two transitions:

$$\{v_i = 0, \ell_i = 1\} \rightarrow \{v_f = 1, \ell_f = 2\}$$

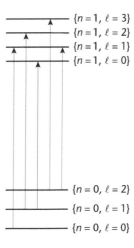

Figure 17.3 Examples of transitions.

and

$$\{v_i = 0, \ell_i = 1\} \rightarrow \{v_f = 1, \ell_f = 0\}$$

In the first, both the vibrational energy and the rotational energy increase, and the transition belongs to the P-branch (see Eq. 17.43). In the second, the vibrational energy increases but the rotational energy decreases, and this transition belongs to the R-branch (see Eq. 17.42).

The rest of the possible initial states have the same behavior. For example, from $\{v_i = 0, \ell_i = 2\}$, I can have the transitions

$$\{v_i = 0, \ell_i = 2\} \rightarrow \{v_f = 1, \ell_f = 3\}$$

and

$$\{v_i = 0, \ell_i = 2\} \rightarrow \{v_f = 1, \ell_f = 1\}$$

The first of these contributes to the R-branch and the second, to the P-branch.

In brief, each molecule in an initial state $\{v_i, \ell_i, m_i\}$ contributes two dips in the spectrum, one in the R-branch and one in the P-branch. If the molecules in the gas are in twenty different initial states, the spectrum will have forty lines.

If I could calculate how many molecules are in a state $\{v_i, \ell_i\}$, I could then determine how these molecules contribute to the spectrum.

§15. *The Infrared Absorption Spectrum of a Gas at a Fixed Temperature.* Let us assume now that we know the probability $\wp(v, \ell, m; T)$ that a molecule in a gas held at the temperature T is in the state $\{v, \ell, m\}$. In this case the probability that molecule undergo the transition

$$i \equiv \{v_i, \ell_i\} \rightarrow f \equiv \{v_i + 1, \ell_i - 1\} \tag{17.44}$$

is

$$p^R(v_i, \ell_i) \equiv \sum_{m_i = -\ell_i}^{\ell_i} \wp(v_i, \ell_i, m_i; T) P^R_{\{v_i, \ell_i\} \rightarrow \{v_i+1, \ell_i-1\}}$$

$$= C \sum_{m_i = -\ell_i}^{\ell_i} \wp(v_i, \ell_i, m_i; T)(v_i + 1)|A_+(\ell_i, m_i)|^2 N(\omega_{if}) \tag{17.45}$$

I used Eq. 17.42 for $P^R_{i \to f} \equiv P^R_{\{v_i,\ell_i\} \to \{v_i+1,\ell_i-1\}}$ and I lumped all the constants together into:

$$C \equiv \frac{\pi}{\epsilon_0 c \hbar^2} \left(\frac{dm(r)}{dr} \right)^2_{r=r_0} \frac{1}{2\beta^2} \tag{17.46}$$

I do not need to keep the Kronecker deltas in the formula since the final state was chosen to make them all equal to 1.

The sum over all possible values of m_i appears for the following reason. A molecule having the quantum number ℓ_i has a fixed rotational energy but it can be in $2\ell_i + 1$ states, corresponding to the possible values of m_i (remember that m_i takes one of the values $-\ell_i$, $-\ell_i + 1$, ..., $\ell_i - 1$, ℓ_i). The value of ℓ_i determines the rotational energy and those of m_i, the orientation of the molecular axis (more precisely, the value of L_z).

All molecules having the state $\{v_i, \ell_i\}$ are allowed to absorb a photon and undergo a transition to $\{v_i + 1, \ell_i - 1\}$, regardless of the value of m_i. Therefore, when we calculate the probability of such a transition, we must sum $\wp(v_i, \ell_i, m_i; T) P^R_{i \to f}$ over all values of m_i compatible with the given value of ℓ_i (i.e. m_i must vary, in increments of 1, from $-\ell_i$ to ℓ_i).

A similar argument gives the probability of photon absorption in the P-branch:

$$p^P(v_i, \ell_i) \equiv \sum_{m_i=-\ell_i}^{\ell_i} \wp(v_i, \ell_i, m_i; T) \, P^P_{\{v_i,\ell_i\} \to \{v_i+1,\ell_i+1\}}$$

$$= C \sum_{m_i=-\ell_i}^{\ell_i} (v_i + 1) \wp(v_i, \ell_i, m_i; T) |A_-(\ell_i, m_i)|^2 N(\omega_{if}) \tag{17.47}$$

for the transition

$$i \equiv \{v_i, \ell_i\} \to f \equiv \{v_i + 1, \ell_i + 1\} \tag{17.48}$$

Eqs 17.45 and 17.47 allow a large number of dips in the absorption spectrum and make the probability of observing such a dip depend on temperature. If we can derive an expression for $\wp(v, \ell, m; T)$, we can use those equations to analyze IR absorption spectra.

§16. *The Probability* $\wp(v, \ell, m; T)$. Statistical mechanics, which you will study after having learned quantum mechanics, tells us that the probability $\wp(v, \ell, m; T)$ that a diatomic molecule is in the state $\{v, \ell, m\}$, in a gas held at the temperature T, is

$$\wp(v, \ell, m; T) = \frac{\exp[-E(v, \ell)/k_B T]}{\sum_{v=0}^{\infty} \sum_{\ell=0}^{\infty} \sum_{m=-\ell}^{\ell} \exp[-E(v, \ell)/k_B T]} \tag{17.49}$$

In this equation, $k_B = 1.3806 \times 10^{-16}$ erg/K (where K is kelvin) is Boltzmann's constant, T is the gas temperature (in kelvin) and $E(v, \ell)$ is the energy of the molecule in the state $\{v, \ell, m\}$.

The probability $\wp(v, \ell, m; T)$ depends on the energy of the molecule, which is independent of the quantum number m; for this reason $\wp$ does not depend on m. This is physically reasonable: m controls the orientation of the angular momentum and there is no reason why in a gas there would be a preferred orientation; therefore, all values of m are equally probable. This means that in the subset of molecules that have a given value for ℓ, the probability that m has a certain magnitude is $1/(2\ell + 1)$, regardless of what the value of m is. This is similar to the fact that, in a throw of an honest die, all results are equally probable and each has the probability 1/6.

These equations allow us to anticipate the temperature dependence of the spectrum. If the temperature is extremely low, $E(v, \ell)/k_B T$ is large and all the molecules in the gas will be in the ground state (see Eq. 17.49). I explained in §13 that in this case the absorption spectrum would have only one line, corresponding to the transition to $\{v = 1, \ell = 1\}$. As the temperature is increased, more and more molecules are in the states $\{v_i = 0, \ell_i = 1, m_i\}$, $\{v_i = 0, \ell_i = 2, m_i = 0\}$, etc. Each such state adds two more lines to the spectrum, one to the R-branch and the other to the P-branch. Since the probability of having a state populated depends exponentially on temperature, the number of peaks in the spectrum and their height is very sensitive to the magnitude of the temperature. Such spectra are excellent "thermometers" for gases that we cannot touch, but from which we can receive radiation. This is how the temperatures of flames or of remote parts of the universe are measured.

We can use Eq. 17.49 in Eqs 17.45 and 17.47 to calculate the probability that the dips corresponding to the transitions Eqs 17.44 and 17.48 are observed in the IR absorption spectrum. Such calculations will allow us to construct the absorption spectrum of the molecule and to study its temperature dependence. Before examining spectra, I take a side trip and perform a few numerical calculations that will help us better understand the quantity $\wp(v, \ell, m; T)$.

§17. *The Probability of Various States in a Gas: a Numerical Study.* To illustrate the behavior of $\wp$ and its connection with the properties of the molecules in the gas, I perform here a numerical study of $\wp$ for CO. Since we are not preoccupied with accuracy I will use Eq. 17.2 for the energy $E(v, \ell)$ and the data provided by Eq. 17.3. Combining them gives

$$E(v, \ell) = 2169.8 \left(v + \tfrac{1}{2}\right) + 1.93127\ell(\ell + 1) \text{ cm}^{-1} \qquad (17.50)$$

Exercise 17.9

Table 16.1 gives the bond length r_0 and the force constant k. Use this information to calculate $\hbar\omega$ and B for CO. Compare your results with the values given in Table 16.1. If your results differ from those given in the table, try to suggest a reasonable explanation for the cause of the discrepancy.

As an example, I calculate the probability of finding in the gas a CO molecule in the state $\{v = 0, \ell = 4, m = 1\}$ when the temperature is $T = 298$ K.

The exponential term in the probability expression (Eq. 17.49) is

$$\exp\left[\frac{-\left(\frac{2169.8}{2} + 4 \times 5 \times 1.93127\right)}{298 \times 1.3806 \times 10^{-16} \times 5.0348 \times 10^{15}}\right]$$

Here $k_B = 1.3806 \times 10^{-16}$ erg/K is Boltzmann's constant and 5.0348×10^{15} converts erg into cm^{-1} (the numerator is the energy of the molecule in cm^{-1}). The sum

$$\sum_{v=0}^{\infty} \sum_{\ell=0}^{\infty} \sum_{m=-\ell}^{+\ell} \exp\left[\frac{-\left(2169.8(v + \tfrac{1}{2}) + \ell(\ell + 1) \times 1.93127\right)}{298 \times 1.3806 \times 10^{-16} \times 5.0348 \times 10^{15}}\right]$$

can be evaluated analytically (see the book *Statistical Mechanics*). Since our purpose here is to use statistical mechanics, not to study it, I use the computer to perform the sum numerically. The exponential gets smaller and smaller as ℓ and v increase, and the sum converges by the time v is 2 or 3 and ℓ reaches 30. Note that the sum over m can be replaced by multiplication by $2\ell + 1$:

$$\sum_{\ell=0}^{\infty} \sum_{m=-\ell}^{+\ell} f(\ell) = \sum_{\ell=0}^{\infty} (2\ell + 1) f(\ell)$$

because the expression under the sum is independent of m.

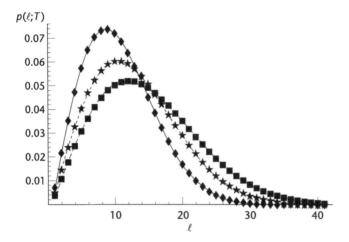

Figure 17.4 The probability $p(\ell; T)$ that a CO molecule in a gas is in the state having $v = 0$ and any one of the $2\ell + 1$ allowed values of m, as a function of ℓ. The three sets of points show results for three temperatures. The graph was made in Workbook QM17.3. ◆ 200 K; ★ 298 K; ■ 400 K.

The probability that the system is in the state having $v = 0$ and any one of the $2\ell + 1$ allowed values of m is plotted in Fig. 17.4 as a function of ℓ, for $T = 200$ K, 298 K, 400 K. As you can see, raising the temperature makes it more likely to find molecules having higher rotational energy.

Exercise 17.10

Show that the probability of finding a molecule with rotational energy much greater than the energy $k_B T$ is very small.

Exercise 17.11

Make a plot showing the probabilities $\wp(v = 0, \ell, m = 0)$, $\wp(v = 1, \ell, m = 0)$ and $\wp(v = 3, \ell, m = 0)$ as a function of ℓ for I_2 (get data from Table 16.1) at $T = 298$ K. Then make the same graph for $T = 60$ K. Describe your finding in words and then try to anticipate how they will affect the absorption spectrum.

Exercise 17.12

Plot the probability $\wp(v, \ell)$ that a molecule is in the state $\{v = 0, \ell\}$ (v and ℓ are specified, but we do not specify a value for m) as a function of ℓ, for CO, at the temperatures 200 K, 298 K, and 400 K. Compare your plots with those obtained in Fig. 17.4. *Hint.* The probability you want is

$$\sum_{m=-\ell}^{\ell} \wp(v, \ell, m) = (2\ell + 1)\wp(v, \ell, m)$$

§18. *Back to the Spectrum: the Relative Intensity of the Absorption Peaks.* I have now all the elements required for calculating the intensities of the peaks in the absorption spectrum.

I assume here—and this is a good assumption—that in the case of CO an overwhelming fraction of the molecules will be in a state with $v = 0$; very few CO molecules are vibrationally excited at room temperature.

Exercise 17.13

Calculate and plot $\wp(v, \ell = 0, m = 0)$ for CO at room temperature and at 800 °C, as a function of v.

Exercise 17.14

Calculate and plot $\wp(v, \ell = 0, m = 0)$ for I_2 and H_2 at room temperature, as a function of v. Explain why the results for H_2 are so different from those for I_2. Get the data you need from Table 16.1.

I have explained earlier that determining the absolute absorption spectrum requires a careful calibration of the light source and detector and also the capability of determining how many photons did not reach the detector for other reasons than absorption. Because such measurements are difficult and are not very informative, it is customary to plot the spectrum as it is delivered by the detector. This is proportional to the real spectrum, but the proportionality constant is unknown. Because of this, I calculate the spectrum in arbitrary units by setting

$$CN(\omega_{\text{if}}) = 1 \tag{17.51}$$

in Eqs 17.45 and 17.47.

§19. *Numerical Analysis.* Now put some numbers into the theory and see how it works. I review here the equations used.

The energy of the photon absorbed in the (R-branch) transition $\{v_i = 0, \ell_i, m_i\} \rightarrow \{v_i = 0, \ell_i + 1, m_i\}$ is

$$\hbar\Omega_R(v_i = 0, \ell_i, m_i) = E(v = 1, \ell + 1) - E(v = 0, \ell)$$
$$= \hbar\omega + 2B(\ell + 1) \tag{17.52}$$

and the height of the absorption peak is proportional to

$$I_R(v_i = 0, \ell_i) = (v_i + 1) \sum_{m_i = -\ell_i}^{+\ell_i} \wp(v_i = 0, \ell_i, m_i) \left[A_+(\ell_i, m_i) \right]^2 \tag{17.53}$$

To derive the last equality in Eq. 17.52, I used the energy given by Eq. 17.2 while Eq. 17.53 is obtained by combining Eq. 17.45 and Eq. 17.51. I dropped from this equation all constants that do not change from one absorption peak to another.

For the P-branch transition $\{v_i = 0, \ell_i, m_i\} \rightarrow \{v_i = 1, \ell_i - 1, m_i\}$, the photon energy is

$$\hbar\Omega_P(\ell_i, m_i) = \hbar\omega - 2B\ell \tag{17.54}$$

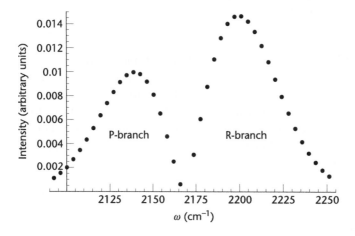

Figure 17.5 The calculated CO absorption spectrum for the transitions $\{v_i = 0, \ell_i, m_i\} \rightarrow \{v_i = 1, \ell_i - 1, m_i\}$. The graph was obtained in Workbook QM17.4.

and the peak intensities are given by

$$I_P(v_i = 0, \ell_i) = (v_i + 1) \sum_{m_i = -\ell_i}^{+\ell_i} \wp(v_i = 0, \ell_i, m_i) \left[A_-(\ell_i, m_i)\right]^2 \qquad (17.55)$$

This equation is obtained by combining Eq. 17.47 with Eq. 17.51.

I have calculated I_R as a function of Ω_R and I_P as functions of Ω_P in Workbook QM17.4. To obtain I_R I started with Eq. 17.53 and used Eq. 17.49 to evaluate $\wp$. The latter is calculated as explained in §17. The data for CO are given in Table 16.1. The evaluation of I_P is similar. A plot of the results, obtained in Workbook QM17.4, is shown in Fig. 17.5. The spectrum strongly resembles the measurements shown in Fig. 17.1, but the intensities are not exactly in the proportion seen in experiment. This is probably due to the use of the harmonic approximation. Had we used numerical methods to calculate the energies and the matrix elements accurately, the results would have been identical to those of the experiment.

18

THE HYDROGEN ATOM

Introduction

§1. In this chapter we study atoms consisting of a nucleus and one electron: e.g. H, D, He$^+$, and Li^{2+}. This is done for three reasons:

- we can solve, exactly, the equations describing these systems and compare the results to highly accurate measurements to test quantum mechanics;

- wave functions similar to those of an electron in the hydrogen atom serve as building blocks in the theory of the chemical bond in molecules;

- this simple system provides an opportunity to practice applying the general rules of quantum mechanics to specific systems.

We start with the theory developed in the previous chapters for a two-particle system. The only difference between a one-electron atom and a diatomic molecule is the form of the potential energy $V(r)$. For the atom, $V(r)$ is the Coulomb interaction energy between electron and nucleus, which you encountered when studying electrostatics. For the diatomic molecule, the interaction energy $V(r)$ is reasonably well-represented by the Morse potential. The properties of these two systems are very different, because the potentials are so different.

The total wave function, in spherical coordinates, depends on the electron–nucleus distance and the angles θ and ϕ. These are used to construct the *atomic orbitals* 1s, 2s, 2p$_x$, ..., 3d$_{x^2-y^2}$, etc., which you must have encountered in your chemistry courses. They form the basis for the qualitative theories of the chemical bond. Here you will learn where these atomic orbitals come from. In later chapters, you will see how they are used to describe the chemical bond. The last section of the chapter describes the absorption spectroscopy of the hydrogen atom.

The Schrödinger Equation for a One-Electron Atom

§2. The hydrogen atom is a two-particle system and its properties are described by the Schrödinger equation (Eq. 15.2):

$$-\frac{\hbar^2}{2\mu r^2}\frac{\partial}{\partial r}r^2\frac{\partial\psi(r,\theta,\phi)}{\partial r} + \frac{\hat{L}^2}{2\mu r^2}\psi(r,\theta,\phi) + V(r)\psi(r,\theta,\phi)$$

$$= E\,\psi(r,\theta,\phi) \tag{18.1}$$

This is valid for any one-electron atom or ion (e.g. H, D, He$^+$, Li^{2+}).

The interaction energy $V(r)$ between the electron and the nucleus is given by the Coulomb formula

$$V(r) = -\frac{e^2 Z}{4\pi\varepsilon_0 r} \tag{18.2}$$

which is familiar from classical electrostatics. This equation is in SI units. The proton charge is $e = 1.602177 \times 10^{-19}$ C (where C stands for coulomb). The electron charge has the same value but opposite sign, and this is why a minus sign appears in Eq. 18.2. The permittivity of the vacuum is $\varepsilon_0 = 8.854187 \times 10^{-12}$ C/N m^2 (N is newton, m is meter). Z is the number of protons in the nucleus: $Z = 1$ for the hydrogen or the deuterium atom, $Z = 2$ for the He$^+$ ion, $Z = 3$ for Li^{2+}, etc. With these values of the constants, and the distance r in meters, Eq. 18.2 gives the electron–nucleus interaction energy in joules.

You have probably encountered the Coulomb interaction energy written as $V(r) = -e^2 Z/r$. This is in the Gaussian system of units, which is not used in this chapter.

Exercise 18.1

1.   Calculate, by using the Coulomb formula, the interaction energy between an electron and a proton when the distance between them is 1 Å and when it is 20 Å. Give the result in eV.

2.   Make a plot of $V(r)$ versus r.

3.   Compare the distance at which the Coulomb interaction is equal to 10^{-3} eV to the distance at which the Morse interaction for CO (see Chapter 16) is equal to 10^{-3} eV.

The reduced mass is

$$\mu = \frac{m_e m_N}{m_e + m_N} \tag{18.3}$$

where $m_e = 9.109389 \times 10^{-31}$ kg is the electron mass and m_N is the mass of the nucleus; in the case of H, the proton mass is $m_N = 1.672623 \times 10^{-27}$ kg. Because the mass of the nucleus is much larger than that of the electron, the reduced mass is very close to the mass of the electron. This is much smaller than the reduced mass of any diatomic molecule.

Exercise 18.2

Calculate the reduced mass for H and for D, in grams.

§3. *Why are the Properties of a One-Electron Atom so Different from Those of a Diatomic Molecule?* It is very striking to see that, as far as the Schrödinger equation is concerned, a one-electron atom differs from a diatomic molecule in only two ways: the nature of the interaction energy and the magnitude of the reduced mass. Given the mathematical similarity between these two systems, it is surprising that their behavior is radically different. Let us try to understand qualitatively where this difference comes from.

The interaction energy between the atoms, in a diatomic molecule, is such that it takes considerable energy to change the interatomic distance; we say that the potential is stiff. This is not the case for the Coulomb potential, which is much softer: very little energy is required to change the distance between the electron and the nucleus. This has two dramatic consequences: neither the harmonic nor the rigid-rotor approximation is adequate for the one-electron atom. This means that it is no longer possible to separate the energy into a vibrational and a rotational contribution (as we did for diatomics).

The atoms in a diatomic molecule repel each other when they are too close, and attract each other when they are too far apart. This is why they tend to undergo small-amplitude oscillations around a certain bond length, which is determined by the position of the minimum in the potential. The electron, however, is attracted by the nucleus at all distances. Because of this, classical physics would say that in the lowest energy state the electron is located on top of the nucleus, collapsing the size of the atom to that of the nucleus. This contradicts a large number of unambiguous experiments that show that the size of the atom is about 10^5 times larger than that of the nucleus. The physical effect that controls the atomic size must be different than the one controlling the size of the diatomic molecule.

When the electron is very close to the nucleus, the mean potential energy V is small but the mean kinetic energy K is large. If the region in which the electron is located increases in size, the kinetic energy is smaller but the potential energy increases. The electron will take the intermediate location for which the total energy $E = K + V$ is minimized.

Finally, the reduced mass of the H atom is about 2000 times smaller than that of a diatomic molecule. We know that the smaller a particle is, the larger the quantum effects. In particular, the zero-point energy is larger and tunneling is more efficient. Because of this, the potential energy is less able to confine a small particle to a small spatial region. The electron in the atom spreads over a much larger region than the interatomic distance r in a diatomic molecule.

Exercise 18.3

Study the difference between the correct solution for the hydrogen atom and the one obtained by applying the harmonic and the rigid-rotor approximations. To apply these approximations, follow the procedure used in Chapter 16, as outlined below.

1.   Try first to expand the Coulomb potential in power series around the point where the potential is smallest. Explain why this expansion would lead to absurd results.

2.   Plot the total potential (i.e. the Coulomb potential plus the centrifugal potential) as a function of r. The centrifugal potential is

$$\frac{\ell(\ell + 1)\hbar^2}{2\mu r^2}$$

3.   Expand the total potential in powers of $r - r_0$, where r_0 is the distance for which the total potential has a minimum. Retain only the terms up to the second order.

4.   Use this expansion to calculate the force constant k, the vibrational frequency ω, and the rotational constant B (see Chapter 16). Note that, in the present case, these quantities will depend on the quantum number ℓ. Explain why this did not happen when we studied the diatomic molecule.

5.   Use the values of ω and B to calculate the energy of the atom (with the harmonic approximation). Note that, unlike the case of a diatomic, the energy of the atom in this crude approximation depends on ℓ (through ω and B).

6.   Use these energies to calculate the frequency of the photons that the system would absorb. Do not worry about selection rules now.

7.   Compare the values you obtained in this way to those obtained in the following section.

The Solution of the Schrödinger Equation for a One-Electron Atom

§4.   *The Approach.* As in the case of the diatomic molecule, I will use the method of separation of variables and attempt to find a solution of the form

$$\psi(r,\theta,\phi) = R(r)\, Y_\ell^m(\theta,\phi) \tag{18.4}$$

Here $Y_\ell^m(\theta,\phi)$ is a spherical harmonic. The spherical harmonics satisfy the equations

$$\hat{L}^2\, Y_\ell^m(\theta,\phi) = \hbar^2 \ell(\ell+1) Y_\ell^m(\theta,\phi) \tag{18.5}$$

and the orthonormalization conditions

$$\langle \ell,m \mid \ell',m' \rangle \equiv \int_0^\pi \sin\theta\, d\theta \int_0^{2\pi} d\phi\, Y_\ell^m(\theta,\phi)^* \, Y_{\ell'}^{m'}(\theta,\phi) = \delta_{\ell,\ell'}\delta_{m,m'} \tag{18.6}$$

($\delta_{i,j} = 1$ if $i = j$ and 0 otherwise.)

The radial wave function $R(r)$ is unknown, but it must be such that ψ given by Eq. 18.4 is a solution of Eq. 18.1. Inserting $\psi(r,\theta,\phi) = R(r)Y_\ell^m(\theta,\phi)$ in Eq. 18.1 and performing a few simplifications gives the following equation for $R(r)$:

$$-\frac{\hbar^2}{2\mu r^2}\frac{\partial}{\partial r}r^2\frac{\partial}{\partial r}R(r) + \frac{\hbar^2 \ell(\ell+1)}{2\mu r^2}R(r) + V(r)R(r) = E\, R(r) \tag{18.7}$$

§5. *The Eigenvalues.* The solutions to the eigenvalue problem Eq. 18.7 are known. The energy eigenvalues are

$$E_n = -\frac{\varepsilon}{n^2} \tag{18.8}$$

with

$$\varepsilon = \frac{\mu e^4 Z^2}{2(4\pi \varepsilon_0)^2 \hbar^2} \tag{18.9}$$

The quantity ε has units of energy and sets an energy scale for this problem. This equation for ε is in SI units.

Exercise 18.4

Prove that ε has units of energy.

The *principal quantum number n* takes the values

$$n = 1, 2, \ldots \tag{18.10}$$

The energy of the atom does not depend on either ℓ or m.

We have noted many times that adding a constant to the energy does not change the physical properties of a system. This happens because we either measure energy differences (for example, the frequency of a photon absorbed by a molecule) or derivatives of the energy (e.g. forces). This means that whenever a formula for energy is given, a decision has been made regarding this arbitrary constant. What decision was made when Eq. 18.8 was derived? Look at $V(r)$ and the rotational energy $\hbar^2 \ell(\ell + 1)/(2\mu r^2)$. They both become zero when $r \to \infty$. This means that we have decided that the energy of the system is zero when the nucleus and the electron are separated by an infinite distance and they have zero kinetic energy (they do not move).

As long as the physics we examine involves only one type of system, any zero of the energy scale is as good (but perhaps not as convenient) as any other one. However, this is not the case if we examine processes involving several systems. In that case, the energy of all systems must be defined on the same energy scale, with the same zero; otherwise they cannot be compared.

The zero of the energy scale chosen here is convenient because it means that Eq. 18.8 is the energy to remove an electron from the atom and place it at an infinite distance from the nucleus. The fact that the energy is negative indicates that the atom is more stable (has lower energy) than the separated electron and nucleus. To ionize an atom that is in the state n, I have to give it the energy E_n.

§6. *The Eigenfunctions.* For each energy eigenvalue E_n, there are several radial eigenfunctions, given by the expression:

$$R_{n,\ell}(r) = \left(\frac{2Z}{na}\right)^{3/2} \Phi_{n,\ell}(\rho) \tag{18.11}$$

with

$$\Phi_{n,\ell}(\rho) \equiv \sqrt{\frac{(n-\ell-1)!}{2n(n+\ell)!}} \left(\frac{2Z\rho}{n}\right)^{\ell} \exp\left[-\frac{2Z\rho}{n}\right] L_{n-\ell-1}^{2\ell+1}\left(\frac{2Z\rho}{n}\right) \tag{18.12}$$

$$\ell = 0, 1, \ldots, n-1 \tag{18.13}$$

$$\rho = \frac{r}{a} \tag{18.14}$$

$$a = \frac{\hbar^2(4\pi\varepsilon_0)}{\mu e^2} \quad \text{(in SI units)} \tag{18.15}$$

Exercise 18.5

Prove that a has units of length.

Since a has units of length, ρ is a dimensionless variable. The units of the wave function is given by the term $(1/a)^{3/2}$ that appears in Eq. 18.11 ($\Phi_{n,\ell}(\rho)$ is dimensionless).

Exercise 18.6

Show that

$$-E_n = \frac{Z^2 e^2}{2(4\pi\varepsilon_0)an^2}$$

The function $L_i^j(r)$ is a *generalized Laguerre polynomial*. You can find its definition in many books. Standard references are Abramovitz and Stegun[a] and Weisstein.[b] The latter is available at **mathworld.wolfram.com**. **Mathematica** provides the function **LaguerreL**[n, ℓ, r] that returns $L_n^\ell(r)$. The definition used by **Mathematica** is the same as the one given by Abramovitz and Stegun.

Most quantum mechanics textbooks give an erroneous formula for $\Phi_{n,\ell}$ (i.e. a formula different from Eq. 18.12), even though they give the correct equations for the particular expressions of $R_{1,0}(r)$, $R_{2,1}(r)$, etc. This should alert you to a simple rule: *never* use an equation or a program unless you understand its derivation and have tested it yourself. An error made in an early text is sometimes copied in subsequent books for decades; many authors copy and only a few doubt.

The Energy

§7. *The Magnitude of the Energies.* The energy of a state, given by Eq. 18.8, is characterized by the *principal quantum number n*. I know the values of all constants in this equation and therefore can calculate the magnitude of the energy.

The reduced mass for the hydrogen atom is

$$\mu = \frac{m_e m_N}{m_e + m_N} = \frac{9.109389 \times 10^{-31} \times 1.672623 \times 10^{-27}}{9.109389 \times 10^{-31} + 1.672623 \times 10^{-27}}$$
$$= 9.10443 \times 10^{-31} \text{ kg}$$

The mass of the nucleus is much larger than that of the electron and the value of μ is very close to that of m_e. One might be tempted to replace μ with m_e in the equation for energy, but that would be a mistake: it will wipe out all dependence on the nuclear mass. If we were to replace μ with m_e, then the energy of the deuterium atom would be the same as that of the hydrogen atom. The experiments show that the spectra of these two atoms are close, but not the same: the peaks in one spectrum are slightly shifted from the peaks of the other. The difference comes from the small differences in the value of μ for the two atoms.

---

[a] *Handbook of Mathematical Functions*, M. Abramovitz and I.A. Stegun, editors, Dover Publications, New York, 1965, Chapter 22.
[b] E.W. Weisstein, *CRC Concise Encyclopedia of Mathematics*, CRC Press, Boca Raton, 1999.

Inserting the values of various constants into Eqs 18.8 and 18.9 gives, for the hydrogen atom,

$$E_n = -\frac{6.2420 \times 10^{18} \times 9.10443 \times 10^{-31} \text{ kg}(1.602177 \times 10^{-19} \text{ C})^4}{2\left(4\pi \times 8.854187 \times 10^{-12}\frac{\text{C}}{\text{N m}^2}\right)^2 (1.054573 \times 10^{-34} \text{ Js})^2}\frac{1}{n^2}$$

$$= -\frac{13.5994}{n^2} \text{ eV} \tag{18.16}$$

Workbook

The calculation is performed in Workbook QM18.1. The factor 6.2420×10^{18} converts the energy from joules to eV. The value of E_n in Eq. 18.16 is valid only for the hydrogen atom, since I used the hydrogen reduced mass and $Z = 1$.

The energy of a photon absorbed in a transition from a state having energy E_{n_i} to one having energy E_{n_f} is given by

$$\hbar\Omega = E(n_f) - E(n_i) = 13.5994\left[\frac{1}{n_i^2} - \frac{1}{n_f^2}\right] \text{ eV} \tag{18.17}$$

If I replace μ with m_e (as is sometimes erroneously done), the energy is

$$E_n = -\frac{13.6068}{n^2} \text{ eV for } \mu \cong m_e \tag{18.18}$$

See if this makes a difference in the spectrum. If we use Eq. 18.16, the frequency Ω of the transition $\{n = 1\} \rightarrow \{n = 2\}$ is determined by

$$\hbar\Omega = E_2 - E_1 = -\frac{13.5994}{4} - \left(-\frac{13.5994}{1}\right) = 10.1995 \text{ eV}$$

Eq. 18.18 predicts a transition at the photon frequency

$$\hbar\Omega = E_2 - E_1 = -\frac{13.6068}{4} - \left(-\frac{13.6068}{1}\right) = 10.2051 \text{ eV}$$

Experiments can easily detect the difference of $10.2051 - 10.1995 = 0.0055$ eV. This difference has been used in some photochemistry experiments to detect D in the presence of H.

Exercise 18.7

A positron and an electron can make a bound species, similar to a hydrogen atom. Calculate the binding energy for $n = 1$ and $n = 2$, and the frequency of the transition $\{n = 1\} \rightarrow \{n = 2\}$. The mass of the positron is equal to that of the electron and its charge is equal to that of the proton.

Exercise 18.8

Calculate the lowest energy of the electron in He^+. Is it lower than that of the electron in H? Give a physical reason for your answer.

§8. *The Energy Scale ε and the Length Scale a.* A one-electron atom is characterized by six constants, m_e, m_N, $\hbar$, e, ε_0 (having various units), and the dimensionless quantity Z. However, in the eigenstates and eigenvalues, these constants appear combined into two quantities: the energy scale ε (Eq. 18.9) and the length a (Eq. 18.15).

a appears in the radial wave function, Eq. 18.11, which depends on the dimensionless quantity $\rho = r/a$. The exponential $\exp[-Zr/na]$ in Eqs 18.11 and 18.12 for $R_{n,\ell}(r)$ controls how far the electron can wander away from the nucleus. If

$$\frac{Zr}{na} \gg 1,$$

the exponential is very small and so is $R_{n,\ell}(r)$. Therefore, the probability of finding the electron at distances from the nucleus that are bigger than na/Z from the nucleus is very small. Roughly speaking, a is the "size" of the atom. ε sets the scale for the magnitude of the energy of the electron.

We have discovered a surprising thing: the six constants characterizing this system have been combined by the theory to define an energy scale ε and a distance scale a. The system is fully characterized by two quantities, not six. This is a great simplification, but also a nuisance: it means that we cannot determine the value of all six constants by experimenting with the hydrogen atom; we can only determine the values of ε and a.

Why are there only two constants in the problem? Why not five or three? The answer to this question must be contained in the Schrödinger equation. In what follows, I use a method called *dimensional analysis* to rewrite the Schrödinger equation for a one-electron atom in a way that shows that the system is described by two quantities, an energy and a length. The idea behind the method is that there is much to gain by scaling the variables in the problem to make the equation dimensionless. Here is how this is done.

I start with the radial Schrödinger equation (combine Eq. 18.7 with the Coulomb formula Eq. 18.2):

$$-\frac{\hbar^2}{2\mu r^2}\frac{\partial}{\partial r}r^2\frac{\partial R}{\partial r} + \frac{\hbar^2\ell(\ell+1)}{2\mu r^2}R - \frac{e^2 Z}{4\pi\varepsilon_0 r}R = E R \qquad (18.19)$$

I divide this equation by $\hbar^2/2\mu$, to obtain

$$-\frac{1}{r^2}\frac{\partial}{\partial r}r^2\frac{\partial R}{\partial r} + \frac{\ell(\ell+1)}{r^2}R - \frac{2Z}{\frac{4\pi\varepsilon_0\hbar^2}{e^2\mu}r}R = \frac{2\mu E}{\hbar^2}R \qquad (18.20)$$

The units of the first term are those of R times $1/\text{length}^2$. This means that the quantity

$$\frac{4\pi\varepsilon_0\hbar^2}{\mu e^2}$$

that appears in the third term of Eq. 18.20 must have units of length so that the units of $\frac{2Z}{\frac{4\pi\varepsilon_0\hbar^2}{e^2\mu}r}R$ are those of R divided by length squared. This length is the length scale a.

Next, I make $r \equiv \rho a$ in Eq. 18.20, which leads to

$$-\frac{1}{a^2\rho^2}\frac{\partial}{a\partial\rho}a^2\rho^2\frac{\partial R}{a\partial\rho} + \frac{\ell(\ell+1)}{a^2\rho^2}R - \frac{2Z}{a(a\rho)}R = \frac{E}{(\hbar^2/2\mu)}R$$

Multiplying this equation by a^2/Z^2 gives

$$-\frac{1}{Z^2\rho^2}\frac{\partial}{\partial\rho}\rho^2\frac{\partial R}{\partial\rho} + \frac{\ell(\ell+1)}{Z^2\rho^2}R - \frac{2}{Z\rho}R = \frac{E}{(\hbar^2 Z^2/2\mu a^2)}R$$

Since ρ is a dimensionless quantity the coefficients multiplying R in the left-hand side are dimensionless. Therefore $E/(\hbar^2 Z^2/2\mu a^2)$ in the right-hand side must also be dimensionless. This means that

$$\frac{Z^2\hbar^2}{2\mu a^2} = \frac{Z^2\hbar^2}{2\mu \left(\frac{\hbar^2\, 4\pi\varepsilon_0}{\mu e^2}\right)^2} = \frac{Z^2\mu e^4}{2\hbar^2(4\pi\varepsilon_0)^2} \tag{18.21}$$

must have units of energy. But this is the energy scale ε (see Eq. 18.9).

From this *dimensional analysis*, we find that the Schrödinger equation and its solutions depend only on a and ε. This "quick and dirty" analysis establishes the scales of the system with practically no calculation. You can use such analysis in cases when an equation is too complicated to solve.

The Radial Wave Functions $R_{n,\ell}$ (r) and the Mean Values of Various Physical Quantities

§9. *Introduction.* We give physical meaning to a wave function $\psi_\alpha(x,y,z)$ by connecting it to the probability $p_\alpha(x,y,z)dxdydz$ of finding the particle in an infinitesimal volume $dxdydz$ centered around the position $\{x,y,z\}$. This probability can be used for calculating the mean values of various observable that depend on $\mathbf{r} = \{x,y,z\}$. Since we study here the radial behavior of the electron, our first task is to write the probability in spherical coordinates and then find the probability $\Pi_\alpha(r)dr$ that when the system is in the state ψ_α, the electron is contained in the space between a sphere of radius r and a sphere of radius $r + dr$. It will turn out that this probability is proportional to the radial wave function squared.

Once the probability is known we can plot it to find the regions where the electron is most likely to "hang out." We can also calculate the average value of any quantity that depends on the coordinate r, such as the mean distance between the electron and the nucleus (i.e. the average value of r), the mean square distance between the electron and the nucleus (i.e. the average value of r^2), the mean Coulomb interaction (i.e. the average value of $V(r)$), the mean centrifugal energy (i.e. the mean value of $\hbar^2\ell(\ell + 1)/(2\mu r^2)$), or the mean value of the radial kinetic energy operator.

None of these averages can be measured at this time. We calculate them because they help us visualize and understand the properties of the electron in the atom. To connect the theory to quantities that can be measured in the laboratory we will have to study the atom's absorption and emission spectra (see Chapter 19).

§10. *A Summary of the Theory.* The wave function of the atom is given by Eq. 18.4:

$$\psi_{n,\ell,m}(r,\theta,\phi) = R_{n,\ell}(r)\, Y_\ell^m(\theta,\phi) \tag{18.22}$$

The spherical harmonics $Y_\ell^m(\theta,\phi)$ have been defined by Eq. 14.34 in Chapter 14 (see Eqs 14.35–14.59 or Workbook QM14.2 for the expressions obtained when ℓ and m take specific values). The general formula for the radial wave function is given by Eqs 18.11–18.15.

The eigenstates of the Hamiltonian must be orthonormal. This means that

$$\langle n,\ell,m \mid n',\ell',m'\rangle \equiv \int_0^\infty r^2 dr \int_0^\pi \sin\theta\, d\theta \int_0^{2\pi} d\phi R_{n,\ell}(r)^* Y_\ell^m(\theta,\phi)^* R_{n',\ell'}(r) Y_{\ell'}^{m'}(\theta,\phi)$$

$$= \int_0^\infty r^2 dr R_{n,\ell}(r)^* R_{n',\ell'}(r) \int_0^\pi \sin\theta d\theta \int_0^{2\pi} d\phi\, Y_\ell^m(\theta,\phi)^* Y_{\ell'}^{m'}(\theta,\phi)$$

$$= \delta_{nn'}\delta_{\ell\ell'}\delta_{mm'} \tag{18.23}$$

The integral is over the whole space and it is performed in spherical coordinates: this is why $\sin(\theta)r^2$ appears in the integrand (we have $dxdydz = r^2\sin(\theta)drd\theta d\phi$). We already know that (see Chapter 14)

$$\int_0^\pi \sin\theta\, d\theta \int_0^{2\pi} d\phi\, Y_\ell^m(\theta,\phi)^* Y_{\ell'}^{m'}(\theta,\phi) \equiv \langle m,\ell \mid m',\ell'\rangle = \delta_{\ell,\ell'}\delta_{m,m'} \tag{18.24}$$

Inserting this in Eq. 18.23 gives

$$\delta_{\ell,\ell'}\delta_{m,m'} \int_0^\infty dr\, r^2 R_{n,\ell}(r)^* R_{n',\ell'}(r) = \delta_{n,n'}\delta_{\ell,\ell'}\delta_{m,m'}$$

which leads to

$$\langle n,\ell \mid n',\ell'\rangle \equiv \int_0^\infty dr\, r^2 R_{n,\ell}(r)^* R_{n',\ell'}(r) = \delta_{n,n'} \tag{18.25}$$

This is the orthonormalization condition for $R_{n,\ell}(r)$. It is easy to use a computer to test that the wave functions derived here satisfy Eq. 18.25 for any particular values of n and ℓ (see Workbook QM18.4).

Exercise 18.9

Verify that Eq. 18.25 is correct, for $n = 0, 1, 2, 3$ and all the allowed values of ℓ.

§11. *The Probability of Finding the Electron at a Certain Distance from the Nucleus.* From the general theory I know that, if a system is in the state $\psi_\alpha(x, y, z)$, the average value $\langle O \rangle_\alpha$ of a dynamical variable O, represented by an operator $\hat{O}$, is

$$\langle O \rangle_\alpha = \langle \alpha \mid \hat{O} \mid \alpha \rangle$$

$$= \int dx \int dy \int dz \, \psi_\alpha(x, y, z)^* \hat{O} \psi_\alpha(x, y, z) \qquad (18.26)$$

In our case the wave function is $\psi_{n,\ell,m}(r, \theta, \phi)$ and Eq. 18.26 becomes

$$\langle O \rangle_{n,\ell,m} = \langle n, \ell, m \mid \hat{O} \mid n, \ell, m \rangle$$

$$= \int_0^\pi \sin(\theta) d\theta \int_0^{2\pi} d\phi \int_0^\infty r^2 dr \, \psi_{n,\ell,m}(r, \theta, \phi)^* \hat{O} \psi_{n,\ell,m}(r, \theta, \phi) \quad (18.27)$$

I have used here the expression $dxdydz = r^2 dr \sin(\theta) d\theta d\phi$ for the volume element in spherical coordinates.

In this section, I refer only to the case when the operator $\hat{O}$ is a function of r. To remind me of this, I will use the notation $\hat{O}(r)$ and reserve $\hat{O}$ for the general case. Using Eq. 18.22 for $\psi_{n,\ell,m}(r, \theta, \phi)$ turns Eq. 18.27 into

$$\langle O(r) \rangle_{n,\ell,m} = \left\{ \int_0^{2\pi} d\phi \int_0^\pi \sin(\theta) d\theta Y_\ell^m(\theta, \phi)^* Y_e^m ll(\theta, \phi) \right\}$$

$$\times \int_0^\infty r^2 dr R_{n,\ell}(r)^* \hat{O}(r) R_{n,\ell}(r)$$

$$\equiv \langle \ell, m \mid \ell, m \rangle \langle n, \ell \mid \hat{O}(r) \mid n, \ell \rangle \qquad (18.28)$$

Using the normalization condition for spherical harmonics (Eq. 18.24) simplifies Eq. 18.28 to

$$\langle O(r) \rangle_{n,\ell,m} = \int_0^\infty r^2 R_{n,\ell}(r)^* \hat{O}(r) R_{n,\ell}(r) \, dr$$

$$\equiv \langle n, \ell \mid \hat{O}(r) \mid n, \ell \rangle \qquad (18.29)$$

If the operator $\hat{O}(r)$ depends on the coordinate r but not on the radial momentum, then $\hat{O}(r)R_{n,\ell}(r) = R_{n,\ell}(r)O(r)$, where $O(r)$ is the appropriate function of r. This provides a further simplification of Eq. 18.29:

$$\langle O(r)\rangle_{n,\ell,m} = \int_0^\infty dr\, r^2 \left|R_{n,\ell}(r)\right|^2 O(r)$$

$$\equiv \int_0^\infty dr\, \Pi_{n,\ell}(r)O(r) \tag{18.30}$$

The quantity

$$\Pi_{n,\ell}(r)dr \equiv r^2 R_{n,\ell}(r)^* R_{n,\ell}(r)\, dr$$

$$\equiv r^2 \left|R_{n,\ell}(r)\right|^2 dr \tag{18.31}$$

is the probability that the electron is located between a sphere of radius r and a sphere of radius $r + dr$. This follows from the general definition of an average (the average of $f(x)$ is $\int f(x)p(x)dx$ where $p(x)dx$ is the probability that x takes values between x and $x + dx$).

The presence of r^2 in Eq. 18.31 may puzzle you, but it should not. The probability of finding a particle in an infinitesimal region is proportional to the volume of that region. In our case, the region is confined between a sphere of radius r and a sphere of radius $r + dr$ and its volume is proportional to r^2.

Eqs 18.30 and 18.31 will be used to calculate the average values of r, r^2, $V(r)$, $e^2 Z/4\pi\varepsilon_0 r^2$, and $\hbar^2\ell(\ell + 1)/2\mu r^2$. However, these equations cannot be used to calculate the mean radial kinetic energy K_r, which is represented by the operator

$$\hat{K}_r = -\frac{\hbar^2}{2\mu r^2}\frac{\partial}{\partial r}r^2\frac{\partial}{\partial r}$$

To convince yourself that this is the radial kinetic energy, take r to be very large in the radial Schrödinger equation (Eq. 18.19) and make the angular momentum zero (i.e. take $\ell = 0$). The radial Schrödinger equation will become $\hat{K}_r R = ER$. Since the electron no longer interacts with the nucleus (r being very large), the term $\hat{K}_r$ must represent the radial kinetic energy.

Since $\hat{K}_r R_{n,\ell} \neq R_{n,\ell} \hat{K}_r$, we cannot use Eq. 18.30 to calculate the mean values of K_r; for that we must use Eq. 18.27.

Keep in mind that the total kinetic energy operator is $\hat{K} + \hbar^2 \ell(\ell+1)/2\mu r^2$ since this term originates from changing

$$-\frac{\hbar^2}{2\mu}\left[\frac{\partial^2}{\partial x^2} + \frac{\partial^2}{\partial y^2} + \frac{\partial^2}{\partial z^2}\right]$$

to spherical coordinates.

§12. *The Functions $R_{n,\ell}(r)$.* To calculate the mean values defined in §11, we need to use the radial wave functions $R_{n,\ell}(r,\theta,\phi)$ given by Eqs 18.11–18.15. Since most of us do not know what a generalized Laguerre polynomial looks like, Table 18.1 gives the expressions for $\Phi_{n,\ell}(\rho)$, for a few pairs of values of n and ℓ.

Workbook

Table 18.1 Expressions for $\Phi_{n,\ell}(\rho)$, calculated in Workbook QM18.9, from Eqs 18.12–18.15, for a few values of n and ℓ allowed by Eq. 18.13. Here $\rho = r/a$ and a is defined by Eq. 18.15.

n	ℓ	$\Phi_{n,\ell}(\rho)$
1	0	$2\exp[-\rho]$
2	0	$-\dfrac{\exp[-\rho/2](\rho-2)}{2\sqrt{2}}$
2	1	$\dfrac{\exp[-\rho/2]\rho}{2\sqrt{6}}$
3	0	$\dfrac{2\exp[-\rho/3](2\rho^2 - 18\rho + 27)}{81\sqrt{3}}$
3	1	$-\dfrac{2}{81}\sqrt{\dfrac{2}{3}}\exp[-\rho/3]\rho(\rho+6)$
3	2	$\dfrac{2}{81}\sqrt{\dfrac{2}{15}}\exp[-\rho/3]\rho^2$
4	0	$-\dfrac{1}{768}\exp[-\rho/4](\rho^3 - 24\rho^2 + 144\rho - 192)$
4	1	$\dfrac{\exp[-\rho/4]\rho(\rho^2 - 20\rho + 80)}{256\sqrt{15}}$
4	2	$-\dfrac{\exp[-\rho/4]\rho^2(\rho+12)}{768\sqrt{5}}$
4	3	$\dfrac{\exp[-\rho/4]\rho^3}{768\sqrt{35}}$

$\Phi_{n,\ell}(\rho)$ is a polynomial of order $n - 1$, multiplied by $\exp[-2Z\rho/n]$. The wave function decays as the electron–nucleus r increases and is close to zero if

$$\frac{Zr}{an} \gg 1$$

In the state $R_{n,\ell}$ the electron lives in a spherical courtyard of radius $r \cong an/Z$, centered on the center of mass of the two-particle system (which is very close to the position of the nucleus). If the number of protons Z is large, the courtyard is small because a more highly charged nucleus attracts the electron more strongly. If the electron has a large energy (n is large), the courtyard is large, because an energetic electron can move further away from the nucleus. It is comforting to see that common sense works in quantum mechanics, as well as in classical physics!

§13. *Plots of $R_{n,\ell}(r)$.* In Figs 18.1–18.3, I show plots of $R_{n,\ell}(r)$, for the hydrogen atom, versus r. In these figures, r has units of Å and $R_{n,\ell}(r)$ has units of Å$^{-3/2}$.

The wave function has no physical significance. I plot it because it participates in various integrals used to compute observables. It is useful to know what it looks like, if we want to understand the values of the integrals.

It is rather striking that the wave function depends so strongly on ℓ, while the energy depends only on the quantum number n. For the diatomic molecule, in the harmonic and rigid-rotor approximations, the radial function did not depend on ℓ at all. Why this difference? In the case of the diatomic, the distance between the two particles varied very little, because the bond is very rigid. You can see from the plots of the radial probability $\Pi_{n,\ell}(r) = r^2 R_{n,\ell}(r)^2$ that this is not the case for the hydrogen atom: the electron can be found at a variety of distances from the nucleus. This means that we cannot replace $\hbar^2 \ell(\ell + 1)/(2\mu r^2)$ with the constant $\hbar^2 \ell(\ell + 1)/(2\mu r_0^2)$, as we did for the diatomic molecule. Because of this, the radial (vibrational) motion is coupled with the rotational one and the radial wave function depends on ℓ. Moreover, we can no longer decompose the total energy into a vibrational and a rotational part.

The dependence of the radial distribution function $\Pi_{n,\ell}(r) = r^2 \left| R_{n,\ell}(r) \right|^2$ on r is not uniform: there are peaks at certain distances indicating that the electron prefers being located in certain shells around the nucleus. Classical physics cannot generate such a behavior, which is related to the "wave nature" of the electron. We have seen a similar behavior for the particle in a box, and for the vibrational wave function of a diatomic molecule.

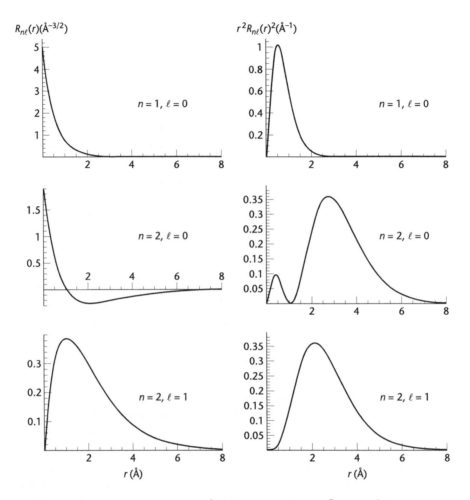

Figure 18.1 Plots of $R_{n,\ell}(r)$ (units $\text{Å}^{-3/2}$) and of $r^2 R_{n,\ell}(r)^2$ (units Å^{-1}) as a function of r (in Å). The values of n and ℓ are shown on the graphs.

§14. *The Mean Values of r, r^2, Coulomb Energy, Centrifugal Energy, and Radial Kinetic Energy.* To understand how electron location depends on the mass of the nucleus, the number of protons in the nucleus, and the quantum numbers n and ℓ, I calculate the mean distance of the electron from the nucleus:

$$\langle r \rangle_{n,\ell} \equiv \langle n, \ell \mid r \mid n, \ell \rangle \tag{18.32}$$

$$= \int_0^\infty r \Pi_{n,\ell}(r)\,dr \tag{18.33}$$

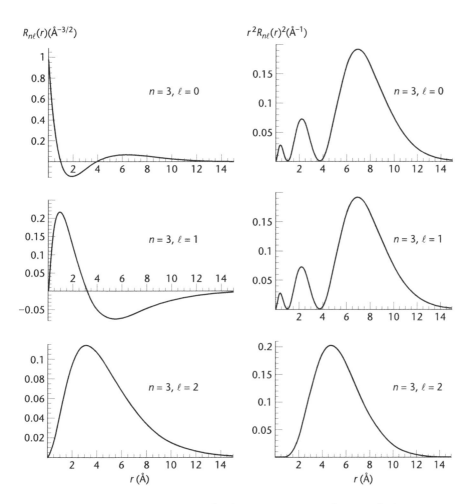

Figure 18.2 Plots of $R_{n,\ell}(r)$ (units $\mathring{A}^{-3/2}$) and of $r^2 R_{n,\ell}(r)^2$ (units $\mathring{A}^{-1}$) as a function of r (in $\mathring{A}$). The values of n and ℓ are shown on the graphs.

$$= \int_0^\infty r^3 R_{n,\ell}(r)^2 dr \qquad (18.34)$$

$$= a \left(\frac{2Z}{n}\right)^3 \int_0^\infty \rho^3 \Phi_{n,\ell}(\rho) d\rho \qquad (18.35)$$

$$= \frac{n^2 a}{Z} \left(1 + \frac{1}{2}\left[1 - \frac{\ell(\ell+1)}{n^2}\right]\right) \qquad (18.36)$$

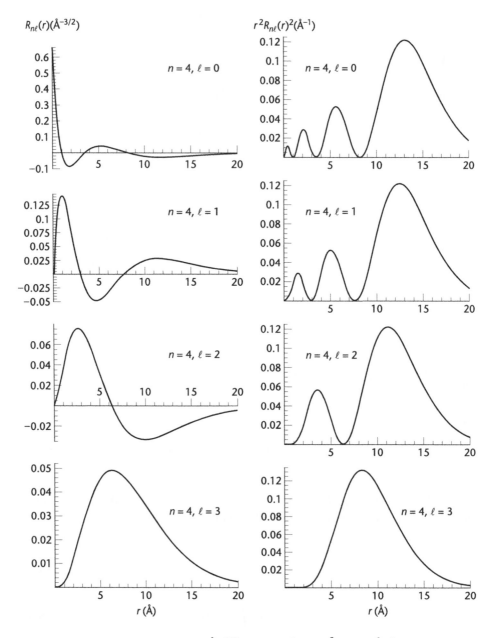

Figure 18.3 Plots of $R_{n,\ell}(r)$ (units $\text{Å}^{-3/2}$) and of $r^2 R_{n,\ell}(r)^2$ (units Å^{-1}) as a function of r (in Å). The values of n and ℓ are shown on the graphs.

Using the dimensionless variable ρ highlights the fact that $\langle r \rangle$ is the length scale a multiplied by a dimensionless factor that depends on n, ℓ, and Z.

The notation $\langle O \rangle$ is often used for the average value of the quantity O. The value of $\langle O \rangle$ depends on the state of the system, but this notation does not tell us what the state is; we are left to deduce it from the context. Here I use the notation $\langle O \rangle_{n,\ell}$ to indicate the quantum numbers of the state used to calculate the average. The symbol $\langle n, \ell \,|\, r \,|\, n, \ell \rangle$ is more explicit and it is often used to denote various integrals. All of these symbols are used frequently in books or articles on quantum mechanics and you need to know them.

Eq. 18.34 follows from Eq. 18.33 and Eq. 18.31, which gives $\Pi_{n,\ell}$. I went from Eq. 18.34 to Eq. 18.35 by changing the integration variable from r to $\rho = r/a$ and by using Eq. 18.11 for $R_{n,\ell}(r)$. Eq. 18.36 was obtained by performing the integral (this was done in I. Waller, Z. f. Phys. **38**, 635 (1926)).

Exercise 18.10

Performing the integral in Eq. 18.36 is rather complicated. **Mathematica** can perform it only if we give numerical values to n and ℓ. Use **Mathematica** or **Mathcad** to perform the integral in Eq. 18.35 for several values of n and ℓ, and check whether the results agree with Eq. 18.36.

Eq. 18.36 shows that $\langle r \rangle_{n,\ell}$ increases with the energy of the electron (i.e. increases with n), decreases with the number of protons in the nucleus, depends on the mass of nucleus (through a which depends on the reduced mass μ), and decreases with the angular momentum squared (whose magnitude is given by ℓ). This last observation is hard to understand, since I would expect intuitively that the electron is further from the nucleus when it has a higher rotational energy. However, if you look at Fig. 18.2 you see that the probability that the electron is far from the nucleus is smaller for $n = 3$ and $\ell = 2$ than it is for $n = 3$ and $\ell = 0$; this is consistent with the expression we found for $\langle r \rangle$.

I have also calculated

- the mean radial kinetic energy

$$\langle K \rangle_{n,\ell} \equiv \langle n, \ell \,|\, -\frac{\hbar^2}{2\mu r^2}\frac{\partial}{\partial r}r^2\frac{\partial}{\partial r} \,|\, n, \ell \rangle$$

$$= \int_0^\infty dr\, r^2 R_{n,\ell}(r)^* \left[-\frac{\hbar^2}{2\mu r^2}\frac{\partial}{\partial r}r^2\frac{\partial R_{n,\ell}(r)}{\partial r} \right] \qquad (18.37)$$

- the mean rotational energy

$$\langle E_r \rangle \equiv \langle n, \ell \mid -\frac{\hbar^2 \ell(\ell+1)}{2\mu r^2} \mid n, \ell \rangle$$

$$= \int_0^\infty dr r^2 R_{n,\ell}(r)^* \left[-\frac{\hbar^2 \ell(\ell+1)}{2\mu r^2} \right] R_{n,\ell}(r)$$

$$= \frac{\hbar^2}{2\mu} \frac{Z^2 \ell(\ell+1)}{a^2 n^3 (\ell+\frac{1}{2})} \tag{18.38}$$

- and the mean Coulomb interaction energy

$$\langle C \rangle \equiv \langle n, \ell \mid -\frac{Ze^2}{4\pi\varepsilon_0 r} \mid n, \ell \rangle$$

$$= \int_0^\infty dr\, r^2 R_{n,\ell}(r)^* \left[-\frac{Ze^2}{4\pi\varepsilon_0 r} \right] R_{n,\ell}(r)$$

$$= -\frac{Ze^2}{4\pi\epsilon_0} \frac{Z}{an^2}$$

$$= -\frac{Z^2 \hbar^2}{2a^2 \mu n^2} \tag{18.39}$$

In the final step here, I used the definition of a given in Eq. 18.15.

These quantities were calculated in Workbook QM18.7 and some of the results are shown in Table 18.2. A few details are worth noting.

All energies are proportional to $\varepsilon = Z^2 e^2 / \mu a^2$, as one would deduce from dimensional analysis.

The radial kinetic energy $\langle K \rangle_{n,\ell}$ is positive and this means that it destabilizes the atom. If we compare $\langle K \rangle_{1,0}$ to $\langle K \rangle_{2,0}$ and to $\langle K \rangle_{3,0}$, we see that the radial kinetic energy becomes smaller as n increases. This is in agreement with a basic quantum mechanical rule that the more delocalized the electron, the smaller its kinetic energy.

Another interesting result is that the mean Coulomb interaction is equal to the total energy divided by 2. This is also true in classical mechanics, for a two-body system interacting through a potential inverse proportional to r (Coulomb or gravity). The total kinetic energy, which is the total energy E_n minus the interaction energy

Table 18.2 Equations for the mean electron–nucleus distance $\langle r \rangle$ (Eq. 18.36), the mean radial kinetic energy $\langle K \rangle$ (Eq. 18.37), the main radial energy $\langle E_r \rangle$ (Eq. 18.38), the mean Coulomb energy $\langle C \rangle$ (Eq. 18.39), and the total energy E_n (Eqs 18.8 and 18.9 combined with Eq. 18.15 for a).

n	ℓ	$\langle r \rangle_{n,\ell}$	$\langle K \rangle_{n,\ell}$	$\langle E_r \rangle_{n,\ell}$	$\langle C \rangle_n$	E_n
1	0	$\dfrac{3a}{2Z}$	$\dfrac{Z^2\hbar^2}{2a^2\mu}$	0	$-\dfrac{Z^2\hbar^2}{a^2\mu}$	$-\dfrac{Z^2\hbar^2}{2a^2\mu}$
2	0	$\dfrac{6a}{Z}$	$\dfrac{Z^2\hbar^2}{8a^2\mu}$	0	$-\dfrac{Z^2\hbar^2}{4a^2\mu}$	$-\dfrac{Z^2\hbar^2}{8a^2\mu}$
2	1	$\dfrac{5a}{Z}$	$\dfrac{Z^2\hbar^2}{24a^2\mu}$	$\dfrac{Z^2\hbar^2}{12a^2\mu}$	$-\dfrac{Z^2\hbar^2}{4a^2\mu}$	$-\dfrac{Z^2\hbar^2}{8a^2\mu}$
3	0	$\dfrac{27a}{2Z}$	$\dfrac{Z^2\hbar^2}{18a^2\mu}$	0	$-\dfrac{Z^2\hbar^2}{9a^2\mu}$	$-\dfrac{Z^2\hbar^2}{18a^2\mu}$
3	1	$\dfrac{25a}{2Z}$	$\dfrac{5Z^2\hbar^2}{162a^2\mu}$	$\dfrac{2Z^2\hbar^2}{81a^2\mu}$	$-\dfrac{Z^2\hbar^2}{9a^2\mu}$	$-\dfrac{Z^2\hbar^2}{18a^2\mu}$
3	2	$\dfrac{21a}{2Z}$	$\dfrac{Z^2\hbar^2}{90a^2\mu}$	$\dfrac{2Z^2\hbar^2}{45a^2\mu}$	$-\dfrac{Z^2\hbar^2}{9a^2\mu}$	$-\dfrac{Z^2\hbar^2}{18a^2\mu}$

(i.e. the Coulomb energy), is then equal to $-E_n/2$. These are surprising regularities, which have esthetic value but no practical consequences.

Exercise 18.11

Calculate the numerical values of $\langle K \rangle$, $\langle E_r \rangle$, and $\langle C \rangle$ for $n = 1$, $\ell = 0$ and for $n = 2$, $\ell = 0, 1$. Give results in eV. Check if $\langle K \rangle + \langle E_r \rangle + \langle C \rangle$ is equal to the energy E_n given by Eq. 18.8. Prove that this must always be the case.

The Angular Dependence of the Wave Function

§15. *Introduction.* So far we have discussed the radial wave function $R_{n,\ell}(r)$. In this section we study the angular dependence of the total wave function

$$\psi_{n,\ell,m}(r,\theta,\phi) = R_{n,\ell}(r)\,Y_\ell^m(\theta,\phi) \tag{18.40}$$

Remember that we started our study of the two-particle, one-electron system by picking a Cartesian coordinate system whose origin is in the center of mass of the system. The coordinate system moves with the center of mass but does not rotate with the electron. The system is described by the vector $\mathbf{r}$ whose Cartesian coordinates are $\{x, y, z\}$ and whose spherical coordinates are $\{r, \theta, \phi\}$.

r is the electron–nucleus distance, and θ and ϕ describe the orientation of the axis connecting the two particles.

The quantum numbers n, ℓ, and m tell us that the total energy is E_n, the square of the angular momentum operator satisfies

$$\hat{L}^2 \psi_{n,\ell,m} = \hbar^2 \ell(\ell + 1)\psi_{n,\ell,m}$$

and the projection of the angular momentum on the OZ axis satisfies

$$\hat{L}_z \psi_{n,\ell,m} = \hbar m \psi_{n,\ell,m}$$

Note that the wave function is degenerate. For a given value of n, the energy is E_n and all wave functions $\psi_{n,\ell,m}$ with $\ell = 0, 1, \ldots n-1$ and $m = -\ell, -\ell+1, \ldots, \ell-1$, ℓ are eigenstates of the energy operator having the same energy. For example, if $n = 2$, then ℓ can be 0 or 1. For $\ell = 0$, $m = 0$, and for $\ell = 1$, m can be -1, 0, or 1. The energy of an atom is the same (i.e. equal to E_n) in any of the states $\psi_{2,0,0}$, $\psi_{2,1,-1}$, $\psi_{2,1,0}$, and $\psi_{2,1,1}$. The degenerate functions differ through the magnitude of the angular momentum squared and/or through its projection on the OZ axis.

The orientation of the Cartesian coordinate system is arbitrary. If the atom is not acted upon by a force, there is no special direction in space and any orientation of the OZ axis is as good as any other. Therefore there is no reason why the atom would prefer one direction of the angular momentum over another direction. In the ground state $\psi_{1,0,0}$ this is not an issue since the angular momentum is zero. Consider next the case of the states $\psi_{2,0,0}$, $\psi_{2,1,-1}$, $\psi_{2,1,0}$, and $\psi_{2,1,1}$. They all have the energy E_2. In the state $\psi_{2,0,0}$, the angular momentum is zero and there is no orientation to worry about. But what about the states $\psi_{2,1,-1}$, $\psi_{2,1,0}$, and $\psi_{2,1,1}$? What is the orientation of the angular momentum vector? It depends on how the atom was forced (excited) into these states. The old-fashioned way was to use an electric discharge or the unpolarized light from an electron-discharge source. In the first case, the atoms were excited by electrons and ions traveling through the gas. Since the electrons traveled in all directions, there is no preferred direction in space. The atoms are excited with equal probability into $\psi_{2,1,-1}$, $\psi_{2,1,0}$, and $\psi_{2,1,1}$; one-third of the atoms excited to have $n = 2$ and $\ell = 1$ will have $m = -1$, one-third will have $m = 0$, and one-third will have $m = 1$. The same is true in the case of unpolarized light: the orientation of the electric field of the light can have any direction; no direction is prefered and each of the values $m = -1, 0, 1$ is equally probably, when the excited state has $\ell = 1$.

Consider now excitation by the polarized light produced by a laser. The electric fields of all the photons have the same direction and now we do have a special direction in space. We can take the OZ axis along the electric field of light. It is now meaningful to ask: what is the magnitude of the projection of the angular momentum on the direction of the electric field? As you will see in the section "Hydrogen atom: absorption and emission spectroscopy" on p. 325, the absorption of a photon will create an excited state with a well defined value of ℓ and m, which depends on the way the light is polarized. The "orbit" of the electron in such a state has a certain orientation which will manifest itself in a preference for emitting a photon in certain directions. Therefore, understanding the orientational information provided by the dependence of $\psi_{n,\ell,m}(r,\theta,\phi)$ on θ and ϕ is of some importance.

However, a more compelling reason for understanding the dependence of $\psi_{n,\ell,m}(r,\theta,\phi)$ on θ and ϕ comes from the use of these functions in the theory of chemical bonding. In that theory, the wave function of an electron in a molecule (a *molecular orbital*) is constructed as a combination of the atomic electronic wave functions (*atomic orbitals*). Since chemical bonds have preferred directions, it is easier to understand them if we construct the molecular orbitals by using atomic orbitals that have well understood spatial orientation. The $2p_x$, $2p_y$, $2p_z$, $3d_{xy}$, etc., atomic orbitals that you have encountered in general chemistry are specific combinations of the wave functions $\psi_{n,\ell,m}$. You will learn here what these combinations are.

A further reason for creating the orbitals $2p_x$, $2p_y$, ... is that they are functions whose values are real numbers; the functions $\psi_{n,\ell,m}(r,\theta,\phi)$ take complex values and are therefore slightly more cumbersome to use on a computer or in analytical work; all other things being equal, real functions are preferable.

§16. *Some Nomenclature.* Spectroscopists have introduced the following nomenclature: the states with $\ell = 0$ are called s-states, those with $\ell = 1$ are p-states, those with $\ell = 2$ are d-states, those with $\ell = 3$ are f-states, those with $\ell = 4$ are g-states, those with $\ell = 5$ are h-states. Furthermore, X-ray spectroscopists call the states with $n = 0$, K states (or they say that an electron in such a state is in the K shell), those with $n = 2$ are L-states, those with $n = 3$ are M-states, those with $n = 4$ are N-states, those with $n = 5$ are O-states. Thus $\psi_{2,1,0}$, $\psi_{2,1,-1}$, $\psi_{2,1,1}$ are all L_p states, or p-states in the L shell.

It is rather hard to defend the urge of scientists to invent unreasonable nomenclature and bizarre units. I find no reason for using K, L, ... and s, p, ... when specifying n, ℓ, and m will do, but these names are too entrenched to be abandoned.

§17. *The s-states.* When $\ell = 0$ (an s-state), m must be equal to zero and the wave function is

$$\psi_{n,0,0}(r,\theta,\phi) = R_{n,0}(r)Y_0^0(\theta,\phi)$$

Eqs 18.11–18.15 give the general formula for $R_{n,\ell}(r)$. Specific examples (for particular values of n and ℓ) are given in Table 18.1. Plots of $R_{n,\ell}(r)$ are shown in Figs 18.1–18.3.

The spherical harmonics $Y_\ell^m(\theta, \phi)$ were described in Chapter 14 (Eq. 14.34 gives the general formula, and specific cases are displayed in Eqs 14.35–14.59). Using this information leads to

$$Y_0^0(\theta, \phi) = \frac{1}{2\sqrt{\pi}} \tag{18.41}$$

This spherical harmonic is independent of θ and ϕ.

The s-wave functions are often denoted by 1s, 2s, ..., or, in general, ns. For hydrogen we have

$$\psi_{n,0,0} \equiv ns(r, \theta, \phi) = \frac{R_{n,0}(r)}{2\sqrt{\pi}} \tag{18.42}$$

For example (use Table 18.1 and Eqs 18.11 and 18.42)

$$2s(r, \theta, \phi) = -\left(\frac{2}{2a}\right)^{3/2} \frac{\exp[-ra/2]\{(r/a) - 2\}}{2\sqrt{2}} \frac{1}{2\sqrt{\pi}} \tag{18.43}$$

Exercise 18.12

Write the formula for 1s and 3s wave functions.

Since $ns(r, \theta, \phi)$ does not depend on θ and ϕ, for a given value of r the function is a sphere (if r is fixed, ns has the same value for all possible values of θ and ϕ).

§18. *The np Orbitals.* There are three wave functions with $\ell = 1$, corresponding to $m = -1, 0, 1$. They are $R_{n,1}(r)Y_1^{-1}(\theta, \phi)$, $R_{n,1}(r)Y_1^0(\theta, \phi)$, and $R_{n,1}(r)Y_1^1(\theta, \phi)$. Two of these are complex-valued functions (use Eqs 14.36–14.38):

$$\psi_{n,1,-1} = R_{n,1}(r)Y_1^{-1}(\theta, \phi) = \frac{1}{2}e^{-i\phi}\sqrt{\frac{3}{2\pi}}\sin(\theta)R_{n,1}(r) \tag{18.44}$$

$$\psi_{n,1,0} = R_{n,1}(r)Y_1^0(\theta, \phi) = \frac{1}{2}\sqrt{\frac{3}{\pi}}\cos(\theta)R_{n,1}(r) \tag{18.45}$$

$$\psi_{n,1,1} = R_{n,1}(r)Y_1^1(\theta, \phi) = -\frac{1}{2}e^{i\phi}\sqrt{\frac{3}{2\pi}}\sin(\theta)R_{n,1}(r) \tag{18.46}$$

These are eigenfunctions of the Hamiltonian, of $\hat{L}^2$, and of $\hat{L}_z$. We can combine them to form the following real wave functions (i.e. their values are real numbers):

$$np_x(r, \theta, \phi) \equiv R_{n,1}(r)\frac{1}{\sqrt{2}}\left[Y_1^{-1} - Y_1^1\right]$$

$$= R_{n,1}(r)\frac{1}{2}\sqrt{\frac{3}{\pi}}\sin(\theta)\cos(\phi) \tag{18.47}$$

$$np_y(r, \theta, \phi) \equiv R_{n,1}(r)\frac{i}{\sqrt{2}}\left[Y_1^1 + Y_1^{-1}\right]$$

$$= R_{n,1}(r)\frac{1}{2}\sqrt{\frac{3}{\pi}}\sin(\theta)\sin(\phi) \tag{18.48}$$

$$np_z(r, \theta, \phi) \equiv R_{n,1}(r)Y_1^0 = R_{n,1}(r)\frac{1}{2}\sqrt{\frac{3}{\pi}}\cos(\theta) \tag{18.49}$$

Very often these expressions are converted to Cartesian coordinates. To do that, we use

$$\cos(\theta) = \frac{z}{r}$$

$$\sin(\theta) = \frac{\sqrt{x^2 + y^2}}{r}$$

$$\sin(\phi) = \frac{y}{\sqrt{x^2 + y^2}}$$

$$\cos(\phi) = \frac{x}{\sqrt{x^2 + y^2}}$$

One can obtain these relationships by using trigonometry and Fig. 13.2. Simple algebra then changes Eqs 18.47–18.49 to

$$np_x(x, y, z) = R_{n,1}(r)\sqrt{\frac{3}{\pi}}\frac{x}{2r} \tag{18.50}$$

$$np_y(x, y, z) = R_{n,1}(r)\sqrt{\frac{3}{\pi}}\frac{y}{2r} \tag{18.51}$$

$$np_z(x, y, z) = R_{n,1}(r)\sqrt{\frac{3}{\pi}}\frac{z}{2r} \tag{18.52}$$

You can now see where the names of these functions come from.

For applications to the theory of chemical bonding, it is useful to know what these functions look like. Since it is not possible to plot functions of three variables, I show in Fig. 18.4 contour plots of $np_x(x, y, z = 0)$. This tells us what these functions look like in the XOY plane (i.e. $z = 0$). The plots were made in Cell 4 of Workbook QM18.9. You can use the program in that cell to make plots for other cuts (e.g. for $np_x(x = 0, y, z)$).

It is not difficult to gain a rough understanding of the shapes seen in Fig. 18.4. $R_{n,1}(r)$ decays exponentially but not monotonically; depending on the value of n,

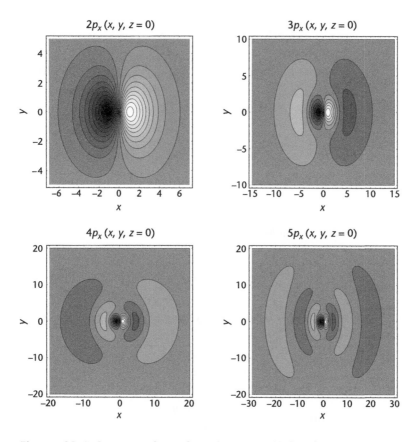

Figure 18.4 Contour plots of $np_x(x, y, z = 0)$ for the hydrogen atom, for $n = 2, \ldots, 5$. The plots were made in Workbook QM18.9, Cell 4. The functions take positive values in the regions where the color is lighter than the background, and negative values where the color is darker.

Table 18.3 Atomic orbitals in spherical and Cartesian coordinates. The last column gives the definitions in terms of spherical harmonics.

Orbital	Spherical coordinates	Cartesian coordinates	Definition
np_x	$\dfrac{1}{2}\sqrt{\dfrac{3}{\pi}}\cos(\phi)\sin(\theta)R_{n,1}(r)$	$\sqrt{\dfrac{3}{\pi}}\dfrac{x}{2r}R_{n,1}(r)$	$R_{n,1}(r)\dfrac{1}{\sqrt{2}}\left[Y_1^{-1}-Y_1^{1}\right]$
np_y	$\dfrac{1}{2}\sqrt{\dfrac{3}{\pi}}\sin(\phi)\sin(\theta)R_{n,1}(r)$	$\sqrt{\dfrac{3}{\pi}}\dfrac{y}{2r}R_{n,1}(r)$	$R_{n,1}(r)\dfrac{i}{\sqrt{2}}\left[Y_1^{1}+Y_1^{-1}\right]$
np_z	$\dfrac{1}{2}\sqrt{\dfrac{3}{\pi}}\cos(\theta)R_{n,1}(r)$	$\sqrt{\dfrac{3}{\pi}}\dfrac{z}{2r}R_{n,1}(r)$	$R_{n,1}Y_1^{0}$
nd_{xy}	$\dfrac{1}{4}\sqrt{\dfrac{15}{\pi}}\sin(2\phi)\sin(\theta)^2R_{n,2}(r)$	$\sqrt{\dfrac{15}{\pi}}\dfrac{xy}{2r^2}R_{n,2}(r)$	$R_{n,2}(r)\dfrac{1}{i\sqrt{2}}\left[Y_2^{2}-Y_2^{-2}\right]$
nd_{yz}	$\dfrac{1}{2}\sqrt{\dfrac{15}{\pi}}\sin(\phi)\cos(\theta)\sin(\theta)R_{n,2}(r)$	$\sqrt{\dfrac{15}{\pi}}\dfrac{yz}{2r^2}R_{n,2}(r)$	$R_{n,2}(r)\dfrac{i}{\sqrt{2}}\left[Y_2^{1}+Y_2^{-1}\right]$
nd_{xz}	$\dfrac{1}{2}\sqrt{\dfrac{15}{\pi}}\cos(\phi)\cos(\theta)\sin(\theta)R_{n,2}(r)$	$\sqrt{\dfrac{15}{\pi}}\dfrac{xz}{2r^2}R_{n,2}(r)$	$R_{n,2}(r)\dfrac{1}{\sqrt{2}}\left[Y_2^{-1}+Y_2^{1}\right]$
$nd_{x^2-y^2}$	$\dfrac{1}{4}\sqrt{\dfrac{15}{\pi}}\cos(2\phi)\sin(\theta)^2R_{n,2}(r)$	$\sqrt{\dfrac{15}{\pi}}\dfrac{x^2-y^2}{4r^2}R_{n,2}(r)$	$R_{n,2}(r)\dfrac{1}{\sqrt{2}}\left[Y_2^{2}+Y_2^{-2}\right]$
nd_{z^2}	$\dfrac{1}{4}\sqrt{\dfrac{5}{\pi}}\left(3\cos(\theta)^2-1\right)R_{n,2}(r)$	$\dfrac{1}{4}\sqrt{\dfrac{5}{\pi}}\left(\dfrac{3z^2}{r^2}-1\right)R_{n,2}(r)$	$R_{n,2}(r)Y_2^{0}$

the function oscillates (and decays) as r increases. By multiplying it by x we make it larger along the x axis, and it will stretch further in that direction. This is why the np_x functions are elongated in the x direction (note that the y and x scales in the figures are not equal). Since the function np_x does not change when we interchange y and z, the plot of $np_x(x, y = 0, z)$ will look identical to the plot of $np_x(x, y, z = 0)$.

There are two more np functions, called np_y and np_z. Their definition and expression in spherical and Cartesian coordinates are collected in Table 18.3. Their properties are very similar to those of np_x, except that the OY (for np_y) or OZ (for np_z) axis plays the role that the OX axis plays for np_x.

§19. *The nd Orbitals.* We can continue using this strategy to obtain real orbitals with $n \geq 3$ and $\ell = 2$ (d-states). In this case m can take the values $m = -2, -1, 0, 1, 2$. The functions $R_{n,2}Y_2^{-2}$, $R_{n,2}Y_2^{-1}$, $R_{n,2}Y_2^{0}$, $R_{n,2}Y_2^{1}$, and $R_{n,2}Y_2^{2}$ are degenerate. All of them are eigenfunctions of the Hamiltonian, the square of the angular momentum $\hat{L}^2$, and the projection $\hat{L}_z$ of the angular momentum on the OZ axis. They correspond to the same eigenenergy E_n and the same eigenvalue of

$\hat{L}^2$ (namely $\hbar^2 \ell(\ell + 1)$), but differ through the eigenvalues of $\hat{L}_z$ (which are $-2\hbar$, $-\hbar$, 0, $\hbar$, $2\hbar$, respectively).

The manner in which these functions are combined to produce five real orbitals is shown in Table 18.3. These expressions were derived in Cells 7–11 of Workbook QM18.9. The names nd_{xy}, nd_{yz}, nd_{xz}, $\mathrm{nd}_{x^2-y^2}$, and nd_{z^2} remind us how the functions depend on x, y, and z.

The contour plots of nd_{xy} orbitals for $n = 3$, 4, 5, and 6, in the XOY plane (e.g. $3\mathrm{d}_{xy}(x, y, z = 0)$) are shown in Figs 18.5 and 18.6. All these orbitals are normalized and orthogonal to each other.

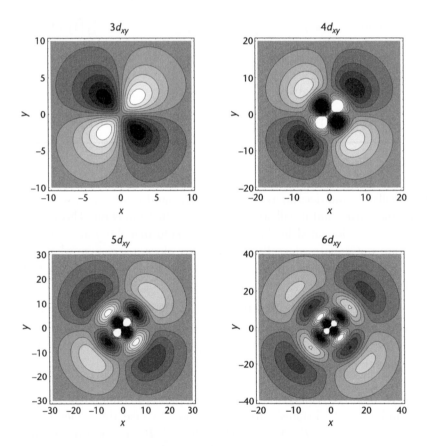

Figure 18.5 Contour plots of $\mathrm{nd}_{xy}(x, y, z = 0)$, for $n = 3, 4, 5, 6$, for the hydrogen atom. The functions are positive in the regions where the graph's color is lighter than the background, and negative where the color is darker.

Printed and bound by CPI Group (UK) Ltd, Croydon, CR0 4YY

17/10/2024

01775697-0001

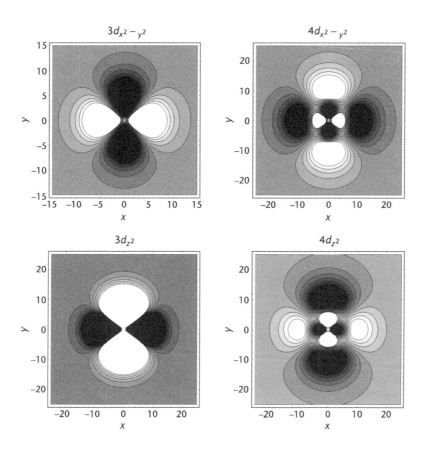

Figure 18.6 The top two figures show $nd_{x^2-y^2}(x, y, z = 0)$ and the bottom two show $nd_{z^2}(x, y, z = 0)$, for the hydrogen atom. The functions are positive in the regions where the color is lighter than the background and negative where the color is darker. The plots were made in Workbook QM18.9.

Workbook

Exercise 18.13

Show that $2p_x$, $3d_{x^2-y^2}$, and $3d_{xy}$ are normalized and orthogonal to each other.

Hydrogen Atom: Absorption and Emission Spectroscopy

§20. *Introduction.* Our study of light absorption and emission by a hydrogen atom uses the theory developed in Chapter 12 and follows closely the pattern that evolved in Chapters 12 and 17. Light absorption by a hydrogen atom causes the excitation of an electron. In this respect our problem is similar to that studied in

Chapter 12, where one electron trapped in a box was excited by light absorption. There are also some similarities to light absorption by a diatomic molecule. These come about because both systems consist of two particles interacting through a central force, and light absorption is strongly influenced by the properties of the angular momentum eigenstate (the spherical harmonics). The differences between the spectroscopy of a hydrogenoid atom and a diatomic molecule exists because the diatomic is not truly a two-body system. Such a molecule has many electrons whose existence leads to the potential energy of the interaction between the atoms. This is so different from the Coulomb interactions that the eigenstates and the energies of the two-atom system differ substantially from those of an electron-nucleus system. In addition, the coupling between the molecule and the light is controlled by the dipole moment of the molecule, not by direct interaction of the light with the atoms. For the hydrogenoid atoms, the light interacts directly with the electron and the nucleus.

As you learned in Chapter 11, §2, the probability that the hydrogen atom absorbs a photon when it is exposed to a pulse of light is

$$P_{i \to f} = \frac{\pi}{\varepsilon_0 c \hbar^2} \left| \langle \psi_f | \hat{\mathbf{m}} \cdot \mathbf{e} | \psi_i \rangle \right|^2 N(\omega_{if}) \tag{18.53}$$

In this formula, I took the line shape function $L(\Omega)$ to be 1 when $\Omega = \omega_{if}$ and zero otherwise.

The dipole moment operator for a hydrogenoid atom is

$$\hat{\mathbf{m}} = q\mathbf{r} - qZ\mathbf{R} \tag{18.54}$$

where q is the electron charge (a negative number), $\mathbf{r}$ is the position of the electron, Z is the number of protons in the nucleus, and $\mathbf{R}$ is the position of the nucleus.

$N(\omega_{if})d\omega$ is the total amount of electromagnetic energy having a frequency between ω and $\omega + d\omega$ that passes through the unit area of a plane perpendicular to the direction of propagation of the light, when the light pulse passes through. $\mathbf{e}$ is the polarization vector of the light, which is a vector of unit length having the same direction as the electric field of the light.

ψ_i and ψ_f are the wave functions of the atom before and after the pulse of light passes through the atom. The transition frequency is

$$\omega_{if} = \frac{E_f - E_i}{\hbar} \tag{18.55}$$

where E_f is the energy of the excited atom and E_i is the energy prior to photon absorption.

In what follows I will use these formulae to calculate the relative intensities of light absorption by a hydrogenoid atom.

Before doing this, I need to warn you that our theory is incomplete. It ignores the effect of electron and nuclear spin and certain relativistic effects. Because of this, our calculations agree with low-resolution experiments, but not with those performed with high accuracy. The latter find two effects that are missing in the theory presented here: some absorption frequencies are slightly shifted from the values predicted by this theory and several peaks appear in some transitions for which this theory predicts only one peak.

It was fortunate that at the time when quantum mechanics was developed, the accuracy of the measurements was too low to detect these effects. As a result, the incomplete theory agreed very well with the experiments. At a time when quantum theory was highly speculative, this agreement was essential in spurring a vigorous development of the theory. Had the experiments been more accurate, they might have delayed the acceptance and the development of the theory.

§21.   *The Transition Dipole.* To evaluate $P_{i \to f}$ we need to calculate the transition dipole $\langle \psi_f \,|\, \hat{\mathbf{m}} \cdot \mathbf{e} \,|\, \psi_i \rangle$. For this we use the electron wave functions (see §10)

$$\psi_{n,\ell,m}(r,\theta,\phi) = R_{n,\ell}(r) Y_\ell^m(\theta,\phi) \tag{18.56}$$

The spherical harmonics $Y_\ell^m(\theta,\phi)$ are given by Eq. 14.34; the specific forms, for various values of m and ℓ are given by Eqs 14.35–14.59 (see also Workbook QM14.2).

The radial wave function $R_{n,\ell}(r)$ is given by Eqs 18.11–18.15; for the forms corresponding to specific values of n and ℓ, see Table 18.1 or Workbook QM18.3.

In the calculations that follow, I will assume that the hydrogenoid atom is in the ground state $\{n = 1, \ell = 0, m = 0\}$ before it interacts with the light. Therefore

$$|\psi_i\rangle \equiv R_{1,0}(r) Y_0^0(\theta,\phi) = \left(\frac{2Z}{a}\right)^{3/2} 2 e^{-r/a} \frac{1}{2\sqrt{\pi}} \tag{18.57}$$

I will further assume that the light is linearly polarized and so I will take the coordinate system to have the OZ direction parallel to the polarization vector $\mathbf{e}$.

This means that

$$\hat{\mathbf{m}} \cdot \mathbf{e} = q\,\hat{\mathbf{R}} \cdot \mathbf{e} - Zq\,\hat{\mathbf{R}} \cdot \mathbf{e} = q\hat{z}\cos(\theta) - Zq\,\hat{\mathbf{R}} \cdot \mathbf{e} \qquad (18.58)$$

The second term in this equation is irrelevant since

$$\langle \psi_f \mid Zq\,\hat{\mathbf{R}} \cdot \mathbf{e} \mid \psi_i \rangle = \int_0^\infty r^2 dr \int_0^\pi \sin(\theta)d\theta \int_0^{2\pi} \psi_f(r,\theta,\phi)^* Zq\hat{\mathbf{R}} \cdot \mathbf{e}\, \psi_i(r,\theta,\phi)d\phi$$

$$= Zq\hat{\mathbf{R}} \cdot \mathbf{e} \int_0^\infty r^2 dr \int_0^\pi \sin(\theta)d\theta \int_0^{2\pi} \psi_f(r,\theta,\phi)^* \psi_i(r,\theta,\phi)d\phi$$

$$= Zq\hat{\mathbf{R}} \cdot \mathbf{e}\langle \psi_f \mid \psi_i \rangle = 0$$

Since the term $Zq\hat{\mathbf{R}} \cdot \mathbf{e}$ is independent of the electron coordinates r, θ, ϕ, I can pull it outside the integral. Then I am left with $\langle \psi_f \mid \psi_i \rangle$, which is zero because ψ_f is orthogonal to ψ_i.

We conclude that the interaction of light with the nucleus does not contribute to the absorption or emission spectrum of H.

The wave functions derived here are for an observer that rides on the center of mass. Therefore we ignore the fact that the atom moves. This motion causes a small shift in the energy of the absorbed photon. The magnitude of this shift is used to measure the velocity of the atom.

§22. *The Selection Rules.* Next we evaluate the transition dipole moment

$$\langle n, \ell, m \mid q\hat{\mathbf{r}} \cdot \mathbf{e} \mid 1,0,0\rangle = q\langle n, \ell, m \mid r\cos(\theta) \mid 1,0,0\rangle \qquad (18.59)$$

for a transition from the state $\{n = 1, \ell = 0, m = 0\}$ to the state $\{n, \ell, m\}$. Using the wave function $\psi_{n,\ell,m}$ (Eq. 18.56) leads to

$$q\langle n, \ell, m \mid r\cos(\theta) \mid 1,0,0\rangle = q\langle n, \ell \mid r \mid 1,0\rangle\langle \ell, m \mid \cos(\theta) \mid 0,0\rangle \qquad (18.60)$$

with

$$\langle n, \ell \mid r \mid 1,0\rangle = \int_0^\infty r^2 dr R_{n,\ell}(r) r R_{1,0}(r) \qquad (18.61)$$

$$\langle \ell, m \mid \cos(\theta) \mid 0,0\rangle = \int_0^\pi \sin(\theta)d\theta \int_0^{2\pi} d\phi Y_\ell^m(\theta,\phi)^* \cos(\theta)Y_0^0(\theta,\phi) \qquad (18.62)$$

The integral over angles is relatively easy to perform. We have (see Eqs 14.35 and 14.37)

$$Y_0^0(\theta, \phi) = \frac{1}{2\sqrt{\pi}}$$

$$\cos(\theta) = 2\sqrt{\frac{\pi}{3}} Y_1^0(\theta, \phi)$$

Therefore

$$\langle \ell, m \mid \cos(\theta) \mid 0, 0 \rangle = \int_0^\pi \sin(\theta)d\theta \int_0^{2\pi} d\phi Y_\ell^m(\theta, \phi)^* \, Y_1^0(\theta, \phi) \, 2\sqrt{\frac{\pi}{3}} \frac{1}{2\sqrt{\pi}}$$

$$= \frac{1}{\sqrt{3}} \langle \ell, m \mid 1, 0 \rangle \qquad (18.63)$$

Since the spherical harmonics are orthonormal,

$$\langle \ell, m \mid 1, 0 \rangle = \delta_{\ell,1}\delta_{m,0}$$

and therefore

$$\langle \ell, m \mid \cos(\theta) \mid 0, 0 \rangle = \frac{1}{\sqrt{3}}\delta_{\ell,1}\delta_{m,0} \qquad (18.64)$$

We have obtained our first selection rule: the absorption of a linearly polarized photon by a hydrogenoid atom in the ground state can take place only to a final state with $\ell = 1$ and $m = 0$.

After photon absorption, the atom is in the state $Y_1^0(\theta, \phi)R_{n,1}(r)$, in a coordinate system whose OZ axis is oriented along the polarization vector of the light. The angular momentum vector $\hat{\mathbf{L}}$ after this excitation has the value $\hbar \times 1(1+1) = 2\hbar$ and its projection on the OZ direction (the direction of linear polarization) is $L_z = 0$. Therefore, the angular momentum vector lies in the XOY plane. A classical interpretation of this result would be that the plane in which the electron orbits is perpendicular to the XOY plane. Of course the story is more complicated in quantum mechanics but the above statement is qualitatively true. Absorption of linearly polarized light prepares H atoms in which electrons orbit with high probability in a plane perpendicular to the polarization vector $\mathbf{e}$ of light. We can test that this is true because these atoms prefer to emit photons in certain directions.

Table 18.4 The matrix elements $\langle n,1 \mid r \mid 1,0 \rangle =$ $\int_0^\infty r^3 R_{n,1}(r)R_{1,0}(r)dr$ calculated in Workbook QM18.10.

n	$\langle n,1 \mid r \mid 1,0 \rangle$
2	$1.29a/Z$
3	$0.517a/Z$
4	$0.305a/Z$
5	$0.209a/Z$
6	$0.155a/Z$
7	$0.121a/Z$
8	$0.0985a/Z$
9	$0.0821a/Z$
10	$0.0698a/Z$

Exercise 18.14

In Chapter 17 §11, Eq. 17.37, I gave you a formula for $\langle \ell_f, m_f \mid \cos(\theta) \mid \ell_i, m_i \rangle$. Use it to determine the selection rules and the matrix elements controlling the emission from $\{n, \ell, m\}$ to $\{n', \ell', m'\}$. Which transitions will be allowed and what are the values of $\langle \ell, m \mid \cos(\theta) \mid \ell', m' \rangle$ for them?

§23. *The Radial Integrals.* The radial integrals, Eq. 18.61, are tedious to calculate analytically but easy to obtain by using a symbolic manipulation program. The results, obtained in Workbook QM18.10, are given in Table 18.4. It is interesting to see that a very large number of transitions are allowed. This is in stark contrast with the diatomic molecule, where only photon absorption by a transition $v \to v+1$ was allowed and only emission due to a transition $v \to v - 1$ was allowed. Again, the difference in the form of the potential energies leads to a dramatic difference in the physical behavior, in spite of the many mathematical similarities between the two systems.

19

THE SPIN OF THE ELECTRON AND ITS ROLE IN SPECTROSCOPY

Introduction

§1. Before quantum mechanics was discovered, the Bohr model described accurately the frequencies of the light emitted or absorbed by the hydrogen atom. However, some discrepancies appeared if the atom emitting (or absorbing) light was exposed to a magnetic field. Where theory predicted one peak in the emission spectrum, experiment found several peaks close to each other. The magnetic field split the predicted spectrum into *multiplets*. This *Zeeman* effect was partly explained by Lorentz, who used a model based on classical electrodynamics to calculate the energy of the electron moving in a magnetic field. There were, however, peaks in the spectrum of some atoms for which Lorentz theory offered no explanation. These were said to show an *anomalous Zeeman* effect.

All this happened before quantum mechanics was invented and you might think that the discrepancy between theory and experiment was due to the ignorance of the theorists. This is not the case. If you take the Schrödinger equation that you know at this point, add to it the energy of interaction between the electron and a magnetic field and solve the resulting equation exactly, you will not obtain the spectrum seen in the anomalous Zeeman effect. This tells us that the equation misses an essential feature, which is not connected to the motion of the electron in the atom when a magnetic field is present (which the equation describes correctly).

After the theory of the hydrogen atom was developed, it was natural to try to understand the structure of more complex atoms. It was reasonable to assume that the ground state of He (i.e. the state of lowest energy) had two electrons in the 1s orbital. While not accurate, this assumption does not lead to any absurd consequences. But, disaster strikes when we consider the lithium atom. Our assumption would require that, in its lowest energy state, the Li atom has three electrons in the 1s orbital. However, this leads to a dramatic conflict with experiment. Consider, for example, the energy I needed for removing one electron from an atom (the ionization potential). Experiments tell us that for hydrogen, $I = 13.598$ eV; for helium, $I = 24.587$ eV; and for lithium, $I = 5.392$ eV. For H, the calculated ionization potential agrees with the measured one. In He each 1s electron interacts with a nucleus of charge $2e$ ($Z = 2$): as a result, the electrons in He are bound more strongly to the nucleus and our simple rule predicts that He has a higher ionization energy than hydrogen. The trend is correct, but the calculated value differs from the experimental one. We have neglected, in this simple model, the interaction between electrons, and this might explain the deviation from the measured value. Such a discrepancy is unpleasant, but it is not catastrophic and one could hope that improvements in the method of calculation will lead to better agreement with measurements. For Li, however, our assumption that it has three electrons in the 1s orbital causes big trouble. If this assumption were correct, the ionization energy of Li would be higher than that of He. But experiment shows that it drops, from 24.587 eV (for He) to 5.392 eV (for Li).

It gets worse when you look at the chemical properties of these elements. Hydrogen is very reactive, helium is inert, and lithium is very reactive. We cannot explain this trend if, in all three atoms, the electrons are in the 1s state. Since the electrons in the lithium 1s orbital are bound more strongly to the nucleus than in the 1s orbital of He, Li should be less reactive than He. It is impossible to explain the chemistry displayed by the periodic table if we keep placing the electrons in the 1s orbital of the atoms.

This dilemma was solved by Wolfgang Pauli, a Swiss graduate student in Munich, who postulated that you cannot put more than two electrons in one orbital. Why two? He had no idea, but his rule resolved all the qualitative puzzles I mentioned and earned him a Nobel Prize. Now you can see why Li was different: the first two electrons went into the 1s orbital. But according to Pauli's rule, there was "no room" for a third electron there. The third electron had to go into a 2s (or $2p_x$, or $2p_y$, or $2p_z$) orbital. The binding energy of the electron in any of these orbitals is much smaller than in a 1s orbital, so Li is easier to ionize than He and its high energy electron is more eager to engage in chemical reactions. Many other facts were qualitatively explained by this rule. But no one knew why, at most, two electrons were allowed in each state. There was also no indication that the multiplets in the spectra and the "not more than two electrons" rule were related.

The first step towards solving these mysteries was taken by two Dutch graduate students, Goudsmit and Uhlenbeck. They proposed that the electron has an *angular momentum*, called *spin*, which can have two states. The "spinning electron" has a magnetic moment (i.e. behaves like a small magnet) and interacts with an external magnetic field. Because of this interaction, the energy of some of the degenerate states in the atom will differ from each other when a magnetic field is present. When excited, these states emit photons having slightly different frequencies, hence the multiplets. The electron spin can only have two states and that fact suggested that the factor of 2 in the Pauli principle was related to spin. Now the principle was formulated differently: no electrons can have identical states in an atom or a molecule. They can occupy the same orbital, hence have the same energy and angular momentum, if they have different spins. Since the spin could have at most two values, at most two electrons can be in the same orbital.

Very quickly spin invaded quantum physics. It was discovered that nuclei have spin and this affects deeply the thermodynamic properties of a gas of homonuclear diatomic molecules and their rotational spectrum. The existence of spin also explained the magnetism of solids and the magnetic properties of molecules. It soon became clear that one could not understand the chemical bond and the electronic spectra of the molecules (i.e. the distinction between fluorescence and phosphorescence) without taking into account the net spin of the electronic state.

The nucleus of a molecule exposed to a magnetic field can have several spin states, having different energies. These states absorb energy from an oscillating magnetic field and this induces transitions between different states of the nuclear spin. Nuclear magnetic resonance spectroscopy (NMR), based on such transitions, has become one of the most important tools available to chemists. Magnetic resonance imaging (MRI), which is now present in any modern hospital, is a close cousin to NMR: it exploits the rate with which nuclear spins lose energy after being excited by an oscillating magnetic field.

In this chapter you will learn how to use the theory of angular momentum to describe the spin states of a particle. The theory is then used to examine how the electron spin affects the emission spectrum of a hydrogen atom that is exposed to a static magnetic field. The spin will appear again, as a major player, in the next chapter where we discuss the chemical bond in the hydrogen molecule. NMR spectroscopy is explained in the last chapter of the book.

§2. *The Spin Operators.* The existence of the electron spin follows naturally from the relativistic quantum mechanical theory of the electron proposed by Dirac. In the non-relativistic quantum theory studied here we need to postulate that the spin is an intrinsic angular momentum of the electron or of a nucleus. This angular momentum has nothing to do with the rotation of the electron around the nucleus (that is the orbital angular momentum $\hat{L}$). Nor does it exist because the electron is a rotating sphere (as was originally believed). The word "intrinsic" tells you that this is a property that the electron has, just like mass or charge.

Since the spin is an angular momentum, it should be described by an angular momentum operator. Unfortunately, the definition of the angular momentum that we used so far involves the rotational motion of the electron, and no such motion exist in the case of spin. We must therefore generalize the definition of the angular momentum to cover this special situation. This will also require a modification of the notation.

The angular momentum operator $\hat{\mathbf{L}}$ was defined by replacing $\mathbf{r}$ and $\mathbf{p}$ in the classical definition

$$\mathbf{L} = \mathbf{r} \times \mathbf{p} \tag{19.1}$$

with the corresponding operators (see Chapter 14). Eq. 19.1 describes the angular momentum of an electron in an atom or the angular momentum of a rotating diatomic molecule.

There is nothing orbiting in the case of electron or nuclear spin and we cannot describe the spin angular momentum by Eq. 19.1. Instead we look for a property of angular momentum that can be generalized to the case when there is no orbiting motion.

In Chapter 14, we saw that the angular momentum operator satisfies the following equations:

$$\left[\hat{L}_x, \hat{L}_y\right] = i\hbar\hat{L}_z \tag{19.2}$$

$$\left[\hat{L}_y, \hat{L}_z\right] = i\hbar\hat{L}_x \tag{19.3}$$

$$\left[\hat{L}_z, \hat{L}_x\right] = i\hbar\hat{L}_y \tag{19.4}$$

Here $[\hat{A}, \hat{B}] \equiv \hat{A}\hat{B} - \hat{B}\hat{A}$ is the commutator of the operators $\hat{A}$ and $\hat{B}$, and $\hat{L}_x$, $\hat{L}_y$, $\hat{L}_z$ are the components of the operator representing the angular momentum vector.

Chapter 14 also mentioned that we can derive all the properties of the angular momentum operator, including its eigenstates and eigenvalues, from the commutation relations, Eqs 19.3–19.4. This means that we can *define* the angular momentum operator as a vector operator $\hat{\mathbf{L}}$ whose components satisfy Eqs 19.3–19.4. This definition is more general than Eq. 19.1 since it does not rely on the existence of a vector $\mathbf{r}$ describing the rotation of the particle.

To deal with spin, we postulate that every particle has an angular momentum represented by a spin operator $\hat{\mathbf{S}}$ that satisfies the commutation relations

$$\left[\hat{S}_x, \hat{S}_y\right] = i\hbar\hat{S}_z \tag{19.5}$$

$$\left[\hat{S}_y, \hat{S}_z\right] = i\hbar\hat{S}_x \tag{19.6}$$

$$\left[\hat{S}_z, \hat{S}_x\right] = i\hbar\hat{S}_y \tag{19.7}$$

§3. *Spin Eigenstates and Eigenvalues.* In Chapter 14, we have seen that the operator $\hat{\mathbf{L}}^2$ has the eigenvalues

$$\hbar^2\ell(\ell+1)$$

with corresponding eigenstates $Y_\ell^m(\theta,\phi)$. This means that these quantities satisfy the eigenvalue equation:

$$\hat{\mathbf{L}}^2 Y_\ell^m(\theta,\phi) = \hbar^2\ell(\ell+1)Y_\ell^m(\theta,\phi) \tag{19.8}$$

In Eq. 19.8, ℓ and m can take the values

$$\ell = 0, 1, 2, \ldots \tag{19.9}$$

$$m = -\ell, -\ell+1, \ldots, \ell-1, \ell \tag{19.10}$$

Furthermore, the eigenfunctions $Y_\ell^m(\theta,\phi)$ of $\hat{\mathbf{L}}^2$ are also eigenfunctions of $\hat{L}_z$:

$$\hat{L}_z Y_\ell^m(\theta,\phi) = \hbar m Y_\ell^m(\theta,\phi) \tag{19.11}$$

We can prove that Eqs 19.8–19.11 follow from the commutation relations Eqs 19.3–19.4. Therefore, similar results must follow from Eqs 19.6–19.7. But we have to be careful. In Eqs 19.8–19.11, θ and ϕ represent the orientation of the vector $\mathbf{r}$. There is no such vector in the case of spin, and our notation must be revised accordingly.

For this, we introduce a new mathematical object, called a *ket* or a *state*, denoted by

$$| \ell_s, m_s \rangle \tag{19.12}$$

This tells us that the state of the spin is such that its angular momentum squared is $\hbar^2 \ell_s(\ell_s + 1)$ and the projection of its angular momentum on the OZ axis is $\hbar m_s$. The abstract object $| \ell_s, m_s \rangle$ satisfies the equations

$$\hat{\mathbf{S}}^2 | \ell_s, m_s \rangle = \hbar^2 \ell_s(\ell_s + 1) | \ell_s, m_s \rangle \tag{19.13}$$

$$\hat{S}_z | \ell_s, m_s \rangle = \hbar m_s | \ell_s, m_s \rangle \tag{19.14}$$

In this way $| \ell_s, m_s \rangle$ is a generalization of $Y_\ell^m(\theta, \phi)$.

Unlike the orbital angular momentum, where ℓ can take any of the values allowed by Eq. 19.9, the values of ℓ_s are fixed. Each particle comes with one value of ℓ_s, just as it comes with its own charge and mass. For example: for an electron $\ell_s = \frac{1}{2}$; for a proton $\ell_s = \frac{1}{2}$; for deuterium $\ell_s = 1$. The *CRC Handbook of Chemistry and Physics* (CRC Press, Boca Raton FL, 62nd edition, 1982) gives, on page B-255, the spins of all nuclei and their isotopes. A small sample is shown in Table 19.1.

Table 19.1 $_a\mathrm{A}^b$ indicates that atom A has a protons and mass b. μ is the magnetic moment of the particle in nuclear magnetons (see text).

Isotope	ℓ_s	μ
$_1^1\mathrm{H}$	1/2	2.79278
$_1^2\mathrm{H}$	1	0.85742
$_2^3\mathrm{He}$	1/2	−2.1275
$_2^4\mathrm{He}$	0	0
$_8^{16}\mathrm{O}$	0	0
$_8^{17}\mathrm{O}$	5/2	−1.8937
$_8^{18}\mathrm{O}$	0	0
$_{13}^{27}\mathrm{Al}$	5/2	3.6414

The quantum number m_s takes the values allowed by Eq. 19.10. An electron, for example, can have

$$m_s = -\frac{1}{2} \quad \text{or} \quad m_s = \frac{1}{2}$$

while for $^{27}_{13}\text{Al}$, m_s can take the values

$$m_s = -\frac{5}{2}, \ -\frac{3}{2}, \ -\frac{1}{2}, \ \frac{1}{2}, \ \frac{3}{2}, \ \frac{5}{2}$$

In general, if a particle has a spin ℓ_s, then m_s can take $2\ell_s + 1$ values.

Thus, the spin of the electron can be in one of two eigenstates

$$| \ell_s = \tfrac{1}{2}, m_s = -\tfrac{1}{2} \rangle \text{or} | \ell_s = \tfrac{1}{2}, m_s = \tfrac{1}{2} \rangle$$

If the context makes it clear that the states refer to an electron spin, giving ℓ_s is superfluous, and these two states are sometimes denoted by $| - \rangle$ and $| + \rangle$, or by $| \alpha \rangle$ and $| \beta \rangle$, or by $| \uparrow \rangle$ and $| \downarrow \rangle$.

If the electron is in state $| \ell_s = \tfrac{1}{2}, m_s = -\tfrac{1}{2} \rangle$ then we know that

$$\hat{\mathbf{S}}^2 | \ell_s = \tfrac{1}{2}, m_s = -\tfrac{1}{2} \rangle = \hbar^2 \tfrac{1}{2} \left(\tfrac{1}{2} + 1 \right) | \ell_s = \tfrac{1}{2}, m_s = -\tfrac{1}{2} \rangle$$

$$\hat{S}_z | \ell_s = \tfrac{1}{2}, m_s = -\tfrac{1}{2} \rangle = -\tfrac{1}{2}\hbar | \ell_s = \tfrac{1}{2}, m_s = -\tfrac{1}{2} \rangle$$

For the state $| \ell_s = \tfrac{1}{2}, m_s = \tfrac{1}{2} \rangle$,

$$\hat{\mathbf{S}}^2 | \ell_s = \tfrac{1}{2}, m_s = \tfrac{1}{2} \rangle = \hbar^2 \tfrac{1}{2} \left(\tfrac{1}{2} + 1 \right) | \ell_s = \tfrac{1}{2}, m_s = \tfrac{1}{2} \rangle$$

$$\hat{S}_z | \ell_s = \tfrac{1}{2}, m_s = \tfrac{1}{2} \rangle = \tfrac{1}{2}\hbar | \ell_s = \tfrac{1}{2}, m_s = \tfrac{1}{2} \rangle$$

Exercise 19.1

Make a list of the states of $^{27}_{13}\text{Al}$ and of the values of $\hat{\mathbf{S}}^2$ and $\hat{S}_z$ for each state.

§4. *The Scalar Product.* You have learned that in order to calculate various observable quantities in quantum mechanics you will calculate certain scalar products. Our theory of the spin states is not complete unless we define a scalar product for them.

In the case of the orbital angular momentum the definition is

$$\langle \ell, m \mid \ell', m' \rangle \equiv \int_0^\pi \sin(\theta) d\theta \int_0^{2\pi} d\phi \, Y_\ell^m(\theta, \phi)^* \, Y_{\ell'}^{m'}(\theta, \phi) = \delta_{\ell\ell'} \delta_{mm'} \qquad (19.15)$$

The first term above is an abstract notation for the integral shown in the second term. The third indicates that the states $Y_\ell^m(\theta, \phi)$ and $Y_{\ell'}^{m'}(\theta, \phi)$ are orthonormal.

In the case of spin, there are no angles to integrate over, so we need to generalize this relationship in a way that preserves its physical consequences, but does not involve angles and integrals. To do this we regard the left-hand-side $\langle \ell, m \mid \ell', m' \rangle$ as the general object, named *scalar product*, and the right-hand-side as a specific embodiment of it.

Therefore we extend the concept of scalar product $\langle \ell, m \mid \ell', m' \rangle$ to the abstract states $|\ell, m\rangle$ and $|\ell', m'\rangle$ by postulating that it has the following properties.

1.

$$\langle \ell, m \mid \ell', m' \rangle = \langle \ell', m' \mid \ell, m \rangle^* \qquad (19.16)$$

where the star indicates complex conjugation (i.e. change $i = \sqrt{-1}$ to $-i$);

2.

$$\langle \ell, m \mid \ell, m \rangle \geq 0 \qquad (19.17)$$

and if this scalar product is zero then $|\ell, m\rangle$ is identically zero;

3.   If $|\eta\rangle = a |\ell', m'\rangle$ then

$$\langle \ell, m \mid \eta \rangle = a \langle \ell, m \mid \ell', m' \rangle \qquad (19.18)$$

4.   If $|\eta\rangle = |\ell', m'\rangle + |\ell'', m''\rangle$ then

$$\langle \ell, m \mid \eta \rangle = \langle \ell, m \mid \ell', m' \rangle + \langle \ell, m \mid \ell'', m'' \rangle \qquad (19.19)$$

It is straightforward to show that the operation defined by the integral in Eq. 19.15 has all of these properties. Moreover, these are the only properties of the integral that turn out to be important in quantum mechanics.

Exercise 19.2

Consider a set of sequences of complex numbers with the property that for any two sequences, say $\{a_1, a_2, a_3 \ldots\}$ and $\{b_1, b_2, b_3 \ldots\}$,

$$\sum_{i \geq 1} a_i^* b_i < \infty$$

Show that if you denote

$$|a\rangle = \{a_1, a_2, a_3, \ldots\}$$

and

$$|b\rangle = \{b_1, b_2, b_3 \ldots\}$$

then the symbol $\langle b \,|\, a\rangle$ defined by

$$\langle b \,|\, a\rangle \equiv \sum_{i \geq 1} b_i^* a_i$$

is a scalar product (it has properties 1–4 above).

By analogy with the theory of orbital angular momentum, we postulate that such a scalar product exists for the states $|\,\ell_s, m_s\rangle$ and that it satisfies

$$\langle \ell_s', m_s' \,|\, \ell_s, m_s \rangle = \delta_{\ell_s' \ell_s} \delta_{m_s' m_s} \qquad (19.20)$$

It turns out that by accepting these generalizations, we can calculate all the properties of the spin states and obtain results that are in perfect agreement with the measurements.

The Emission Spectrum of a Hydrogen Atom in a Magnetic Field: the Normal Zeeman Effect

§5. *Introduction.* One of the earliest measurements that could not be explained by ignoring spin was the emission spectrum of an atom interacting with a magnetic field. The rest of this chapter is taken up with a description of the results obtained in such measurements and with their explanation by quantum mechanics.

The first measurements of an emission spectrum in the presence of a magnetic field were made by Zeeman and were explained promptly by a classical theory developed by Lorentz. It so happened that spin played no role in the emission by the Na atoms Zeeman worked with. However, soon after Zeeman's discovery, other people, working with different atoms, obtained results that disagreed with Lorentz theory. This disagreement was qualitative: the measured spectrum had more peaks than theory predicted. This was called the *anomalous Zeeman effect* in contrast to the one that agreed with Lorentz theory, which was called the *normal Zeeman* effect (or sometimes the Zeeman effect).

You will see shortly that the normal Zeeman effect can be explained fully by the interaction between the magnetic field and the motion of the electron in the atom. The anomalous Zeeman effect can only be understood if we accept the assumption that the electron has a spin $\ell_s = 1/2$.

In this section I explain the normal Zeeman effect and leave the anomalous one for the next. As far as I know, this effect is of little practical importance and is used only in laboratory experiments. However, understanding it will familiarize you with the physics of spin in the presence of magnetic fields, which is essential for understanding NMR (a method of great practical importance).

§6.   *The Experiment.* Zeeman's experiments intended to investigate whether light can be affected by magnetism. Faraday has shown that a magnetic field can cause the plane of polarization of light to rotate. He also performed measurements that concluded that magnetic fields have no influence on the light emitted by a flame. This erroneous conclusion was reinforced by Maxwell, who believed that no force produced in the laboratory can alter even slightly the charge oscillations, which, in his theory, were responsible for light emission.

It is easier to explain Zeeman's result if I remind you a certain nomenclature. An emission spectrum is a plot of the intensity of the emitted light as a function of its frequency. For an atom, this spectrum consists of very narrow peaks, present at distinct frequencies. These are caused by transitions from an atomic state of high energy, to one of low energy.

In his measurements Zeeman put a sodium salt in a flame. The high temperature causes the salt to dissociate and produce excited sodium atoms, which emit light. In the absence of the magnetic field, the emission spectrum of this sodium flame has two lines. Zeeman's experiments focused on one of them and ignored the other. Therefore, in the absence of the magnetic field there is only one peak in his measurements. Moreover, the frequency of the emitted photons and their polarization is the same, regardless of the location of the detector.

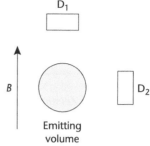

Figure 19.1 The sodium atoms contained in the gray region emit photons in the presence of the magnetic field. Two photon detectors are used, one in the direction parallel to the field (D_1) and the other in the direction perpendicular to the field (D_2).

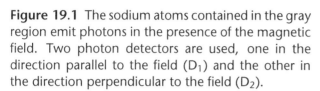

When a magnetic field is present, it is conceivable that the properties of the emitted light are not the same in all directions, because the field breaks the isotropy of space. To test whether this happens, Zeeman performed two experiments. In one, he detected photons traveling parallel to the magnetic field (the detector D_1 in Fig. 19.1); in the other, those traveling in a direction perpendicular to it (the detector D_2 in Fig. 19.1). In this way, he could probe whether the magnetic field affected the directional properties of the emission.

Here are the facts that Zeeman discovered. The atoms emitted, in direction perpendicular to the magnetic field, photons of three frequencies, Ω_+, Ω_0, and Ω_-. The frequency Ω_0 was identical to that emitted in the absence of the field; Ω_- was smaller and Ω_+ was larger. We say that the magnetic field splits the emission spectrum into a triplet. The difference $\Delta = \Omega_+ - \Omega_0$ was equal to $\Omega_0 - \Omega_-$; it is called the splitting of the emission line by the magnetic field, or the level splitting.

The magnetic field splits the spectrum of the light emitted along the magnetic field into a doublet (the spectrum has two peaks) of frequencies Ω_+ and Ω_-.

Zeeman also found that the magnitude of the level splitting is proportional to the strength of the magnetic field **B**. He made a clever use of this observation. He measured the spectrum of sunlight and found that it had multiplets; from the magnitude of the splitting he calculated the magnetic field in the sun. This was the first time that anyone proved that such a field exists and measured it.

Back to Earth and Zeeman's experiment. He also found that the photons emitted in direction perpendicular to the field are linearly polarized. The polarization of the photons of frequency Ω_+ and Ω_- is perpendicular to **B** and the photons of frequency Ω_0 are polarized parallel to it.

The photons emitted along the field are circularly polarized, with the electric-field vector rotating in opposite directions. As you can imagine, interesting things happen when you move the detectors around.

Exercise 19.3

Describe what might happen to the emission spectrum, as you rotate the detector from the direction perpendicular to the magnetic field to the direction parallel to it.

This *normal Zeeman effect* is observed only when the spin of the electron does not affect the emission spectrum. It was Zeeman's and Lorentz's good luck that the first experiments were performed on such an emission peak. Subsequent experiments, with other atoms, revealed that the modifications produced by the magnetic field differed from those predicted by Lorentz theory. The *anomalous Zeeman effect* was thus born.

§7. *A Modern (but Oversimplified) Version of the Lorentz Model.* To calculate the effect that the magnetic field has on the electron in the hydrogen atom, we add to the Hamiltonian the energy of the interaction between the electron and the magnetic field. Then we solve the Schrödinger equation corresponding to this Hamiltonian, to calculate the energy of the atom in the presence of the field. Since the energy of the interaction between the electron and the field is small, compared to the energy of the atom in the absence of the field, we can solve Schrödinger equation by perturbation theory.

The operator corresponding to the energy of an electron in a magnetic field is

$$\hat{\mathcal{H}} = \frac{e}{2m_e} \hat{\mathbf{L}} \cdot \mathbf{B} \tag{19.21}$$

B is the *magnetic flux density* vector, e is the proton charge, and $\hat{\mathbf{L}}$ is the orbital angular momentum operator. In this chapter I will use m_e for the mass of the electron, since the letter m is needed for the quantum number describing the eigenvalues of $\hat{L}_z$. The unit of **B** in the SI system is a tesla ($kg/(s^2 A)$).

This equation is obtained by the procedure used throughout this textbook. We take the expression for the energy, given by classical electrodynamics, and replace the classical quantities with the corresponding operators. Eq. 19.21 is obtained by replacing the angular momentum, in the classical expression used by Lorentz, with the angular momentum operator. The magnetic field **B** is treated classically (nothing new is obtained if we quantize this quantity).

You will find in the literature that your predecessors have been busy introducing a great variety of pointless notation and nomenclature. Unfortunately, they are widely used and you need to know what they mean.

The *magnetic dipole moment* of the moving electron is

$$\hat{\mu} \equiv \gamma_e \hat{\mathbf{L}} \equiv -\frac{e}{2m_e}\hat{\mathbf{L}} \tag{19.22}$$

while

$$\gamma_e = -\frac{e}{2m_e} \tag{19.23}$$

is the *magnetogyric* ratio of the electron. With this notation, the interaction energy (Eq. 19.21) becomes

$$\hat{\mathcal{H}} = -\hat{\mu} \cdot \mathbf{B} \tag{19.24}$$

This energy is very small compared to the total energy of the electron in the absence of the field.

Eq. 19.21 can be understood from a qualitative argument, based on classical electrodynamics. From your study of electricity, you know that if you pass an electric current through a loop (formed by a wire) placed into a magnetic field, a force acts on the wire. The loop behaves as if it is a magnet. Let us assume that the behavior of an electron rotating around the nucleus must be the same as that of an electron moving through a wire loop. Then, it too should behave as a magnet and should interact with a magnetic field. Classical theory can be used to show that the energy of this interaction is given by Eq. 19.21 (but the magnetic moment **L** is the classical angular momentum). In a more advanced presentation of quantum mechanics, the interaction of electrons with electric and magnetic fields is derived by a more rigorous procedure.

§8. *The Energy of the Hydrogen Atom in a Magnetic Field.* The Hamiltonian of the hydrogen atom in a magnetic field is

$$\hat{H} = \hat{H}_0 + \frac{e}{2m_e}\hat{\mathbf{L}} \cdot \mathbf{B} \tag{19.25}$$

Here $\hat{H}_0$ is the Hamiltonian when the magnetic field is absent. This approximation is correct only when the magnetic field is large enough to allow us to neglect terms in the Hamiltonian that are not discussed here.

Since the additional term in Eq. 19.25 is much smaller than $\hat{H}_0$, we assume that the wave function of the hydrogen atom is not greatly disturbed by the presence of the field. Therefore, the wave function in the presence of the field is approximately the same as that in its absence:

$$\psi_{n,\ell,m}(r,\theta,\phi) \equiv R_{n,\ell}(r)Y_\ell^m(\theta,\phi) \equiv |n,\ell,m\rangle \qquad (19.26)$$

We can now calculate the energy of the electron when a magnetic field is present from

$$E_{n,\ell,m} = \langle n,\ell,m \,|\, \hat{H}_0 \,|\, n,\ell,m \rangle + \langle n,\ell,m \,|\, \frac{e}{2m_e}\hat{\mathbf{L}}\cdot\mathbf{B} \,|\, n,\ell,m \rangle$$

$$= -\frac{\mu e^4 Z^2}{2(4\pi\varepsilon_0)^2 n^2\hbar^2} + \frac{e}{2m_e}\langle n,\ell,m \,|\, \hat{\mathbf{L}} \,|\, n,\ell,m \rangle \cdot \mathbf{B} \qquad (19.27)$$

The first term is the energy of the hydrogen atom in the absence of the magnetic field $\mathbf{B}$. I did not set $Z = 1$ (as I should, for the hydrogen atom) just in case that you want to use this equation for He^+, for which $Z = 2$.

The second term is easily calculated by choosing the OZ coordinate axis along the direction of $\mathbf{B}$. This means that the components of $\mathbf{B}$ are $\{0,0,B\}$ (where B is the length of the vector $\mathbf{B}$) and so

$$\hat{\mathbf{L}}\cdot\mathbf{B} = \hat{L}_z B$$

Since

$$\hat{L}_z \,|\, n,\ell,m \rangle = \hat{L}_z R_{n,\ell}(r)\, Y_\ell^m(\theta,\phi)$$

$$= m\hbar R_{n,\ell}(r)Y_\ell^m(\theta,\phi) = m\hbar \,|\, n,\ell,m \rangle$$

and

$$\langle n,\ell,m \,|\, n,\ell,m \rangle = 1,$$

Eq. 19.27 becomes

$$E_{n,\ell,m} = -\frac{\mu e^4 Z^2}{2(4\pi\varepsilon_0)^2 n^2\hbar^2} + \frac{e\hbar}{2m_e}mB \qquad (19.28)$$

The quantity

$$\mu_B \equiv \frac{e\hbar}{2m_e} \qquad (19.29)$$

has units of a magnetic dipole (i.e. energy/magnetic field) and it is called a *Bohr magneton*. μ_B is often used as a magnetic-dipole unit in molecular physics.

Eq. 19.28 shows that when $B = 0$, the energy of the hydrogen atom is independent of ℓ and m. States having the same value of n, but different values of ℓ and m are *degenerate* (have the same energy). This is no longer true if $B \neq 0$.

Consider what happens to the 2s and 2p- states (which have $n = 2$) when a magnetic field is present. The 2s state has $\ell = 0$ and therefore m must be equal to zero. Because $m = 0$, the energy of this state *is not affected* by the magnetic field (see Eq. 19.28). The atom has three 2p states since $\ell = 1$ and m can take the values -1, 0, or 1. In the absence of the magnetic field these states are degenerate. Therefore the emission spectrum corresponding to a transition from any one of these states to the ground state, has one peak.

When a magnetic field is present *the degeneracy is lifted* and these four states have three different energies (the energy levels are shown in Fig. 19.2). The energy of the 2s state and that of the 2p state (having $m = 0$) is unaffected by the field; these two states remain degenerate even when the magnetic field is present. The energy of the 2p state having $m > 0$ increases with the field, and that of the 2p state having $m < 0$ decreases. Now you see why, in the early times of quantum mechanics, m was called the magnetic quantum number.

§9. *The Emission Frequencies.* Since in the presence of the magnetic field the atom has three excited states with distinct energies, it can emit photons of three frequencies. One frequency is unaffected by the field and we identify it with Ω_0 (see Fig. 19.2). Ω_+ is the frequency of the photon emitted from the level $\{n = 2, \ell = 1, m = 1\}$ and Ω_- is the one from $\{n = 2, \ell = 1, m = -1\}$. These frequencies, and their dependence on the strength of the magnetic field, are in excellent agreement with those obtained by experiment.

Theory also predicts that the splitting of the level is:

$$\Delta = \Omega_+ - \Omega_0 = \Omega_0 - \Omega_- = \frac{e\hbar B}{m_e} \tag{19.30}$$

This formula was derived by Lorentz, by using classical theory, and he used it with Zeeman's data to calculate the ratio e/m_e. He found a value that agreed with that obtained by Thomson for the charged particles in cathode rays (electrons) and concluded that light is emitted by oscillating electrons in the atom. There is a lesson to be learned here: Lorentz obtained correct results by using an incorrect theory (classical electrodynamics and mechanics); the extremely good agreement with the

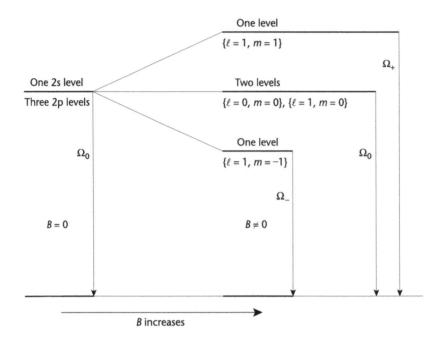

Figure 19.2 This figure shows what happens to the 2s and 2p energy levels of a hydrogen atom placed in a magnetic field. When there is no field ($B = 0$), the electron could be in one of the four degenerate levels with $n = 2$. When the field is present, the energy levels having $m \neq 0$ are shifted symmetrically above and below the energy level of the electron when the field was absent. The vertical arrows show the frequencies of the photons emitted when $B = 0$ and when $B \neq 0$.

experiment was accidental. The anomalous Zeeman effect dispelled any illusion that classical theory can explain atomic physics.

Exercise 19.4

Make an energy-level diagram, like the one shown in Fig. 19.2, for the energies of the $n = 3$ levels.

Exercise 19.5

Calculate the magnitude of the energies in the diagrams shown in Fig. 19.2, in units of eV, for a magnetic field B of 1 T. Calculate the magnitude of $\Omega_+ - \Omega_0$ and $\Omega_0 - \Omega_-$ in units of cm^{-1}.

§10. *The Spectrum.* What you have learned so far does not provide a complete explanation of Zeeman effect. We have said nothing about the polarization of the photons nor have we explained why the spectrum has two peaks for photons traveling parallel to the magnetic field and three when they travel perpendicular to it. To explain these observations we need to use a theory that calculates the probability that an atom emits a photon of a given polarization, in a given direction. We do not have time to do justice to this theory here. However, I can give you a qualitative explanation for the reason why the spectra measured in different directions are so different from each other.

The quantum number m describes the orientation of the angular momentum with respect to the OZ axis. When $m = 1$, the angular momentum of the electron in the atom is oriented along the magnetic field (which is the direction of the OZ axis). When $m = 0$, $\hat{\mathbf{L}}$ is perpendicular to $\mathbf{B}$, and when $m = -1$, the direction of $\hat{\mathbf{L}}$ is opposite to that of $\mathbf{B}$. From classical physics you know that the orbital angular momentum is perpendicular to the plane in which the electron rotates. Therefore, different values of m correspond to different orientations of the electronic orbit. If you think of the rotating electron as a current loop emitting electromagnetic radiation, it is not surprising that the intensity of the radiation emitted depends on loop orientation. There are some directions in which a certain loop does not emit radiation. This is why in some directions we see only two emission peaks and in others we see three.

You are right to be suspicious of such classical explanation, but this one happens to be correct. The quantum theory does not make use of it, but calculates the probability that an excited atom emits radiation of a certain frequency and polarization in a certain direction. This depends on the direction of emission and become zero in certain directions. This is why two photons are emitted in a certain direction and three in other.

The polarization of emission can be vaguely understood from the conservation of the angular momentum: the angular momentum of the electron before emission must be equal to the angular momentum of the emitted photons plus the angular momentum of the electron after emission. Photons of different polarization have different angular momenta and this controls, to some extent the polarization of the emission. This rule is not imposed arbitrarily but arises naturally when the matrix elements are calculated.

I do not expect you to understand these statements in any detail, but I hope that they stimulate your curiosity to read more about light emission.

The Role of Spin in Light Emission by a Hydrogen Atom: the Anomalous Zeeman Effect

§11. *The Interaction between Spin and a Magnetic Field.* We have seen that an electron rotating around the nucleus has an orbital angular momentum $\hat{\mathbf{L}}$ and interacts with an external magnetic field according to Eq. 19.21. We can generalize this equation and postulate that the energy of a particle having spin $\hat{\mathbf{S}}$, in the presence of a magnetic field, is

$$\hat{H}_s = -\gamma \hat{\mathbf{S}} \cdot \mathbf{B} \tag{19.31}$$

This formula is valid for either the electron or the nuclear spin. The *gyromagnetic ratio γ* of nucleus depends on the nuclear structure and its magnitude is determined experimentally. The values of γ for a few nuclei are given in Table 19.2.

For an electron, the gyromagnetic ratio is

$$\gamma_e = -\frac{g_e e}{2m_e} \tag{19.32}$$

where the *electronic g-factor* is $g_e = 2.002$.

§12. *The Energies of the Electron in the Hydrogen Atom: the Contribution of Spin.* We assume that the Hamiltonian of the electron in the hydrogen atom is:

$$\hat{H} = \hat{H}_0 + \frac{e}{2m_e}\hat{\mathbf{L}} \cdot \mathbf{B} + \frac{g_e e}{2m_e}\hat{\mathbf{S}} \cdot \mathbf{B} \tag{19.33}$$

Table 19.2 The gyromagnetic ratio of several nuclei (1 tesla = 10^4 gauss).

Nucleus	γ (rad/s gauss)
^{1}H	26,753
^{2}H	4,107
^{13}C	6,728
^{17}O	−3,628
^{27}Al	6,971
^{29}Si	−5,319

Here $\hat{H}_0$ is the Hamiltonian of the hydrogen atom in the absence of the magnetic field. The second term in Eq. 19.33 is the interaction energy between the orbiting electron and the magnetic field and the last one is the interaction of the field with the spin of the electron. This Hamiltonian neglects a few other terms that are not essential for the phenomena analyzed here.

Furthermore, we assume that the magnetic field does not modify the states of the electron substantially and therefore the eigenstates in the presence of the field are (to a good approximation)

$$| n, \ell, m, s, m_s \rangle = R_{n,\ell}(r) Y_\ell^m(\theta, \phi) | s, m_s \rangle \qquad (19.34)$$

Since $s = \frac{1}{2}$ for the electron, m_s can take the values $-\frac{1}{2}$ and $+\frac{1}{2}$.

The energy of the atom is

$$\begin{aligned}
E_{n,\ell,m,s,m_s} &= \langle n, \ell, m, s, m_s | \hat{H} | n, \ell, m, s, m_s \rangle \\
&= \langle n, \ell, m, s, m_s | \hat{H}_0 | n, \ell, m, s, m_s \rangle \\
&+ \langle n, \ell, m, s, m_s | \frac{e}{2m_e} \hat{\mathbf{L}} \cdot \mathbf{B} | n, \ell, m, s, m_s \rangle \\
&+ \langle n, \ell, m, s, m_s | \frac{g_e e}{2m_e} \hat{\mathbf{S}} \cdot \mathbf{B} | n, \ell, m, s, m_s \rangle \qquad (19.35)
\end{aligned}$$

We take the OZ axis along $\mathbf{B}$, which means that $\mathbf{B} = \{0, 0, B\}$ and $\hat{\mathbf{L}} \cdot \mathbf{B} = \hat{L}_z B$.

The first term in Eq. 19.35 is the energy of the hydrogen atom when there is no magnetic field. This is given by (see Chapter 18)

$$-\frac{m_e e^4 Z^4}{2(4\pi\varepsilon_0)^2 n^2 \hbar^2} \qquad (19.36)$$

The second term is calculated by using

$$\hat{L}_z | n, \ell, m, s, m_s \rangle = \hbar m | n, \ell, m, s, m_s \rangle$$

and the third, is evaluated by taking into account that

$$\hat{S}_z | n, \ell, m, s, m_s \rangle = \hbar m_s | n, \ell, m, s, m_s \rangle$$

These equations, together with

$$\langle n, \ell, m, s, m_s \, | \, n, \ell, m, s, m_s \rangle = 1$$

turn Eq. 19.35 into

$$E_{n,\ell,m,s,m_s} = -\frac{m_e e^4 Z^4}{2(4\pi\varepsilon_0)^2 n^2 \hbar^2} + \frac{e}{2m_e}\left(m + g_e m_s\right) B \qquad (19.37)$$

Because of the presence of the spin, the energy now depends on m_s, in addition to the quantum numbers n and m. Furthermore, the energy depends of ℓ and s, since their values limit the values m and m_s can take. In the case of the normal Zeeman effect ($m_s = 0$), the energy level having $m = 0$ was not affected by the magnetic field, no matter how large the field was. This is no longer true when the interaction of the spin with the field is taken into account. Since m_s cannot be zero, the energy depends on **B** regardless of what value m takes. This results in an additional lifting of degeneracy than in the case of the normal Zeeman effect. The energy level generated by Eq. 19.37 are shown in Fig. 19.3, for the case when $n = 1$ and $n = 2$.

Exercise 19.6

Calculate the energies of the 1s, 2s, and 2p states of the hydrogen atom in eV, in the presence of a magnetic field of 1 T .

The appearance of the quantum number m_s, in addition to n and m, considerably increases the number of levels. There are now five distinct energies for $n = 2$ and two for $n = 1$. This means that there are more peaks in the emission spectrum when the spin is taken into account.

§13. *Comments and Warnings.* I presented here a simplified description of the Hamiltonian of a particle with spin and of the calculation of the energy levels of the electron in the hydrogen atom exposed to a magnetic field. While incomplete, this presentation gives you an idea of the role of spin in generating multiplets in spectra and of some of the physical factors involved in the "splitting" of the photon frequencies.

Here are some of the things left out. Imagine that you are traveling with the electron: it would appear to you that the nucleus moves around you. A moving

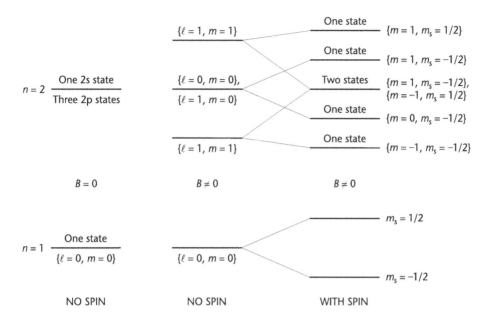

Figure 19.3 The energy levels of the electron in the hydrogen atom. The left-hand column shows the 1s, 2s, and 2p levels in the absence of a magnetic field. The middle column shows the levels that would be obtained if the spin did not exist but the magnetic field is present. The right-hand column shows the levels when $B \neq 0$ and the spin is taken into account.

charge creates a magnetic field, which acts on the spin of the electron. The energy corresponding to this interaction is

$$H_{so} = \frac{e^2}{2m_e^2 c^2} \frac{1}{r^3} \hat{\mathbf{L}} \cdot \hat{\mathbf{S}}$$

where c is the speed of light and r is the electron–nucleus distance. This term in the Hamiltonian is called the *spin–orbit* coupling. This has nothing to do with the presence of an external magnetic field. It couples the orbital angular momentum of the electron to its own spin. It provides a mechanism through which changes in orbital motion can cause a change in the spin state. This term is large in organometallic compounds containing heavy atoms and plays an important role in the spectroscopy and the photochemistry of those molecules.

If the nuclear spin is not zero, the nucleus has a magnetic moment. This produces a magnetic field that acts on the spin of the electron. This affects the energy of the electron just like an external magnetic field would. I will not write down the expression for the energy of this interaction but note that it has two terms: one depends on $\hat{\mathbf{L}} \cdot \hat{\mathbf{S}}_N$ and the other on terms containing products of $\hat{\mathbf{S}}_N$ and $\hat{\mathbf{S}}$. Here $\hat{\mathbf{S}}_N$ is the nuclear spin operator and $\hat{\mathbf{S}}$ is the electron spin operator. The first term makes it possible to couple the orbital motion to the nuclear spin, so that changing the state of the rotating electron can be accompanied by a change of the nuclear spin. The other term couples the electron spin to the nuclear spin. These terms affect the spectra of the atoms and molecules (inducing more splittings, call the hyperfine splitting) and play a role in NMR spectroscopy.

We have introduced spin and its interactions through a succession of guesses. We have assumed that the spin is an intrinsic angular momentum of the particle which is accompanied by a magnetic moment. In some cases we used classical equations for energy and replaced certain quantities in them with the corresponding operators. Dirac developed a relativistic theory of the electrons from which all these terms follow mathematically, including the existence of spin. The theory generates two additional terms in the Hamiltonian of the hydrogen atom. One appears because the velocity of the electron is sufficiently close to the speed of light to require a relativistic correction to the kinetic energy. The other, called the Darwin term, does not have a simple physical interpretation. These terms were not included in our analysis. Dirac's equation also predicted the existence of positron and introduced for the first time the concept of antimatter.

20

THE ELECTRONIC STRUCTURE OF MOLECULES: THE H₂ MOLECULE

Introduction

§1. In this chapter, we study the properties of the electrons in the hydrogen molecule. Our intention is to use quantum mechanics to understand why molecules are stable and to calculate their energy and other electronic properties. This is the most complex subject in quantum mechanics and the current state of this field is too advanced to be presented in an introductory textbook. Nevertheless, the fundamental principles can be learned by examining the bond in the hydrogen molecule. Because of the complexity of this chapter an outline of the content of each section is given below.

On p. 355 we start with the Born–Oppenheimer (BO) approximation which says that the energy $E(R)$ of the electrons should be calculated while keeping distance R between the atom fixed. Born and Oppenheimer have also shown that the electronic energy $E(R)$ is the potential energy controlling the movement of the nuclei. By calculating $E(R)$ for a variety of inter-nuclear distances, we determine the *potential energy surface* of the molecule. This controls the bond length R_e (the value of R for which $E(R)$ has a minimum), the binding energy (the difference between the energy of the molecule at $R = R_e$ and the energy of the separated atoms) and the vibrational motion of the nuclei. This approximation has a shortcoming: no radiationless transition (i.e. a transition that is not accompanied by photon emission or absorption) between two electronic states is possible. The approximation breaks down when the molecule reaches a nuclear distance for which two electronic states have nearly equal energy.

On pp. 362–369 the energy eigenstates and eigenvalues of the electrons (within the BO approximation) are calculated by using a *variational principle*. This states that the expression $F[\Phi] = \langle \Phi \,|\, \hat{H}_{BO} \,|\, \Phi \rangle$, which depends on the normalized wave function Φ of the electrons, has a minimum when Φ is equal to the exact wave function Ψ. Moreover, $F[\Psi]$ is the exact energy of the electrons in the molecule. If we guess that $\Phi(a, b, c, \ldots)$ has a certain functional form, which depends on the parameters $a, b, c, \ldots$, we can minimize $F[\Phi(a, b, c, \ldots)]$ with respect to the parameters. If $\bar{a}, \bar{b}, \bar{c}, \ldots$ are the parameter values for which $F[\Phi(a, b, c, \ldots)]$ has a minimum, then $\Phi(\bar{a}, \bar{b}, \bar{c}, \ldots)$ is the most accurate wave function having the guessed functional form. In addition, $F[\Phi(\bar{a}, \bar{b}, \bar{c}, \ldots)]$ is the most accurate energy that the functional form can produce.

On pp. 369–372 we deal with the many-body wave function. Quantum mechanics of a system of N particles is made extremely complicated by the fact that the wave function depends on $3N$ coordinates. All other fields of physics that study an N-body problem deal with N functions of three coordinates. Since all observables are calculated by performing integrals involving wave functions, the theory needs to evaluate $3N$-dimensional integrals. This is beyond the capability of existing computers. Because of this, we represent the wave function as sums of products of functions of three variables called *molecular orbitals* (MO). This allows a reduction of the $3N$-dimensional integrals to six-dimensional integrals, whose evaluation is manageable on the existing computers.

A fundamental principle of quantum mechanics requires that the wave function of the electrons changes sign when we permute the coordinates of two electrons. Because of this, we must make sure that the products of molecular orbitals that represent the wave function of the electrons are *antisymmetrized* (pp. 372–387).

To implement these ideas we must construct appropriate molecular orbitals, as discussed on pp. 383–387. Often they are constructed as a *linear combination of atomic orbitals* (LCAO). In our study of the hydrogen molecule we use only two atomic orbitals. In this case we can have only two molecular orbitals, σ_g and σ_u and can determine them by using the symmetry of the molecule.

On pp. 387–394 we discuss the requirement that the electronic wave function must contain information about the spin of the electrons and it must be an eigenstate of the spin operators $\hat{\mathbf{S}}^2$ and S_z. To make sure that this is the case the products of molecular orbitals, which are used to construct the electronic wave functions, must be multiplied with the appropriate spin states. The electronic wave function $\Psi(1, 2)$ constructed on the basis of this idea has the form

$$\Psi(1, 2) = \sum_{n=1}^{m} C_n \Phi(1, 2) \qquad (20.1)$$

where $\Phi(1, 2)$ consists of products of molecular orbitals multiplied by the appropriate spin states. The functions $\Phi(1, 2)$ are known (we constructed them) and the coefficients C_n are determined by using the variational principle. This procedure is called the *configuration interaction* (CI) method and the wave functions $\Phi(1, 2)$ are called configurations.

To implement the variational principle we need to evaluate a great variety of integrals. The results of these evaluations are given on pp. 394–405.

Before implementing the CI method, on pp. 405–409 I examine the results obtained by using perturbation theory. This is an old method, which is often presented in the textbooks, but is no longer in use. We will see that it leads to qualitatively erroneous results that are substantially improved by the CI method. We study it because it dramatizes the need for using CI.

Finally, on pp. 409–416, we use the CI method to calculate the ground state energy and wave function of the electrons in the H_2 molecule.

The calculations presented here illustrate most of the conceptual difficulties encountered while solving electronic structure problems by more advanced methods. I hope that I made this problem as simple as possible, but not simpler.

The Born–Oppenheimer Approximation

§2. To calculate any property of a molecule, we must solve the Schrödinger equation for nuclei *and* electrons. This is a difficult calculation which has been attempted only recently and the methodology is still under development. All computer codes performing electronic structure calculations start with an approximation, proposed by Born and Oppenheimer. They showed that, under most circumstances, one can solve the problem in two stages.

First, write the Schrödinger equation for the electrons, with fixed nuclear positions. Nuclei that do not move have zero kinetic energy. Because of this, the kinetic energy operator for the nuclei does not appear in the Hamiltonian. This means that the Born–Oppenheimer Hamiltonian of the electrons in the hydrogen molecule is:

$$\hat{H}(\mathbf{r}_1, \mathbf{r}_2; R) = -\frac{\hbar^2}{2m}\nabla^2_{\mathbf{r}_1} - \frac{\hbar^2}{2m}\nabla^2_{\mathbf{r}_2} \tag{20.2}$$

$$-\frac{e^2}{4\pi\varepsilon_0 r_{1A}} - \frac{e^2}{4\pi\varepsilon_0 r_{2A}} \tag{20.3}$$

$$-\frac{e^2}{4\pi\varepsilon_0 r_{1B}} - \frac{e^2}{4\pi\varepsilon_0 r_{2B}} \tag{20.4}$$

$$+\frac{e^2}{4\pi\varepsilon_0 r_{12}} \tag{20.5}$$

$$+\frac{e^2}{4\pi\varepsilon_0 R} \tag{20.6}$$

We use here the following notation for the coordinates (see Fig. 20.1):

- $\mathbf{r}_i$, $i = 1, 2$ is the vector giving the position of electron i,
- $r_{i\alpha}$ $i = 1, 2$ and $\alpha = A, B$ is the distance of electron i to the nucleus α,
- r_{12} is the distance between the electrons,
- R is the distance between the nuclei.

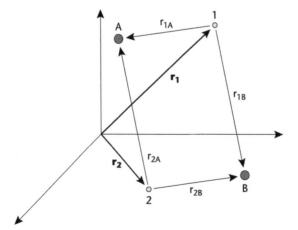

Figure 20.1 The vectors and the distances used in the Hamiltonian of H_2.

The terms in the Hamiltonian Eq. 20.6 have the following physical meaning:

- $-\frac{\hbar^2}{2m}\nabla_{\mathbf{r}_i}^2$ is the kinetic energy operator of electron i,

- $-\frac{e^2}{4\pi\varepsilon_0 r_{i\alpha}}$ is the Coulomb interaction between electron i and nucleus α,

- $\frac{e^2}{4\pi\varepsilon_0 r_{12}}$ is the Coulomb interaction between the electrons,

- $\frac{e^2}{4\pi\varepsilon_0 r_{AB}}$ is the Coulomb interaction between the nuclei.

The fact that in the BO approximation the nuclei are fixed has two consequences. (1) The kinetic energy operator of the nuclei is no longer present in the Hamiltonian. (2) The nuclear coordinates are no longer operators but numerical parameters, on a par with the mass or the charge of the particles. This is why in the symbol $\hat{H}(\mathbf{r}_1,\mathbf{r}_2;R)$, the distance R between nuclei is given a separate place in the argument list.

The Schrödinger equation

$$\hat{H}(\mathbf{r}_1,\mathbf{r}_2;R)\,\Psi(\mathbf{r}_1,\mathbf{r}_2;R) = E(R)\,\Psi(\mathbf{r}_1,\mathbf{r}_2;R) \tag{20.7}$$

provides the wave function and the energy of the electrons in the molecule, when the internuclear distance is fixed at R. Note that in the electronic wave function $\Psi(\mathbf{r}_1,\mathbf{r}_2;R)$ the nuclear coordinates are parameters. We obtain different electronic states for different values of R.

As with all eigenvalue problems, Eq. 20.7 has many solutions, which we label $E_0(R)$, $E_1(R),\ldots$ and $\Psi_0(\mathbf{r}_1,\mathbf{r}_2;R)$, $\Psi_1(\mathbf{r}_1,\mathbf{r}_2;R),\ldots$. $E_0(R)$ is the energy of the *ground electronic state* and $\Psi_0(\mathbf{r}_1,\mathbf{r}_2;R)$ is the wave function of the electrons in the ground state; $E_1(R)$ is the potential energy surface of the *first excited electronic state* and $\Psi_1(\mathbf{r}_1,\mathbf{r}_2;R)$ is the corresponding electronic wave function. We can solve Eq. 20.7 repeatedly, for various values of R, and generate the functions $E_0(R)$, $E_1(R),\ldots$. These are called the ground state *potential energy surface* or the first excited state potential energy surface, respectively. Different electronic states generate different potential energy surfaces.

These features are general. For an arbitrary molecule the Hamiltonian will contain the kinetic energies of all electrons, the Coulomb interaction between all electrons, the Coulomb interaction of all electrons with all nuclei and the interaction between all nuclei. If the coordinates of the N nuclei in the molecule are labeled $\mathbf{R} \equiv \{\mathbf{R}_1, \mathbf{R}_2, \ldots, \mathbf{R}_N\}$, then the potential energy surfaces $E_0(\mathbf{R})$, $E_1(\mathbf{R}),\ldots$ are functions of the $3N$ coordinates of all nuclei. The electronic wave functions depend on the coordinates of all electrons and all nuclei. It is the presence of these multi-dimensional functions in the theory that makes solving the Schrödinger equation extremely difficult. All other branches of physics deal with functions of three variables, which are much easier to handle.

§3. *The Electronic Energies $E_n(\mathbf{R})$ are the Potential Energies for the Nuclear Motion.* We started with the full Hamiltonian and then discarded the kinetic energy operator of the nuclei. This provided a Hamiltonian for the electrons, in which the nuclei no longer move and their positions are parameters in the electronic wave functions and energies. However, most physical phenomena occur because of nuclear motion. We could not have studied the spectroscopy of the diatomic molecules without information about their vibrational and rotational motion. How is this information obtained in the BO approximation?

Born and Oppenheimer have shown that the electronic energies $E_i(\mathbf{R}_1, \ldots, \mathbf{R}_m)$, where $\mathbf{R}_1, \ldots, \mathbf{R}_m$ are the nuclear coordinates, are the potential energies controlling nuclear motion. For example, the Schrödinger equation for the diatomic molecule AB, in the electronic ground state, is

$$
-\frac{\hbar^2}{2M_A}\nabla^2_{\mathbf{R}_A}\phi(\mathbf{R}_A, \mathbf{R}_B) - \frac{\hbar^2}{2M_B}\nabla^2_{\mathbf{R}_B}\phi(\mathbf{R}_A, \mathbf{R}_B) + E_0(R)\phi(\mathbf{R}_A, \mathbf{R}_B)
$$
$$
= E\,\phi(\mathbf{R}_A, \mathbf{R}_B) \tag{20.8}
$$

The vectors $\mathbf{R}_A$ and $\mathbf{R}_B$ specify the position of the nuclei of A and B, M_A and M_B are the masses of the atoms A and B, and E is the total energy of the nuclei (the eigenvalue). In this equation, the electronic energy $E_0(R)$ is the *potential energy of the nuclei*. This is why $E_0(R)$ is called the potential energy surface of the ground electronic state. In previous chapters, this quantity was denoted by $V(R)$.

Furthermore, if the electrons are in the state $\Psi_1(\mathbf{r}_1, \mathbf{r}_2; \mathbf{R}_A, \mathbf{R}_B)$, the Schrödinger equation for the nuclei is the same as Eq. 20.8, except that $E_0(R)$ is replaced by $E_1(R)$.

For most internuclear distances the energy difference between E_1 and E_0 is much larger than the energies of the collisions between molecules at room temperature. Unless we heat a system to very high temperatures, the molecules are in the ground electronic state. Why then bother with the excited states? Molecules can be excited into the state Ψ_1, or $\Psi_2, \ldots$ by exposing them to visible or ultraviolet light of appropriate frequency, or to a beam of energetic electrons. The evolution of the nuclei, after light absorption and excitation to an electronically excited state Ψ_i, is described by the nuclear Schrödinger equation, Eq. 20.8, with $E_0(R)$ replaced by $E_i(R)$. This evolution determines the absorption spectrum and whether or not the molecule dissociates into fragments. The latter phenomenon is the essence of photochemistry.

§4. *Why is this an Approximation and what are we Missing?* I described the recipe recommended by Born and Oppenheimer, but I did not explain why and when it can be used safely and what happens when it breaks down.

In the original work, the approximation was based on the large difference between the mass of the electron and that of the nuclei. Because the kinetic energy terms in the Hamiltonian are divided by the mass of the particle, the kinetic energy of the nuclei is much smaller than that of the electron. This means that we can treat the kinetic energy of the nuclei as a small perturbation. To obtain the approximation explained here, Born and Oppenheimer expanded the equation and its solutions in powers of m/M and kept only the low-order terms. The approximation is general and can be used whenever the system consists of very light and very heavy particles: treat the light ones the way Born and Oppenheimer treated the electrons and the heavy ones the way they treated the nuclei.

This is, however, not the whole story: sometimes the small terms do matter and their presence in the correct equation leads to new phenomena. One consequence of the Born–Oppenheimer approximation is that the approximate Eqs 20.2 through 20.8 contain no coupling between different electronic states. If these equations were correct, then once the system is excited to an electronic state Ψ_i it will remain in that state until it emits a photon. If photon emission is forbidden by the selection rules, the molecule will remain forever in the state Ψ_i. Experiments show that in most cases this is not true: a molecule excited to an electronic state Ψ_i manages to change its electronic state without photon emission or absorption. If the Born–Oppenheimer procedure was correct such *radiationless transitions* would not take place.

To understand why this happens, we need to abandon the idea that the ratio of the electron mass to the nucleus mass is the only important factor. It is often the case that the difference between the ground-state energy E_0 and the excited-state energies E_1, E_2, etc. is much larger than the energy of the collisions between the molecules in a gas or a liquid. In this case the collisions do not have enough energy to excite the molecules, and they will stay in the state Ψ_0 for ever. The existence of the other excited states will not matter and the fact that the ground state is not coupled to them is inconsequential. In this situation the Born–Oppenheimer approximation works. There are exceptions, however. The energy required to excite the electrons in a metal is infinitesimally small. Because of this, when a molecule collides with the surface of a metal, there is a finite probability that the metal will be electronically excited. This excitation energy is taken from the energy of the incident molecule. Because of this energy loss, the molecule is no longer able to overcome the attraction to the metal surface and ends up "glued" to the surface (the technical term is that the molecule is adsorbed). This process violates the BO approximation. The extent to which these electronic excitations affect surface-molecule processes, such as adsorption or catalysis, is a topic of much current research. My colleagues Alec Wodtke and Eric McFarland are leaders in that field.

Behavior in disagreement with the BO approximation is also common in photo-chemistry. Let us examine what happens to a molecule which has been excited to an upper electronic state Ψ_i (by absorption of a photon or by collision with an electron). If the BO approximation is correct, the molecule can change its state only by emiting a photon. There are many cases, however, where the molecule does not emit a photon, but manages to return to the ground electronic state. How do we know that? We monitor photon absorption and photon emission and find that photons have been adsorbed and have not been emitted. The sample also heats up: the energy of the adsorbed photons ends up as heat and the molecules end up in the ground electronic state. We say that the molecules have undergone *radiation-less transitions* that changes their state from Ψ_i to the ground state without photon emission.

It is not difficult to understand this phenomenon qualitatively. If the energy of the state Ψ_i is very different from the energy of the other electronic states Ψ_j, then it is very difficult to have a radiationless transition from Ψ_i to Ψ_j. The reason is simple: the transition does not conserve energy. In this case, the molecule will either emit a photon or will stay in the state Ψ_i for a very long time, as the Born–Oppenheimer approximation predicts. However, this is not the whole story. The potential energies $E_i(R)$ of the states Ψ_i depend on the internuclear distance. For some values of R the energy $E_i(R)$ can become equal to the energy $E_j(R)$ of another state. When this happens we say that the two states cross. When this crossing occurs our argument that a radiationless transition, from the state Ψ_i to the state Ψ_j, does not conserve energy, is no longer operative. When the molecule is in the state Ψ_i and the nuclei reach a position R for which $E_i(R) = E_j(R)$, a transition from Ψ_i to Ψ_j can occur without violating energy conservation. The transition rate is determined by the terms that the Born–Oppenheimer approximation threw away, on account of being small. When a crossing occurs, we cannot neglect these terms and their effects: the BO approximation breaks down.

Are such crossings common or is this some exotic notion invented by people trying to give Born and Oppenheimer a hard time? You can decide for yourself, after

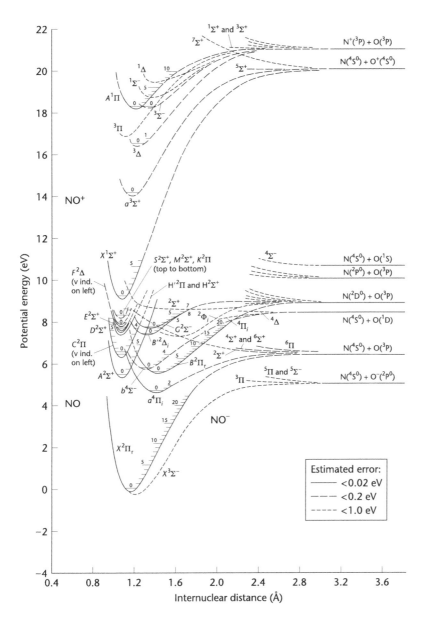

Figure 20.2 Potential energy diagram for NO, from J.I. Steinfeld, *Molecules and Radiation: An Introduction to Modern Molecular Spectroscopy*, The MIT Press, 1981. That work cites: F.R. Gilmore, *Potential Energy Curves for N₂, NO, O₂, and corresponding ions*, RAND Corporation Memorandum R-4034-PR, June 1964.

you take a look at Fig. 20.2 which shows the potential energy surfaces of the NO molecule (these "surfaces" are curves, since they depend only on the distance between the O and N atoms). This infamous molecule, a common pollutant in the smog of Los Angeles, has a large number of electronic states, many of which cross. To understand the chemistry of NO formation and decomposition in atmosphere we need to know these states and their crossings. Without that, we cannot undestand smog photochemistry.

The Variational Principle

§5. If we want to understand the properties of electrons in a molecule, we must solve the Schrödinger equation with the Born–Oppenheimer Hamiltonian. For the H_2 molecule, this means solving Eq. 20.7 with the Hamiltonian given by Eqs 20.2–20.6. It is easy to generalize this equation for an arbitrary molecule: just include in Eqs 20.2–20.6 the kinetic energies of all electrons, the interaction energies between all electrons, the interaction energies of all electrons with all nuclei, and the interaction energies between all nuclei.

One of the most popular methods for solving the Schrödinger equation uses the variational principle explained below. This works with the "object"

$$
E[\Psi; R] \equiv \langle \Psi \mid \hat{H} \mid \Psi \rangle
$$

$$
= \int d\mathbf{r}_1 \cdots d\mathbf{r}_N \Psi^*(\mathbf{r}_1, \ldots, \mathbf{r}_N; R) \hat{H}(\mathbf{r}_1, \ldots, \mathbf{r}_N; R) \Psi(\mathbf{r}_1, \ldots, \mathbf{r}_N; R)
$$

$$(20.9)$$

Here $\mathbf{r}_1, \ldots, \mathbf{r}_N$ are the coordinates of the N electrons in the molecule and R is a symbol for all nuclear coordinates. $\hat{H}(\mathbf{r}_1, \ldots, \mathbf{r}_N; R)$ is the Hamiltonian of the electrons in the Born–Oppenheimer approximation.

Think of the symbol $E[\Psi; R]$ as a function of Ψ, which also depends on the nuclear coordinates R. In this calculation, the nuclear coordinates are held fixed and you should think of them as parameters, not as variables. The argument Ψ of $E[\Psi; R]$ is *a function* (not a number) and the value of $E[\Psi; R]$ is a number. Because of this, $E[\Psi; R]$ is called a *functional* of Ψ (not a function of Ψ.)

The variational theorem says that if you insert in Eq. 20.9 all conceivable functions Ψ that satisfy the normalization condition

$$
\langle \Psi \mid \Psi \rangle \equiv \int d\mathbf{r}_1 \cdots d\mathbf{r}_N \Psi^*(\mathbf{r}_1, \ldots, \mathbf{r}_N; R) \Psi(\mathbf{r}_1, \ldots, \mathbf{r}_N; R) = 1, \qquad (20.10)
$$

you get the smallest values for $E[\Psi; R]$ if, and only if, Ψ is the lowest energy eigenstate of $\hat{H}$. In addition, this lowest value of $E[\Psi; R]$ is the ground-state energy of the system having the Hamiltonian $\hat{H}$.

This theorem is valid for any Hamiltonian in quantum mechanics, not just for that of the electrons in the molecule.

§6. *How to use the Variational Principle.* Many methods for solving differential equations are based on guessing the form of the solution and then varying the parameters in that form to make that guess as accurate as possible. Some of these guesses are based on mathematical theorems, and some on physical arguments. Occasionally one picks, in desperation, a form that seems flexible enough to describe the solution.

No matter how the choice is made, the selected form must contain some parameters $a, b, c, \ldots$ that can be adjusted to make the guess as accurate as possible. The variational principle is used to make this adjustment. Here is how this is done.

Assume that you have chosen a functional form $\Psi(\mathbf{r}_1, \ldots, \mathbf{r}_N; R; a, b, c)$ that contains three parameters a, b, c, whose numerical values are unknown. In addition the function satisfies Eq. 20.10. Then

$$E[\Psi; R; a, b, c] = \int d\mathbf{r}_1 \cdots d\mathbf{r}_N \Psi^*(\mathbf{r}_1, \ldots, \mathbf{r}_N; R; a, b, c)$$
$$\times \hat{H}(\mathbf{r}_1, \ldots, \mathbf{r}_N; R)\Psi(\mathbf{r}_1, \ldots, \mathbf{r}_N; R; a, b, c) \qquad (20.11)$$

depends on a, b, and c; as I change the value of a or b or c, the value of $E[\Psi; R; a, b, c]$ changes. The variational principle tells me that the values of a, b, c for which $E[\Psi; R; a, b, c]$ has a minimum give the best representation of the ground-state eigenfunction of $\hat{H}$ that the guess $\Psi(\mathbf{r}_1, \ldots, \mathbf{r}_N; R; a, b, c)$ is capable of. Moreover, the minimum value of $E[\Psi; R; a, b, c]$ is the best energy that the guess can produce. A bad guess gives bad results, but the variational principle ensures that you obtain the best result your bad guess can give.

§7. *Application to the Harmonic Oscillator.* To make this concrete, I present two examples. In the first, I calculate the ground state of a harmonic oscillator (representing the vibrational motion of a diatomic molecule). My guess for the wave function is

$$\phi(x) = Ae^{-\alpha x^2} \qquad (20.12)$$

where

$$x = r - r_0, \tag{20.13}$$

α is an unknown real number, and x is the difference between the bond length r and the equilibrium bond length r_0. I base this guess on my intuition that the most probable bond length should be $r = r_0$ (i.e. $x = 0$) and that the probability of finding a bond stretched a distance x away from r_0 decays rapidly with the magnitude of x.

The function $\phi(x)$ must be normalized and therefore

$$1 = \int_{-\infty}^{+\infty} \phi^*(x)\phi(x)dx = A^2 \int_{-\infty}^{+\infty} \left(\exp\left[-\alpha x^2\right]\right)^2 dx \tag{20.14}$$

Workbook

From this equation I obtain the prefactor A. The integral is performed in Cell 1 of Workbook QM20.1. It is finite and equal to $\sqrt{(\frac{\pi}{\alpha})}$ only if α is greater than zero. If this condition is not fulfilled the guess Eq. 20.12 cannot be a state of a quantum system. This value for the integral leads to (use Eq. 20.14)

$$A = \pm \left(\frac{2\alpha}{\pi}\right)^{1/4} \tag{20.15}$$

Since all equations for observable quantities involve the square of the wave function, the two solutions given by Eq. 20.15 are physically equivalent; I pick the positive one.

The functional is then

$$E[\phi; \alpha] = \int_{-\infty}^{+\infty} \phi^*(x)\hat{H}\phi(x)dx$$

$$= \left(\frac{2\alpha}{\pi}\right)^{1/2} \int e^{-\alpha x^2}\left[-\frac{\hbar^2}{2\mu}\frac{d^2}{dx^2} + \frac{\mu\omega}{2}x^2\right]e^{-\alpha x^2}dx \tag{20.16}$$

The quantity in the brackets is the Hamiltonian of the harmonic oscillator.

This integral is performed in Cell 4 of Workbook QM20.1 and has the value

$$E[\phi; \alpha] = -\frac{\mu\omega^2}{8\alpha} + \frac{\alpha\hbar^2}{2\mu} \tag{20.17}$$

To find the minimum with respect to α, I solve

$$\frac{\partial E[\phi; \alpha]}{\partial \alpha} = 0$$

which gives (Cell 5 of Workbook QM20.1)

$$\alpha = \pm \frac{\mu\omega}{2\hbar} \tag{20.18}$$

As I have already explained, a negative solution is not acceptable. This means that the best wave function of the form I have chosen is

$$\phi(x) = \left(\frac{2\alpha}{\pi}\right)^{1/4} \exp[-\alpha x^2] = \left(\frac{\mu\omega}{\pi\hbar}\right)^{1/4} \exp\left[-\frac{\mu\omega x^2}{2\hbar}\right] \tag{20.19}$$

The best energy is (insert α from Eq. 20.18 into Eq. 20.17)

$$E\left[\phi; \alpha = \frac{\mu\omega}{2\hbar}\right] = \frac{\hbar\omega}{2} \tag{20.20}$$

Because I made a good guess of the form of the wave function, these results are exact (see Chapter 16, §13). When given the correct functional form, the variational principle finds the parameter values that give the exact solution.

Exercise 20.1

Try the guess

$$\phi(x) = A e^{-\alpha|x|}$$

This functional form is a bad guess because it has a cusp at the origin and the first derivative is discontinuous. Quantum mechanics does not allow such solutions. Nevertheless, you could use it in the variational principle to see what happens. You need to think carefully how to handle the point $x = 0$. Compare your results to the exact solution.

Exercise 20.2

Use the variational principle to find the ground state of the hydrogen atom.

§8. *An Application of the Variational Principle that uses a Basis Set.*
Here is another example of using the variational principle for solving the
Schrödinger equation. I apply it to a three-dimensional problem in which a particle
moves under the influence of a potential.

As my guess for the wave function, I use

$$\Psi(\mathbf{r}) = \sum_{n=1}^{N} a_n \phi_n(\mathbf{r}) \tag{20.21}$$

Here $\mathbf{r}$ is the position of the particle. This equation expresses $\Psi(\mathbf{r})$ as a *linear
combination* of the *basis set* $\{\phi_n(\mathbf{r})\}_{n=1}^{N}$. We know the basis-set functions $\phi_n(\mathbf{r})$ but
not the coefficients a_n. We determine them by calculating $E[\Psi; a_1, \ldots, a_N]$ and
choosing $a_1, \ldots, a_N$ to minimize E.

Inserting this form of Ψ into Eq. 20.9 for E gives

$$E[\Psi; a_1, \ldots, a_N] = \int_{-\infty}^{+\infty} d\mathbf{r} \left(\sum_{n=1}^{N} a_n \phi_n(\mathbf{r}) \right)^* \hat{H} \left(\sum_{m=1}^{N} a_m \phi_m(\mathbf{r}) \right)$$

$$= \sum_{n=1}^{N} \sum_{m=1}^{N} a_n^* a_m H_{nm} \tag{20.22}$$

with

$$H_{nm} = \int_{-\infty}^{+\infty} d\mathbf{r} \phi_n^*(\mathbf{r}) \hat{H} \, \phi_m(\mathbf{r}) \tag{20.23}$$

The complex numbers H_{nm} are called the *matrix elements of $\hat{H}$ in the basis set*
$\{\phi_n(\mathbf{r})\}_{n=1}^{N}$. Since we know the basis-set functions and the Hamiltonian, we can
calculate the values of these matrix elements.

$E[\Psi; a_1, \ldots, a_N]$ depends on both a_n, $n = 1, \ldots, N$, and a_n^*, $n = 1, \ldots, N$. But these
parameters are not independent: if we know a_n then we know a_n^* (and vice versa).
I choose to consider $E[\Psi; a_1, \ldots, a_N]$ to be a function of $a_1^*, \ldots, a_N^*$.

To determine the values of $a_1, \ldots, a_N$ that minimize $E[\Psi; a_1, \ldots, a_N]$, I use the N equations

$$\frac{\partial E}{\partial a_v^*} = \sum_{m=1}^{N} H_{vm} a_m = 0 \qquad (20.24)$$

The first equality follows from Eq. 20.22; the second is the condition that $E[\Psi; a_1, \ldots, a_N]$ has a minimum with respect to the coefficients a_n.

However, Eq. 20.24 is not good enough! I must also ensure that the coefficients $\{a_1, \ldots, a_N\}$ are determined so that the function Ψ, given by Eq. 20.21, is normalized. This means that $a_1, \ldots, a_N$ must be such that

$$1 = \int \Psi^* \Psi d\mathbf{r} = \sum_{n=1}^{N} \sum_{m=1}^{N} a_n^* a_m S_{nm} \qquad (20.25)$$

with

$$S_{nm} \equiv \int_{-\infty}^{+\infty} d\mathbf{r} \phi_n^*(\mathbf{r}) \phi_m(\mathbf{r}) \qquad (20.26)$$

This equation is obtained by replacing Ψ in Eq. 20.25 with the expression given by Eq. 20.21. S_{nm} is called the *overlap of* ϕ_n *with* ϕ_m. The matrix having the elements S_{nm} is called the *overlap matrix of the basis set*.

To ensure that $a_1, \ldots, a_N$ minimize E and also satisfy Eq. 20.25, we use the method of *Lagrange multipliers*. This tells us to minimize

$$E[\Psi; a_1, \ldots, a_N] + \lambda \left(\int_{-\infty}^{+\infty} \Psi^* \Psi d\mathbf{r} - 1 \right)$$

$$= \sum_{n=1}^{N} \sum_{m=1}^{N} a_n^* a_m [H_{nm} + \lambda S_{nm}] - \lambda \qquad (20.27)$$

The parameter λ is called a Lagrange multiplier and its value is unknown, at this point. You will see shortly how λ is determined.

Taking the derivatives of this expression with respect to each a_ν^* and setting them to zero gives us the desired equations for $a_1, \dots, a_N$. They are

$$\sum_{m=1}^{N} H_{\nu m} a_m = \lambda \sum_{m=1}^{N} S_{\nu m} a_m, \quad \nu = 1, \dots, N \tag{20.28}$$

If you have studied matrix algebra you realize that $H_{\nu m}$ and $S_{\nu m}$ are the elements of two $N \times N$ matrices $\overset{\leftrightarrow}{H}$ and $\overset{\leftrightarrow}{S}$, and $\{a_m\}_{m=1}^{N}$ are the components (i.e. the "coordinates") of an N-dimensional vector $\vec{a}$. We can therefore write Eq. 20.28 as

$$\overset{\leftrightarrow}{H} \cdot \vec{a} = \lambda \overset{\leftrightarrow}{S} \cdot \vec{a} \tag{20.29}$$

If the basis-set functions are orthonormal, then

$$S_{nm} = \begin{cases} 1 & \text{if } n = m \\ 0 & \text{otherwise} \end{cases}$$

and Eq. 20.29 becomes

$$\overset{\leftrightarrow}{H} \cdot \vec{a} = \lambda \vec{a} \tag{20.30}$$

Workbook

This is an eigenvalue problem for the matrix $\overset{\leftrightarrow}{H}$. Eq. 20.29 can be viewed as a generalization of Eq. 20.30 and for this reason it is called a generalized eigenvalue problem. This problem can be solved by using programs provided by various computer manufacturers. The manner in which **Mathematica** performs matrix algebra and solves eigenvalue problems is explained in Workbook QM20.2.

The unknown Lagrange multiplier λ turns out to be the generalized eigenvalue of $\overset{\leftrightarrow}{H}$ and it is determined when a computer program is used to solve Eq. 20.29. If the basis set is orthogonal, the Lagrange multiplier is the eigenvalue of the Hamiltonian matrix $\overset{\leftrightarrow}{H}$.

As you know from linear algebra, an $N \times N$ matrix has N eigenvalues and N eigenvectors. If we denote the eigenvalues by $\lambda_1, \dots, \lambda_N$, then for each eigenvalue λ_α the computer returns an eigenvector $\vec{a}^\alpha$. I denote the components of this vector by

$\{a_1^\alpha, a_2^\alpha, \ldots, a_N^\alpha\}$. The wave function corresponding to the eigenvector $\vec{a}^\alpha$ is

$$\Psi_\alpha = \sum_{n=1}^N a_n^\alpha \phi_n(\mathbf{r}) \tag{20.31}$$

Exercise 20.3

Show that no matter what the basis set $\{\phi_1, \ldots, \phi_N\}$ is, the matrix element of the Hamiltonian satisfies the equation $H_{ij}^* = H_{ji}$.

Exercise 20.4

For the orthonormal basis set $\{\phi_1(r), \phi_2(r), \phi_3(r)\}$ the Hamiltonian matrix is

$$\overset{\leftrightarrow}{H} = \begin{pmatrix} 2.1 & 3.2 & 1.7 \\ 3.2 & 3.6 & 1.8 \\ 1.7 & 1.8 & 0.9 \end{pmatrix}$$

Use **Mathematica** or **Mathcad** to calculate the three eigenvectors and eigenvalues of $\overset{\leftrightarrow}{H}$. Write an explicit expression of the three wave eigenfunctions as linear combinations of the functions in the basis set.

Exercise 20.5

Show that you can derive Eq. 20.28 by the following procedure. Write the wave function $\Psi(r)$ as

$$\Psi(r) = \sum_{n=1}^N a_n \phi_n(r),$$

where $\phi_n(r)$ are known functions (a basis set). Introduce this expression in $\hat{H}\Psi(r) = E\Psi(r)$. Multiply with $\phi_j(r)^*$ and integrate over r.

The Many-Body Wave Function as a Product of Orbitals

§9. *The Curse of Multi-Dimensionality.* The variational principle is not the only method for solving the Schrödinger equation, but it is one of the most popular.

No matter which method we use, to connect to experiments we have to calculate integrals such as

$$E[\Psi] = \int \Psi^*(\mathbf{r}_1, \ldots, \mathbf{r}_N)\, \hat{H}\, \Psi(\mathbf{r}_1, \ldots, \mathbf{r}_N) d\mathbf{r}_1 \ldots d\mathbf{r}_N,$$

where $\hat{H}$ is the Hamiltonian or another operator. Any method of integration that uses a computer amounts to evaluating a sum. For example, a two-dimensional integral requires the evaluation of the sum

$$\int dx \int dy f(x,y) \cong \sum_{i=1}^{N} \sum_{j=1}^{N} w_{ij} f(x_i, y_j)$$

where $f(x_i, y_j)$ is the value of the function at the *grid point* $\{x_i, y_j\}$ and w_{ij} is a "weight" whose value depends on the method used for picking the values of x_i and y_j. If we use a 10-point grid for each variable, which is very modest, a two-dimensional integral requires 10^2 evaluations of the integrand $w_{ij} f(x_i, y_j)$.

Consider now what happens if we want to calculate the electronic wave function of a molecule having thirty electrons. The wave function will depend on thirty electron positions, which means ninety coordinates. To evaluate the energy functional we will have to perform a 90-dimensional integral. If we take ten grid points for each variable we must perform 10^{90} evaluations of the integrand. If we plan to calculate other observable quantities, we also need to keep in memory 10^{90} values of the wave function at the grid points. If a computer performs 10^{12} wave function evaluations per second, the calculation of the integral will take 10^{78} s. If we assume that memorizing one value of the wave function will take 12 atoms, then the memory needed to store the wave function will weigh approximately $10^{70} \times 12/(6 \times 10^{23}) \approx 10^{47}$ mol. Even taking the weight of 1 mol to be 1 g, anything weighing 10^{49} g is a hefty chunk of matter, which is not going to be available. These absurdly optimistic assumptions tell us that a brute force solution of the 90-dimensional electronic eigenvalue problem is not possible. We need an approximation that reduces the dimensionality of the problem, while providing a reasonable accuracy.

§10. *The Wave Function as a Product of Orbitals.* Consider now what happens to these forbidding integrals if we write the wave function as

$$\Psi(\mathbf{r}_1, \ldots, \mathbf{r}_N) = \sum_{i_1=1}^{M_1} \sum_{i_2=1}^{M_2} \cdots \sum_{i_N=1}^{M_N} C(i_1, i_2, \ldots, i_N)\phi_{i_1}(\mathbf{r}_1)\phi_{i_2}(\mathbf{r}_2)\cdots \phi_{i_N}(\mathbf{r}_N) \quad (20.32)$$

For example, for two electrons we can use

$$\Psi(\mathbf{r}_1, \mathbf{r}_2) = C(1,1)\phi_1(\mathbf{r}_1)\phi_1(\mathbf{r}_2) + C(1,2)\phi_1(\mathbf{r}_1)\phi_2(\mathbf{r}_2)$$
$$+ C(2,1)\phi_2(\mathbf{r}_1)\phi_1(\mathbf{r}_2) + C(1,3)\phi_1(\mathbf{r}_1)\phi_3(\mathbf{r}_2) + \cdots$$

The set $\{\phi_1(\mathbf{r}), \phi_2(\mathbf{r}), \ldots, \phi_N(\mathbf{r})\}$ consists of known functions that depend on the position of one electron only. In what follows, I call them *molecular orbitals*, even though the literature reserves this name for a special class of one-electron functions. The success of this approximation will depend on our ability to choose well the functional forms of these orbitals. The symbols $C(i_1, i_2, \ldots, i_N)$ are unknown constants, to be determined by using the variational principle.

Eq. 20.32 simplifies tremendously the multi-dimensional integrals. You can see why, if we look at what happens when we evaluate, as an example, the integral over the Coulomb interaction between electrons.

By expressing the N-electron wave function $\Psi(\mathbf{r}_1, \ldots, \mathbf{r}_N)$ in terms of the products $\phi_{i_1}(\mathbf{r}_1) \ldots \phi_{i_N}(\mathbf{r}_N)$ the integrals containing the N-electron wave function are reduced to a simpler form. For example, the interaction energy between electrons 1 and 2 becomes:

$$\int \phi_{i_1}(\mathbf{r}_1)^* \cdots \phi_{i_N}^*(\mathbf{r}_N) \frac{e^2}{4\pi\varepsilon_0|\mathbf{r}_1 - \mathbf{r}_2|}\phi_{\epsilon_1}(\mathbf{r}_1) \cdots \phi_{\epsilon_N}(\mathbf{r}_n)d\mathbf{r}_1 \ldots \mathbf{r}_N$$

$$= \frac{e^2}{4\pi\varepsilon_0} \int d\mathbf{r}_1 d\mathbf{r}_2 \phi_{i_1}(\mathbf{r}_1)\phi_{i_2}(\mathbf{r}_2)\frac{1}{|\mathbf{r}_1 - \mathbf{r}_2|}\phi_{\epsilon_1}(\mathbf{r}_1)\phi_{\epsilon_2}(\mathbf{r}_2)$$

$$\times \int d\mathbf{r}_3 \phi_{i_3}(\mathbf{r}_3)\phi_{\epsilon_3}(\mathbf{r}_3) \times \cdots \times \int d\mathbf{r}_N \phi_{i_N}(\mathbf{r}_N)\phi_{\epsilon_N}(\mathbf{r}_N) \quad (20.33)$$

If I picked an orthonormal set of orbitals, then

$$\int d\mathbf{r}_i \phi_\alpha(\mathbf{r}_i)\phi_\beta(\mathbf{r}_i) = \begin{cases} 1 & \text{if } \alpha = \beta \\ 0 & \text{otherwise} \end{cases}$$

and the integral in Eq. 20.33 is zero unless $i_3 = \epsilon_3, \ldots, i_N = \epsilon_N$. This reduces the N-dimensional integral further, to the six-dimensional integral of the form

$$\frac{e^2}{4\pi\varepsilon_0} \int d\mathbf{r}_1 \int d\mathbf{r}_2 \frac{\phi_\alpha(\mathbf{r}_1)\phi_\beta(\mathbf{r}_2)\phi_\lambda(\mathbf{r}_1)\phi_\mu(\mathbf{r}_2)}{|\mathbf{r}_1 - \mathbf{r}_2|}$$

We can handle this on today's computers, although doing it efficiently is a challenge. A modern computer program performing electronic-energy calculations spends most of the time evaluating these six-dimensional integrals.

The approximation defined by Eq 20.32 solves the problem of highly dimensional integration. How about storing the wave function? We know the orbitals, and define them to have simple forms that can be evaluated rapidly. Therefore, we only need to memorize the coefficients $C(i_1, \ldots, i_N)$ to be able to reconstruct rapidly the wave function Eq 20.32.

§11. *How to Choose Good Orbitals.* The quality of the results and the efficiency of the calculation based on Eq. 20.32 are strongly affected by the orbitals used. Many choices are possible and we defer discussion of this until later sections.

The Electron Wave Function must be Antisymmetric

§12. *Introduction.* Before implementing the scheme described above, we must make a further modification. One of the principles of quantum mechanics, which we have not discussed yet, demands that the wave function of a system of electrons must change sign when the coordinates of any two electrons are exchanged. In symbols,

$$\Psi(\mathbf{r}_1, \ldots, \mathbf{r}_i, \ldots, \mathbf{r}_j, \ldots, \mathbf{r}_N) = -\Psi(\mathbf{r}_1, \ldots, \mathbf{r}_j, \ldots, \mathbf{r}_i, \ldots, \mathbf{r}_N)$$

This requirement has no analog in classical physics and has profound consequences. It is no exaggeration to say, that all chemistry and biochemistry would be radically different if this requirement did not exist.

This section explains how this principle is implemented and explores some of its consequences. Its effects will be felt throughout this chapter, and I will comment on them in the appropriate places.

§13. *Indistinguishable Particles.* I confine this discussion to the case of two particles, but it should be clear that the principle is general. Two particles are indistinguishable if their Hamiltonian has the property

$$\hat{H}(\mathbf{r}_1, \mathbf{r}_2) = \hat{H}(\mathbf{r}_2, \mathbf{r}_1) \tag{20.34}$$

The Born–Oppenheimer Hamiltonian of the two electrons in the hydrogen molecule is

$$\hat{H}(\mathbf{r}_1, \mathbf{r}_2) = -\frac{\hbar^2}{2m}\nabla^2_{\mathbf{r}_1} - \frac{\hbar^2}{2m}\nabla^2_{\mathbf{r}_2} + \frac{e^2}{4\pi\varepsilon_0|\mathbf{r}_2 - \mathbf{r}_1|}$$

$$- \frac{e^2}{4\pi\varepsilon_0|\mathbf{r}_1 - \mathbf{R}_A|} - \frac{e^2}{4\pi\varepsilon_0|\mathbf{r}_1 - \mathbf{R}_B|} - \frac{e^2}{4\pi\varepsilon_0|\mathbf{r}_2 - \mathbf{R}_A|}$$

$$- \frac{e^2}{4\pi\varepsilon_0|\mathbf{r}_2 - \mathbf{R}_B|} + \frac{e^2}{4\pi\varepsilon_0|\mathbf{R}_A - \mathbf{R}_B|} \qquad (20.35)$$

Here $\mathbf{r}_1$ and $\mathbf{r}_2$ are the position vectors of the electrons, $\mathbf{R}_A$ and $\mathbf{R}_B$ are the position vectors of the nuclei, and $|\mathbf{r}_1 - \mathbf{r}_2|$, $|\mathbf{r}_1 - \mathbf{R}_A|$, $|\mathbf{r}_1 - \mathbf{R}_B|$, $|\mathbf{r}_2 - \mathbf{R}_A|$, $|\mathbf{r}_2 - \mathbf{R}_B|$, and $|\mathbf{R}_A - \mathbf{R}_B|$ are the distances between the corresponding particles.

Now permute the labels 1 and 2 in $\mathbf{r}_1$ and $\mathbf{r}_2$. You can easily convince yourself that the Hamiltonian does not change. Because of this, we say that the two electrons in the hydrogen molecule are indistinguishable. It is easy to see that any two electrons in any molecule, are indistinguishable.

Exercise 20.6

Are the protons in the H_2 molecule indistinguishable?

What does this property mean for the wave function? Your first reaction might be that the wave function must also be symmetric ($\Psi(\mathbf{r}_1, \mathbf{r}_2) = \Psi(\mathbf{r}_2, \mathbf{r}_1)$): interchanging the labels makes no difference in the laboratory, so it should make no difference in the wave function.

Surprizingly, this statement is wrong. Two wave functions that describe the same physical situation can differ by a special factor, called a *phase factor*. Indeed, since all physical observables $\langle \hat{O} \rangle$ are of the form

$$\langle \hat{O} \rangle \equiv \int \Psi(\mathbf{r}_1, \mathbf{r}_2)^* \, \hat{O} \, \Psi(\mathbf{r}_1, \mathbf{r}_2) d\mathbf{r}_1 d\mathbf{r}_2$$

the wave functions $\Psi(\mathbf{r}_1, \mathbf{r}_2)$ and $\alpha\Psi(\mathbf{r}_1, \mathbf{r}_2)$ give the same result for $\langle \hat{O} \rangle$ as long as

$$\alpha^*\alpha = 1 \qquad (20.36)$$

(since $\int (\alpha\Psi)^* \, \hat{O} \, (\alpha\Psi) d\mathbf{r}_1 d\mathbf{r}_2 = \alpha^*\alpha \int \Psi^* \, \hat{O} \, \Psi d\mathbf{r}_1 d\mathbf{r}_2$).

A complex number α satisfying Eq. 20.36 can always be written as

$$\alpha = e^{i\phi} \tag{20.37}$$

where ϕ is a real number. This is why α is called a phase factor.

To repeat: two wave functions representing physically identical situations can differ by a phase factor. As a consequence, the fact that permuting the labels does not change the physical situation implies that the wave function must satisfy

$$\Psi(\mathbf{r}_2, \mathbf{r}_1) = e^{i\phi}\Psi(\mathbf{r}_1, \mathbf{r}_2) \tag{20.38}$$

which is more general than the equality of the wave functions. A more general condition is likely to be richer in physical consequences.

This argument tells us that permuting the labels is equivalent to multiplying the wave function with a phase factor. To determine the value of this phase factor, I introduce the permutation operator $\hat{P}$, defined by

$$\hat{P}\Psi(\mathbf{r}_1, \mathbf{r}_2) \equiv \Psi(\mathbf{r}_2, \mathbf{r}_1) \tag{20.39}$$

Using Eq. 20.38 gives

$$\hat{P}\Psi(\mathbf{r}_1, \mathbf{r}_2) = e^{i\phi}\Psi(\mathbf{r}_1, \mathbf{r}_2) \tag{20.40}$$

This tells me that $e^{i\phi}$ is an eigenvalue of $\hat{P}$ and $\Psi(\mathbf{r}_1, \mathbf{r}_2)$ is its eigenfunction. This still does not pin down ϕ. But look what happens when $\hat{P}$ is again applied to Eq. 20.40:

$$\hat{P}\hat{P}\Psi(\mathbf{r}_1, \mathbf{r}_2) = e^{i\phi}\hat{P}\Psi(\mathbf{r}_1, \mathbf{r}_2) = e^{i\phi}e^{i\phi}\Psi(\mathbf{r}_1, \mathbf{r}_2)$$

On the other hand, Eq. 20.39 can be used twice to obtain

$$\hat{P}\hat{P}\Psi(\mathbf{r}_1, \mathbf{r}_2) = \hat{P}\Psi(\mathbf{r}_2, \mathbf{r}_1) = \Psi(\mathbf{r}_1, \mathbf{r}_2)$$

Since these two equations both calculate $\hat{P}^2\Psi \equiv \hat{P}\hat{P}\Psi$, I must have

$$\left(e^{i\phi}\right)^2 = 1$$

and therefore

$$e^{i\phi} = \pm 1 \qquad (20.41)$$

Which value should I take?

Logic can no longer help here; we must appeal to experiment. It turns out that if the two indistinguishable particles have half-integer spin (1/2, 3/2, etc) then we must take the negative value and

$$\Psi(\mathbf{r}_1, \mathbf{r}_2) = -\Psi(\mathbf{r}_2, \mathbf{r}_1) \qquad (20.42)$$

Such particles are called *fermions*. Particles having integer spin are called *bosons* and their wave function must instead satisfy $\Psi(\mathbf{r}_1, \mathbf{r}_2) = \Psi(\mathbf{r}_2, \mathbf{r}_1)$.

This result can be easily generalized to systems having N indistinguishable particles. If they are fermions, their wave function must change sign when we permute any two coordinates (or the labels of any two electrons):

$$\Psi(\mathbf{r}_1, \ldots, \mathbf{r}_i, \ldots, \mathbf{r}_j, \ldots, \mathbf{r}_N) = -\Psi(\mathbf{r}_1, \ldots, \mathbf{r}_j, \ldots, \mathbf{r}_i, \ldots, \mathbf{r}_N) \qquad (20.43)$$

Exercise 20.7

You want to do a variational calculation of the wave function and the energy of two indistinguishable fermions in one dimension. Which of these is a better guess for the wave function?

$$x_1^2 \sin(x_1 x_2^3) \exp[-(x_1 - x_2)^2]$$

or

$$\left[x_1^2 \sin(x_1 x_2^3) - x_2^2 \sin(x_2 x_1^3) \right] \exp[-(x_1 - x_2)^2]$$

If the two particles are bosons, which of the two is the better choice? Explain your answers.

§14. *The Antisymmetrization of a Product of Orbitals.* We return now to the representation of the wave function given by Eq. 20.32. Let us examine it in the case of two electrons, such as the H_2 molecule or the He atom. Assume that the orbitals are the eigenfunctions of a one-electron Hamiltonian

$$\hat{H}(\mathbf{r})\phi_i(\mathbf{r}) = \varepsilon_i\phi_i(\mathbf{r}) \qquad (20.44)$$

I will not discuss now how $\hat{H}(\mathbf{r})$ is chosen, or how this equation is solved to provide $\phi_1(\mathbf{r}), \dots, \phi_N(\mathbf{r})$. For now, assume that we know them.

The index i in $\phi_i(\mathbf{r})$, $i = 1, 2, \dots, N$, labels the state of the electron, and it is telling me the energy (ε_i in this case) and also *the spin* of the electron. Two orbitals can differ because they have different energies or because they have different spin quantum number m_s. The electrons having the same energy ε and different spins (i.e. one has $m_s = \frac{1}{2}$ and the other $m_s = -\frac{1}{2}$) are in different states, which are represented by different orbitals. The orbitals that include information about the spin state of the electron are called *spin orbitals*.

We know now that the wave function

$$\Psi(\mathbf{r}_1, \mathbf{r}_2) = c_{11}\phi_1(\mathbf{r}_1)\phi_1(\mathbf{r}_2) + c_{12}\phi_1(\mathbf{r}_1)\phi_2(\mathbf{r}_2) + c_{22}\phi_2(\mathbf{r}_1)\phi_2(\mathbf{r}_2)$$
$$+ c_{13}\phi_1(\mathbf{r}_1)\phi_3(\mathbf{r}_2) + \cdots,$$

which was recommended in §10, is not acceptable since it is not antisymmetric. It must be modified to satisfy the symmetry requirements.

A product of two spin-orbitals becomes antisymmetric if we combine them as follows:

$$\hat{\mathcal{A}}\phi_i(\mathbf{r}_1)\phi_j(\mathbf{r}_2) \equiv \frac{1}{\sqrt{2}}\left[\phi_i(\mathbf{r}_1)\phi_j(\mathbf{r}_2) - \phi_i(\mathbf{r}_2)\phi_j(\mathbf{r}_1)\right] \qquad (20.45)$$

This equation defines the *antisymmetrization operator* $\hat{\mathcal{A}}$. It is clear that this function changes sign when I exchange the labels of the particles (i.e. when $\mathbf{r}_1$ and $\mathbf{r}_2$ are interchanged).

When writing this form, I have assumed that the set of spin-orbitals $\{\phi_1, \phi_2, \dots, \phi_N\}$ is orthonormal:

$$\langle \phi_i \mid \phi_j \rangle \equiv \int \phi_i(\mathbf{r})^*\phi_j(\mathbf{r})d\mathbf{r} = \begin{cases} 1 & \text{if } i = j \\ 0 & \text{if } i \neq j \end{cases}$$

The factor $1/\sqrt{2}$ is introduced to normalize $\hat{A}\phi_i(\mathbf{r}_1)\phi_j(\mathbf{r}_2)$. Indeed, it is easy to show that

$$\int d\mathbf{r}_1 d\mathbf{r}_2 \left[\hat{A}\phi_i(\mathbf{r}_1)\phi_j(\mathbf{r}_2)\right]^* \left[\hat{A}\phi_i(\mathbf{r}_1)\phi_j(\mathbf{r}_2)\right] = 1$$

To antisymmetrize $\Psi(\mathbf{r}_1, \mathbf{r}_2)$, we apply the antisymmetrization operator $\hat{A}$ to each term in the sum:

$$\hat{A}\Psi(\mathbf{r}_1, \mathbf{r}_2) = c_{11}\hat{A}\phi_1(\mathbf{r}_1)\phi_1(\mathbf{r}_2) + c_{12}\hat{A}\phi_1(\mathbf{r}_1)\phi_2(\mathbf{r}_2) + c_{22}\hat{A}\phi_2(\mathbf{r}_1)\phi_2(\mathbf{r}_2)$$

$$+ c_{13}\hat{A}\phi_1(\mathbf{r}_1)\phi_3(\mathbf{r}_2) + \cdots \tag{20.46}$$

This test function satisfies the symmetry requirements.

§15. *How do we Generalize to More than Two Electrons?* This section is optional. It is easy to create an antisymmetric wave function for a two-electron system. How can we do this for molecules with more than two electrons? An elegant answer was given by Slater, who pointed out that an antisymmetric wave function can be obtained by forming determinants with the spin-orbitals. Here is how this is done, for a molecule with three electrons. Suppose $\phi_1, \ldots, \phi_N$ are the spin-orbitals that I plan to use to form the electronic wave function. First I use them to form determinants such as

$$\Psi_{1,3,4}(\mathbf{r}_1, \mathbf{r}_2, \mathbf{r}_3) \equiv \frac{1}{3!} \begin{vmatrix} \phi_1(\mathbf{r}_1) & \phi_1(\mathbf{r}_2) & \phi_1(\mathbf{r}_3) \\ \phi_3(\mathbf{r}_1) & \phi_3(\mathbf{r}_2) & \phi_3(\mathbf{r}_3) \\ \phi_4(\mathbf{r}_1) & \phi_4(\mathbf{r}_2) & \phi_4(\mathbf{r}_3) \end{vmatrix}$$

Remember that interchanging two columns in a determinant causes its sign to change. This is why the function $\Psi_{1,3,4}(\mathbf{r}_1, \mathbf{r}_2, \mathbf{r}_3)$ is antisymmetric.

Exercise 20.8

Show that

$$\frac{1}{2!} \begin{vmatrix} \phi_i(\mathbf{r}_1) & \phi_i(\mathbf{r}_2) \\ \phi_j(\mathbf{r}_1) & \phi_j(\mathbf{r}_2) \end{vmatrix}$$

is equal to $\hat{A}\phi_i(\mathbf{r}_1)\phi_j(\mathbf{r}_2)$ given by Eq. 20.45.

Exercise 20.9

Use the properties of determinants to show that $\Psi_{1,3,4}(\mathbf{r}_1, \mathbf{r}_2, \mathbf{r}_3)$ defined above is equal to

$$\frac{1}{3!} [\phi_1(\mathbf{r}_1)\phi_3(\mathbf{r}_2)\phi_4(\mathbf{r}_3) - \phi_1(\mathbf{r}_1)\phi_4(\mathbf{r}_2)\phi_3(\mathbf{r}_3) - \phi_3(\mathbf{r}_1)\phi_1(\mathbf{r}_2)\phi_4(\mathbf{r}_3)$$

$$+ \phi_4(\mathbf{r}_1)\phi_1(\mathbf{r}_2)\phi_3(\mathbf{r}_3) + \phi_3(\mathbf{r}_1)\phi_4(\mathbf{r}_2)\phi_1(\mathbf{r}_3) - \phi_4(\mathbf{r}_1)\phi_3(\mathbf{r}_2)\phi_1(\mathbf{r}_3)]$$

The test wave function for the three-electron system will therefore be

$$\Psi(\mathbf{r}_1, \mathbf{r}_2, \mathbf{r}_3) = C_{1,2,3}\Psi_{1,2,3}(\mathbf{r}_1, \mathbf{r}_2, \mathbf{r}_3) + C_{1,2,4}\Psi_{1,2,4}(\mathbf{r}_1, \mathbf{r}_2, \mathbf{r}_3)$$

$$+ C_{2,3,4}\Psi_{2,3,4}(\mathbf{r}_1, \mathbf{r}_2, \mathbf{r}_3) + C_{2,3,5}\Psi_{2,3,5}(\mathbf{r}_1, \mathbf{r}_2, \mathbf{r}_3) + \cdots \quad (20.47)$$

This function is a sum of the determinants $\Psi_{i,j,k}$ formed with the spin-orbitals. The coefficients $C_{i,j,k}$ are determined by using the variational principle, which gives the best ground-state wave function and energy that a function of the form of Eq. 20.47 is capable of producing.

Each determinant in Eq. 20.47 is called a configuration. The method of calculation based on Eq. 20.47 is called *configuration interaction*. Its accuracy depends on how we choose the spin-orbitals, on how many of them we use, and on how many determinants we form with them. If we do everything right, we can obtain practically exact results for small molecules. For large molecules we are limited by insufficient computer power.

The simplest approximation, in this context, uses only one determinant and considers the spin-orbitals to be unknown functions. The variational principle is then used to derive equations for the orbitals. In this way it is possible to obtain the best spin-orbitals that a one-determinant approximation can produce. This is called the *Hartree–Fock* (HF) method, and the orbitals produced in this way are the HF orbitals.

The Hartree–Fock method gives useful results for the structure of the molecule and its vibrational frequencies, but has difficulties with the strength of the chemical bond. To obtain reliable bond energies, we use the HF orbitals in a configuration interaction calculation.

There are many other methods, whose accuracy and computer-power demands fall between those of the Hartree–Fock method and the configuration interaction method. All of them are included in a variety of software packages.

§16. *The Pauli Principle.* This section concludes by examining a few physical consequences of the antisymmetry requirement. Let us see what happens if we place two electrons in the same spin-orbital. For a two-electron system, this means that we take $i = j$ in Eq. 20.45. But this makes the two-electron wave function $\hat{A}\phi_i(\mathbf{r}_1)\phi_i(\mathbf{r}_2)$ equal to zero. Therefore, the probability of having two electrons in the same orbital is zero. This is the famous Pauli exclusion principle, which says that no two electrons can occupy the same orbital. This principle was formulated before the discovery of spin, and Pauli postulated that one can have only two electrons in each orbital. We now know that these two electrons have different spin-states, one has "spin up" and the other "spin down." Therefore the two electrons in the same orbital are not identical.

Let us see how this works in the case of Li. The spin-orbitals we chose to work with are those of the wave function of the ion Li^{2+}. They have the same form as the orbitals of the hydrogen atom, except that the nuclear charge is $+3$. The spin-orbitals we choose in our analysis are $\phi_1(\mathbf{r}) = 1s_+(\mathbf{r})$, $\phi_2(\mathbf{r}) = 1s_-(\mathbf{r})$, and $\phi_3(\mathbf{r}) = 3s_+(\mathbf{r})$. The notation "$1s_+(\mathbf{r})$" tells me that the electron has $n = 1$, $\ell = 0$, $m = 0$, and $m_s = \frac{1}{2}$. Here n is the radial quantum number, and the 1 in $1s_+$ tells us that $n = 1$. Since $n = 1$, the quantum number ℓ must be 0 (ℓ takes the values $0, 1, \ldots, n - 1$) and this is what the s in $1s_+(\mathbf{r})$ tells me. Since $\ell = 0$, we must have $m = 0$ (m takes the values $-\ell, -\ell + 1, \ldots, \ell - 1, \ell$). Finally, the $+$ in $1s_+$ tells me that the spin quantum number is $m_s = \frac{1}{2}$. $1s_-$ is a state with $n = 1$, $\ell = 0$, $m = 0$, and $m_s = -\frac{1}{2}$, and $2s_+$ is a state with $n = 2$, $\ell = 0$, $m = 0$, and $m_s = \frac{1}{2}$.

Exercise 20.10

What are the quantum numbers of $2p_{x,+}$ and of $3d_{x^2-y^2,+}$?

The lowest energy determinant is

$$
\begin{vmatrix}
1s_+(\mathbf{r}_1) & 1s_+(\mathbf{r}_2) & 1s_+(\mathbf{r}_3) \\
1s_-(\mathbf{r}_1) & 1s_-(\mathbf{r}_2) & 1s_-(\mathbf{r}_3) \\
2s_+(\mathbf{r}_1) & 2s_+(\mathbf{r}_2) & 2s_+(\mathbf{r}_3)
\end{vmatrix}
$$

If we put two electrons in the same orbital, as in

$$
\begin{vmatrix}
1s_+(\mathbf{r}_1) & 1s_+(\mathbf{r}_2) & 1s_+(\mathbf{r}_3) \\
1s_-(\mathbf{r}_1) & 1s_-(\mathbf{r}_2) & 1s_-(\mathbf{r}_3) \\
1s_+(\mathbf{r}_1) & 1s_+(\mathbf{r}_2) & 1s_+(\mathbf{r}_3)
\end{vmatrix}
$$

the determinant is zero (it has two identical rows). Such a state cannot exist.

Why did Pauli introduce such a principle? Imagine that we do not know about the principle and do the following. The hydrogen atom has an electron in a 1s orbital. The lowest energy of the He atom would place two electrons in the 1s orbital of a "hydrogen" atom with nuclear charge +2. Such a state neglects the interaction between electrons. We assume that this will lead to quantitative, but not qualitative, errors. The energy to ionize the He atom is much higher than that required to ionize the H atom, since in He the electron is removed from an 1s orbital with nuclear charge 2. This crude calculation tells us that ionizing He takes more energy than ionizing H, which is in agreement with the observations. The value of the ionization energy is not correct, but we expect this because we neglected electron–electron interaction. A catastrophy occurs, however, when we consider Li. The lowest energy state would have all three electrons in the 1s orbital of a Li^{2+} ion. Hence, for Li this will give an ionization energy much larger than that of He. Unfortunately, for this theory, the ionization potential of Li is very much lower than that of He. We arrive at a qualitative discrepancy with the facts. A more detailed examination of other properties and other atoms reveals more discrepancies. Pauli solved all these contradictions by postulating that no more than two electrons can occur in a given orbital. This implies that the state of Li is two electrons in the 1s orbital and one in a 2s orbital. Since the energy of the 2s orbital is much higher than that of 1s, the ionization energy of Li takes much less energy than the ionization of He. This is now in agreement with the facts.

Exercise 20.11

Use the theory developed in Chapter 18 to calculate the energy required to ionize a He atom, in the case that the electron–electron interaction can be neglected. Explain whether adding the electron–electron interaction will increase or decrease the ionization energy as compared to that calculated in the absence of the interaction.

§17. *Which Electrons should be Antisymmetrized?* How do I calculate the electronic energy of a gas of 10^{24} hydrogen atoms? The gas has 10^{24} electrons, and it seems that the ground-state wave function should be a $10^{24} \times 10^{24}$ determinant. This neglects the electrons in the wall of the container so, perhaps, they should also be included in the antisymmetrization. If this is necessary it seems that we should antisymmetrize the wave function of all electrons in the universe.

However, the experiments tell me that the properties of the H atoms (e.g. the light-absorption spectrum, the ionization energy) in the gas are the same as the properties of an isolated H atom. Therefore, each atom in the gas can be treated as if it is alone in the universe; it should not be necessary to antisymmetrize the wave function of all the electrons in the gas. Indeed, this is the correct procedure.

To understand why this procedure is legitimate, let us look at the case of two hydrogen atoms, labeled A and B. To describe this system I use a limited basis set, consisting of two molecular orbitals. One, denoted $1s_A$ is the $1s$ orbital of an electron located on the hydrogen atom A. The other one is $1s_B$ and it is located around the nucleus B. Adding more orbitals to the basis set will increase the accuracy of the calculations, but it will not affect the conclusion we are about to reach. Also, taking the spin into account will not change our results, and therefore I ignore it.

Examine now two descriptions of the system. One,

$$\Phi(\mathbf{r}_1, \mathbf{r}_2) = 1s_A(1)1s_B(2) \tag{20.48}$$

does not antisymmetrize the two electrons. In essence, it considers that two electrons, each located on a different atom, are distinguishable. This wave function ignores the symmetry principle.

The other,

$$\hat{\mathcal{A}}\Phi(\mathbf{r}_1, \mathbf{r}_2) = \frac{1}{\sqrt{2}} \left[1s_A(\mathbf{r}_1)1s_B(\mathbf{r}_2) - 1s_A(\mathbf{r}_2)1s_B(\mathbf{r}_1) \right] \tag{20.49}$$

takes into account the symmetry principle.

There is no doubt that, strictly speaking, Eq. 20.49 is correct and Eq. 20.48 is not. What I will show here is that if we have a physical situation which can be described as two H atoms (rather than a H_2 molecule), then the wave function that is not symmetrized is as good as the antisymmetrized one. In other words, by using Eq. 20.48 we make an error, but the error is extremely small. As you will see later in this chapter, if the two hydrogen atoms are close together and form a H_2 molecule, then only the antisymmetrized wave function gives correct results; the one that is not antisymmetrized, is worthless.

Here is how these statements are supported. Any observable quantity, for this two-hydrogen atoms system, is given by

$$\langle \Phi \mid \hat{O} \mid \Phi \rangle \equiv \int \Phi(\mathbf{r}_1, \mathbf{r}_2)^* \, \hat{O} \, \Phi(\mathbf{r}_1, \mathbf{r}_2) \, d\mathbf{r}_1 d\mathbf{r}_2$$

or

$$\langle \hat{A}\Phi \,|\, \hat{O} \,|\, \hat{A}\Phi \rangle \equiv \int \hat{A}\Phi(\mathbf{r}_1, \mathbf{r}_2)^* \, \hat{O} \, \hat{A}\Phi(\mathbf{r}_1, \mathbf{r}_2) \, d\mathbf{r}_1 d\mathbf{r}_2,$$

depending on which wave function we use. Here $\hat{O}$ is the operator of interest (e.g. the Hamiltonian, if we want to calculate the energy). If you use Eqs 20.48 and 20.49 for Φ and $\hat{A}\Phi$, you will find that $\langle \hat{A}\Phi \,|\, \hat{O} \,|\, \hat{A}\Phi \rangle$ differs from $\langle \Phi \,|\, \hat{O} \,|\, \Phi \rangle$ through integrals of the form

$$\int d\mathbf{r}_1 d\mathbf{r}_2 1s_A(\mathbf{r}_1) 1s_B(\mathbf{r}_2) \, \hat{O} \, 1s_A(\mathbf{r}_2) 1s_B(\mathbf{r}_1) \qquad (20.50)$$

If these integrals are zero, then $\langle \hat{A}\Phi \,|\, \hat{O} \,|\, \hat{A}\Phi \rangle$ is equal to $\langle \Phi \,|\, \hat{O} \,|\, \Phi \rangle$; this means that the antisymmetrized wave function $\hat{A}\Phi$ gives the same physical results as Φ, and we need not bother antisymmetrizing Φ. When does this happen?

Look how the product $1s_A(\mathbf{r}_1) 1s_B(\mathbf{r}_1)$, which appears in the integrand of Eq. 20.50, behaves when the distance R between the nuclei is large. The electron in $1s_A$ is located near nucleus A and that in $1s_B$ is located near nucleus B. Because of this, when R is large $1s_A(\mathbf{r}_1) 1s_B(\mathbf{r}_1)$ is zero: if $\mathbf{r}_1$ is close to A then $1s_B(\mathbf{r}_1)$ is zero; if $\mathbf{r}_1$ is close to B then $1s_A(\mathbf{r}_1)$ is zero; when $\mathbf{r}_1$ is between the nuclei both orbitals are zero.

How large should R be for this to happen? The 1s orbital decays exponentially on a length scale equal to the Bohr radius. This is of order 0.5 Å. So, if the two hydrogen atoms are separated by about 2 Å, then $1s_A(\mathbf{r}_1) 1s_B(\mathbf{r}_2) \approx 0$, the integrals of the form Eq. 20.50 are zero, and $\langle \Phi \,|\, \hat{O} \,|\, \Phi \rangle$ is equal to $\langle \hat{A}\Phi \,|\, \hat{O} \,|\, \hat{A}\Phi \rangle$; we need not antisymmetrize the two electrons and can treat the two H atoms as two independent entities. It is a relief to find out that to calculate the properties of one hydrogen atom, we do not need to antisymmetrize its electrons with all the electrons in the universe. The same argument holds for any atoms or molecules that are separated by a distance greater than that of a chemical bond.

We can reformulate our conclusion to say that if two electrons are localized in orbitals $\phi_1(\mathbf{r})$ and $\phi_2(\mathbf{r})$ that do not overlap, then we can treat them as if they are distinguishable particles: the wave function $\phi_1(\mathbf{r}_1)\phi_2(\mathbf{r}_2)$ can be used instead of $\hat{A}\phi_1(\mathbf{r}_1)\phi_2(\mathbf{r}_2)$. This does not mean that the principle becomes invalid when the electrons become spatially separated. It only means that its effects are very small and could be neglected.

Exercise 20.12

If the wave function $\Psi(\mathbf{r}_1, \mathbf{r}_2)$ describes a two-electron system, the probability that electron 1 is at $\mathbf{r}$ is

$$p_1(\mathbf{r}) = \int d\mathbf{r}_2 \Psi^*(\mathbf{r}, \mathbf{r}_2) \Psi(\mathbf{r}, \mathbf{r}_2)$$

The probability that electron 2 is at $\mathbf{r}$ is

$$p_2(\mathbf{r}) = \int d\mathbf{r}_1 \Psi^*(\mathbf{r}_1, \mathbf{r}) \Psi(\mathbf{r}_1, \mathbf{r})$$

Use these definitions to calculate $p_1(\mathbf{r})$ and $p_2(\mathbf{r})$ for the states $\Phi(\mathbf{r}_1, \mathbf{r}_2)$ and $\hat{A}\Phi(\mathbf{r}_1, \mathbf{r}_2)$ given by Eqs 20.48 and 20.49. Give a physical interpretation of these results.

The Molecular Orbitals in a Minimal Basis Set: σ_u and σ_g

§18.   *Outline.* To calculate the properties of the electrons in the H_2 molecule we will construct the configuration interaction (CI) wave function

$$\Psi(\mathbf{r}_1, \mathbf{r}_2) = C_{12}\hat{A}\phi_1(\mathbf{r}_1)\phi_2(\mathbf{r}_2) + C_{13}\hat{A}\phi_1(\mathbf{r}_1)\phi_3(\mathbf{r}_2)$$
$$+ C_{23}\hat{A}\phi_2(\mathbf{r}_1)\phi_3(\mathbf{r}_2) + \cdots \qquad (20.51)$$

This will be used, as a test function, in the variational principle. Insert this wave function in the energy functional and express it as a function of the coefficients $C(i,j)$. The best values of these coefficients are those that minimize the energy functional. Before this can be done, we have a lot of preparatory work to do. In this section the molecular orbitals $\phi_1(\mathbf{r})$, $\phi_2(\mathbf{r}), \ldots$ are determined by using a *minimal basis set* of two *atomic orbitals*.

§19.   *The MO-LCAO Method.* Nowadays, the molecular orbitals are chosen to be the eigenfunction of the Hartree–Fock Hamiltonian, which we mentioned briefly in an earlier section. Explaining the Hartree–Fock theory would take too much space and I will not deal with it here.

In the early days of quantum chemistry the molecular orbitals were taken to be the energy eigenfunctions of a Hamiltonian in which the electrons in the molecule did not interact with each other; each electron behaved as if it was alone in the molecule. This methodology will be used here.

In most methods used in quantum chemistry the molecular orbitals $\phi_1(\mathbf{r})$, $\phi_2(\mathbf{r})$,... are represented by some approximate form. A very popular one uses a *linear combination of atomic orbitals* (LCAO):

$$\phi_i(\mathbf{r}) = a_{i1} 1s_A(\mathbf{r}) + a_{i2} 1s_B(\mathbf{r}) + a_{i3} 2s_A(\mathbf{r})$$
$$+ a_{i4} 2s_B(\mathbf{r}) + a_{i5} 2p_x(\mathbf{r}) + \cdots \tag{20.52}$$

Here $1s_A(\mathbf{r})$ is the 1s atomic orbital of atom A, $1s_B(\mathbf{r})$ is the 1s atomic orbital of atom B, $2s_A(\mathbf{r})$ is the 2s atomic orbital of atom A, etc. The coefficients a_α are determined by forcing ϕ_i to be the eigenfunction of a Hamiltonian in which the electron interacts with the nuclei, but not with the other electrons.

§20. *The Minimal Basis Set.* An additional simplification is obtained by assuming that sufficient insight into the chemical bond can be obtained if it is assumed that the molecular orbitals can be written as a combination of only two atomic orbitals. In this case we can construct the molecular orbitals for H$_2$ by using the symmetry properties of the Hamiltonian. The use of this *minimal basis set* means that we keep only the first two terms in Eq. 20.52:

$$\phi(\mathbf{r}) = a_1 1s_A(\mathbf{r}) + a_2 1s_B(\mathbf{r}) \tag{20.53}$$

This choice, made in the late 1920s, was forced upon the early researchers by lack of computing power. To some extent it is based on common sense. When the independent electron (i.e. not interacting with the other electron in the molecule) is near nucleus A, and far from B, its wave function should be similar to $1s_A$. When it is close to nucleus B, and far from A, it should resemble $1s_B$. When it is between nuclei, it should be a combination of the two atomic orbitals.

It turns out, however, that the use of a minimal basis set in the calculation limits severely the accuracy of the potential energy surfaces. Fortunately, it maintains the mathematical structure of the theory and the qualitative features of the chemical bond. It is therefore useful in pedagogical work.

§21. *Determine the Molecular Orbitals by Using Symmetry.* There is no reason why an electron should prefer one atom over the other. By using this condition alone, the values of the coefficients a_1 and a_2 can be determined. This is possible because the molecule is symmetric and the basis set contains only two atomic orbitals.

The probability that an electron in the state ϕ given by Eq. 20.53 is in the state $1s_A$ is proportional to a_1^2; the probability that it is in the state $1s_B$ is proportional to a_2^2. Because of the symmetry of the molecule these probabilities must be equal and therefore

$$a_1^2 = a_2^2 \tag{20.54}$$

This equation has two solutions:

$$a_1 = a_2 \tag{20.55}$$

and

$$a_1 = -a_2 \tag{20.56}$$

This generates two orbitals:

$$\phi_1 = a_1(1s_A - 1s_B) \tag{20.57}$$

and

$$\phi_2 = a_1(1s_A + 1s_B) \tag{20.58}$$

§22. *The Molecular Orbitals are Normalized.* Strictly speaking the molecular orbitals used in the calculation do not have to be normalized. However, normalizing them simplifies the calculations. Normalization has an additional beneficial effect: it determines the value of a_1 in Eqs 20.57 and 20.58. The normalization of ϕ_1 requires that:

$$\langle \phi_1 \mid \phi_1 \rangle \equiv \langle a_1(1s_A - 1s_B) \mid a_1(1s_A - 1s_B) \rangle$$

$$= a_1^2 \{ \langle 1s_A \mid 1s_A \rangle + \langle 1s_B \mid 1s_B \rangle - 2\langle 1s_A \mid 1s_B \rangle = 1$$

But $\langle 1s_A \mid 1s_A \rangle = \langle 1s_B \mid 1s_B \rangle = 1$, because the atomic orbitals belonging to the same atom are normalized. Therefore

$$\langle \phi_1 \mid \phi_1 \rangle = 1 = 2a_1^2(1 - S) \tag{20.59}$$

where

$$S \equiv \langle 1s_A \mid 1s_B \rangle = \int d\mathbf{r}\, 1s_A(\mathbf{r}) 1s_B(\mathbf{r}) \tag{20.60}$$

is the *overlap integral* between the atomic orbitals.

Eq. 20.59 is used to calculate a_1:

$$a_1 = \pm \frac{1}{\sqrt{2(1-S)}} \tag{20.61}$$

Since these two solutions differ through a phase factor, they are physically equivalent and I use the positive one.

Introducing this in Eq. 20.57 leads to

$$\phi_1 \equiv \sigma_u(\mathbf{r}) = \frac{1s_A - 1s_B}{\sqrt{2(1-S)}} \tag{20.62}$$

From now on I abandon the notation ϕ_1 in favor of σ_u. This is a traditional notation that indicates that the electron participates in a σ-bond.

A second orbital, called σ_g,

$$\phi_2 \equiv \sigma_g(\mathbf{r}) = \frac{1s_A + 1s_B}{\sqrt{2(1+S)}}, \tag{20.63}$$

is obtained from ϕ_2, by using the procedure that converted ϕ_1 to σ_u.

Exercise 20.13

Verify that $\hat{L}_z \sigma_g = 0$ and $\hat{L}_z \sigma_g = 0$.

Constructing these two orbitals was easy because we used a basis set of two atomic orbitals and the molecule is symmetric. This procedure would not work if the molecule lacked symmetry (e.g. HLi^+) or if we had more atomic orbitals in the basis set.

§23. *The Symmetry of σ_g and σ_u.* The hydrogen molecule remains unchanged when we perform a *reflection* in a plane that is perpendicular to the bond and cuts it through the middle. This reflection amounts to placing nucleus A where B was located, and placing B at the original location of A. In other words the nuclei are permuted. The subscript g comes from the German word "gerade", which means even, and is telling us that σ_g remains unchanged when we interchange nuclei A and B. The u in σ_u comes from 'ungerade' and indicates that σ_u changes to $-\sigma_u$ when that interchange is made.

Exercise 20.14

Make contour plots of σ_u and σ_g in a plane that contains the molecular axis, for $R = 0.4$, and $R = 2.5$ Å. To calculate the overlap integral $S(R)$ see §42.

Exercise 20.15

Show that the symmetry arguments used above fail to determine the values of the coefficients a_i for the case when the basis set is $\{1s_A, 1s_B, 2s_A, 2s_B\}$.

Exercise 20.16

Show that the symmetry arguments used above fail to determine the values of the coefficients a_i for the case when the basis set is $\{1s_A, 1s_B\}$ and the nuclei A and B are not identical.

The Antisymmetrized Products used in the Configuration Interaction Wave Functions must be Eigenfunctions of $\hat{S}^2$ and $\hat{S}_z$

§24. *Outline.* So far we have ignored the spin of the electrons, and this is a very bad idea. The total spin operator of the two electrons,

$$\hat{\mathbf{S}} = \hat{\mathbf{S}}(1) + \hat{\mathbf{S}}(2), \tag{20.64}$$

is the sum of the spin $\hat{\mathbf{S}}(1)$ of electron 1 and the spin $\hat{\mathbf{S}}(2)$ of electron 2. This operatore *commutes* with the Hamiltoniam of the H_2 molecule. Because of this, the eigenstates of the Hamiltonian must also be eigenstates of $\hat{S}^2$ and $\hat{S}_z$. This means that the antisymmetrized functions, used to generate the CI wave function, must be forced to satisfy this condition. This section shows how this is done.

As explained in Chapter 19, an electron can have two spin states. The one denoted $|\alpha\rangle$ has $m_s = 1/2$ and satisfies

$$\hat{S}_z|\alpha\rangle = \frac{\hbar}{2}|\alpha\rangle$$

The state $|\beta\rangle$ has $m_s = -\frac{1}{2}$ and

$$\hat{S}_z|\beta\rangle = -\frac{\hbar}{2}|\beta\rangle$$

With two molecular orbitals σ_u and σ_g and two spin states $|\alpha\rangle$ and $|\beta\rangle$, we can form four *spin-orbitals*:

$$\phi_1(\mathbf{r}) = \sigma_u(\mathbf{r})\,|\alpha\rangle \tag{20.65}$$

$$\phi_2(\mathbf{r}) = \sigma_u(\mathbf{r})\,|\beta\rangle \tag{20.66}$$

$$\phi_3(\mathbf{r}) = \sigma_g(\mathbf{r})\,|\alpha\rangle \tag{20.67}$$

$$\phi_1(\mathbf{r}) = \sigma_g(\mathbf{r})\,|\beta\rangle \tag{20.68}$$

These spin-orbitals are to be used to form antisymmetric two-electron wave functions. They must be constructed to be eigenstates of $\hat{S}_z = \hat{S}_z(1) + \hat{S}_z(2)$ and of $\hat{\mathbf{S}}^2 = (\hat{\mathbf{S}}(1) + \hat{\mathbf{S}}(2))^2$. With the basis set used here, we can make six such functions. They are:

$$\Phi_1(1,2) \equiv {}^1\Sigma_{g+}(1,2) = \sigma_g(1)\sigma_g(2)\,|0,0\rangle \tag{20.69}$$

$$\Phi_2(1,2) \equiv {}^1\Sigma_{g-}(1,2) = \sigma_u(1)\sigma_u(2)\,|0,0\rangle \tag{20.70}$$

$$\Phi_3(1,2) \equiv {}^3\Sigma_{u,-1} = \frac{\sigma_g(1)\sigma_u(2) - \sigma_g(2)\sigma_u(1)}{\sqrt{2}}|1,-1\rangle \tag{20.71}$$

$$\Phi_4(1,2) \equiv {}^3\Sigma_{u,0} = \frac{\sigma_g(1)\sigma_u(2) - \sigma_g(2)\sigma_u(1)}{\sqrt{2}}|1,0\rangle \tag{20.72}$$

$$\Phi_5(1,2) \equiv {}^3\Sigma_{u,1} = \frac{\sigma_g(1)\sigma_u(2) - \sigma_g(2)\sigma_u(1)}{\sqrt{2}}|1,1\rangle \tag{20.73}$$

$$\Phi_6(1,2) \equiv {}^1\Sigma_{u,0} = \frac{\sigma_g(1)\sigma_u(2) + \sigma_g(2)\sigma_u(1)}{\sqrt{2}}|0,0\rangle \tag{20.74}$$

These equations are a handful and it would take the remainder of the section to explain them. They contain new notation and also some new physical ideas. Before I explain them, I remind you that these functions are used to construct the CI test function

$$\Psi(1,2) = \sum_{n=1}^{6} C_n \Phi_n(1,2) \tag{20.75}$$

This will be used in the variational principle to determine the coefficients C_n and the energies of various states of the system (see §5 on p. 362).

§25. *The Strategy for Constructing the Functions* $\Phi_i(1,2)$. The functions $\Phi_i(1,2)$ must be

• products of two orbitals and two-spin states, one for each electron;

• each such product must change sign when we permute the coordinates 1 and 2 of the electrons;

• each product must be an eigenstate of $\hat{S}^2$ and $\hat{S}_z$.

I will show next that the functions $\Phi_1(1,2)\ldots\Phi_6(1,2)$ given by Eqs 20.69–20.74 satisfy these conditions.

§26. *The Spin States.* The symbols $|0,0\rangle$, $|0,1\rangle$ etc. tell us the total spin state of the two electrons in the molecule. I will first state that they have certain properties and then show how these properties are proven.

The spin states are given by:

$$|0,0\rangle \equiv \frac{|\alpha,1\rangle\,|\beta,2\rangle - |\alpha,2\rangle\,|\beta,1\rangle}{\sqrt{2}} \tag{20.76}$$

$$|1,1\rangle \equiv |\alpha,1\rangle\,|\alpha,2\rangle \tag{20.77}$$

$$|1,0\rangle \equiv \frac{|\alpha,1\rangle\,|\beta,2\rangle + |\alpha,2\rangle\,|\beta,1\rangle}{\sqrt{2}} \tag{20.78}$$

$$|1,-1\rangle \equiv |\beta,1\rangle\,|\beta,2\rangle \tag{20.79}$$

The state $|\alpha,1\rangle$ tells me that the spin of the electron 1 is α (i.e. $m_s = 1/2$) and $|\beta,2\rangle$ says that the electron 2 is in state β (i.e. $m_s = -1/2$), etc. The product

$|\alpha, 1\rangle \, |\beta, 2\rangle$ indicates the spin state of the two electrons: 1 is in state α and 2 is in the state β, etc.

The general form of the symbols $|0,0\rangle$, $|1,0\rangle$ etc is $|\ell_s, m_s\rangle$. The quantum number ℓ_s gives the eigenvalues of $\hat{\mathbf{S}}^2$ and m_s those of $\hat{S}_z$. The state $|0,0\rangle$ has $\ell_s = 0$ and $m_s = 0$. This state is not degenerate and because of this it is called a *singlet*. The three states $|1,-1\rangle$, $|1,0\rangle$ and $|1,1\rangle$ have $\ell_s = 1$ and differ through the values of m_s; this must be an integer between $-\ell_s$ and ℓ_s. These states satisfy the following eigenvalue equations

$$\hat{\mathbf{S}}^2 \, | \, \ell_s, m_s\rangle = \hbar^2 \ell_s(\ell_s + 1) \, | \, \ell_s m_s\rangle \tag{20.80}$$

and

$$\hat{S}_z \, | \, \ell_s, m_s\rangle = \hbar m_s \, | \, \ell_s m_s\rangle \tag{20.81}$$

The length of the spin angular momentum vector of two electrons in the three states $|1, m_s\rangle$, $m_s = -1, 0, 1$ is the same. These states differ only through the orientation of this vector (see Chapter 19), one orientation for each value of m_s.

§27. *The Singlet State* $|0,0\rangle$. I will show here that the state $|0,0\rangle$, defined by Eq. 20.76, is an eigenfunction of $\hat{\mathbf{S}}^2$ and $\hat{S}_z$.

When the Z-component of the total spin operator $\hat{\mathbf{S}}$ (see Eq. 20.64) is applied to the two-electron spin state, we obtain

$$\hat{S}_z|0,0\rangle = \hat{S}_z(1)|0,0\rangle + \hat{S}_z(2)|0,0\rangle$$

$$= \frac{1}{\sqrt{2}}\left(\frac{\hbar}{2}|\alpha, 1\rangle \, |\beta, 2\rangle - \left(-\frac{\hbar}{2}\right)|\alpha, 2\rangle \, |\beta, 1\rangle\right)$$

$$+ \frac{1}{\sqrt{2}}\left(-\frac{\hbar}{2}|\alpha, 1\rangle \, |\beta, 2\rangle - \frac{\hbar}{2}|\alpha, 2\rangle \, |\beta, 1\rangle\right)$$

$$= 0 \tag{20.82}$$

In this calculation the spin operator $\hat{\mathbf{S}}(1)$ of the electron 1, applied to a state such as $|\alpha, 1\rangle \, |\beta, 2\rangle$, acts on the spin state of the electron 1. Therefore, $\hat{S}_z(1)|\alpha, 1\rangle \, |\beta, 2\rangle = \left(\hat{S}_z(1)|\alpha, 1\rangle\right)|\beta, 2\rangle = \frac{\hbar}{2}|\alpha, 1\rangle \, |\beta, 2\rangle$, $\hat{S}_z(2)|\alpha, 1\rangle \, |\beta, 1\rangle$

$= \alpha, 1 \rangle \left(\hat{S}_z(1) | \beta, 2 \rangle \right) = - \frac{\hbar}{2} | \alpha, 1 \rangle | \beta, 2 \rangle$, etc. Putting all these together we find that the total spin along the OZ-axis is zero.

A similar, but more tedious, calculation shows that

$$\hat{\mathbf{S}}^2 | 0, 0 \rangle = 0. \tag{20.83}$$

The total spin angular momentum has zero length.

§28. *The Triplet States* $| 1, m_s \rangle$. By using the method described above it can be shown that the states $| 1, m_s \rangle$, $m_s = -1, 0, 1$ have the properties (it is a good exercise to try to prove the results given below):

$$\hat{S}_z | 1, 1 \rangle = \frac{\hbar}{2} | 1, 1 \rangle \ \ (\text{i.e. } m_s = 1) \tag{20.84}$$

$$\hat{S}_z | 1, 0 \rangle = 0 | 1, 0 \rangle \ \ (\text{i.e. } m_s = 0) \tag{20.85}$$

$$\hat{S}_z | 1, -1 \rangle = -\frac{\hbar}{2} | 1, -1 \rangle \ \ (\text{i.e. } m_s = -1) \tag{20.86}$$

and

$$\hat{\mathbf{S}}^2 | 1, m_s \rangle = \hbar^2 1(1 + 1) | 1, m_s \rangle \ \ \text{for } m_s = -1, 0, 1 \tag{20.87}$$

The latter equation means that $\ell_s = 1$. The state with $\ell_s = 1$ is triply degenerate (i.e. $2l_s + 1 = 3$): the total spin vector can have three different OZ-components.

§29. *Pairing up the Spin and the Orbital Functions to Create Antisymmetric Configurations.* Finally, I can now explain how we combine the orbital wave functions with the spin states to create antisymmetric configurations: Such combinations can be made in two ways:

• combine an antisymmetric orbital wave functions with a symmetric spin state;

• combine a symmetric orbital with an antisymmetric spin function.

§30. *The States* Φ_1, Φ_2, *and* Φ_6. These functions are defined by Eqs 20.69, 20.70, and 20.74. They all contain the spin state $| 0, 0 \rangle$, which is antisymmetric.

Exercise 20.17

Start with the definition Eq. 20.76 and show that $|0,0\rangle$ changes sign when you permute the labels 1 and 2.

This antisymmetric spin state must be multiplied with a symmetric orbital function. With the two molecular orbitals σ_g and σ_u at our disposal we can make only three symmetric functions. They are:

$$\sigma_g(1)\sigma_g(2) \tag{20.88}$$

$$\sigma_u(1)\sigma_u(2) \tag{20.89}$$

$$\frac{\sigma_g(1)\sigma_u(2) + \sigma_g(2)\sigma_u(1)}{\sqrt{2}} \tag{20.90}$$

As you can see in Eqs 20.69, 20.70, and 20.74 these three functions are multiplied with $|0,0\rangle$ to create Φ_1, Φ_2, and Φ_6. By doing this we ensure that the product of the orbital functions with the spin states is antisymetric.

§31. *The States* Φ_3, Φ_4 *and* Φ_5. To create these states we have multiplied the antisymmetric orbital function $(1/\sqrt{2})\{\sigma_g(1)\sigma_u(2) - \sigma_g(2)\sigma_u(1)\}$ with the three symmetric spin functions $|1,-1\rangle$, $|1,0\rangle$ and $|1,1\rangle$. Such products generate antisymmetric states, which is what we want.

§32. *The Notations* $^1\Sigma_{g+}$, $^1\Sigma_{g-}$, $^3\Sigma_{u,-1}$, $^3\Sigma_{u,0}$, $^3\Sigma_{u,1}$, $^1\Sigma_{u,0}$. It is now time to explain the meaning of this notation. The letter Σ tells us that these states describe σ-bonds. The upper script at the left of Σ indicates whether the state is a singlet (e.g. $^1\Sigma$) or a triplet (e.g. $^3\Sigma$). The numerical value of the upper script is the degeneracy $2\ell_s + 1$ of the spin state.

As explained in §23 the symbols u and g tell us whether the state is symmetric or antisymmetric with respect to a reflection through a plane perpendicular to the bond and cutting it in the middle. The gerade states do not change sign when such a reflection is performed and the ungerade ones do. The subscript $+$ and $-$ in $^1\Sigma_{g+}$ and $^1\Sigma_{g-}$ indicate that the first state is constructed with two gerade MOs and the second with two ungerade MOs. The numbers 1, 0 or -1 in $^3\Sigma_{u,-1}$, $^3\Sigma_{u,0}$, $^3\Sigma_{u,1}$ $^1\Sigma_{u,0}$ indicate the value of m_s of the two-electron spin state.

You may wonder why do we need this barbaric notation. In the early days of quantum mechanics, when very little computing power was available, symmetry played a very important role because it allows us to determine whether certain matrix elements are zero, without performing any calculations. For example, it can be shown that the absorption of a linearly polarized photon through a transition from $^1\Sigma_{g+}$ to $^3\Sigma_{u,0}$ is forbidden. Spectroscopists were very happy when theorists gave them information about the symmetry of electronic states. As computing power increased, this information lost its importance. After all, a computer can evaluate integrals very rapidly, so we know which ones are zero. Nevertheless, substantial savings of computer power can be obtained if we can tell in advance that certain integrals in the theory are zero, and symmetry is still used in the computer codes. A branch of mathematics called group theory is very useful for studing the symmetry properties of molecules and solids.

§33. *The Configurations in Terms of Atomic Orbitals: Physical Interpretation.* In Eqs 20.69–20.74 we have expressed the configuration wave functions $\Phi_1 \dots \Phi_6$ in terms of the molecular orbitals σ_u and σ_g. To discern the physical meanings of these functions, and also for performing some of the calculations, it is useful to have them in terms of the atomic orbitals $1s_A$ and $1s_B$. To obtain such expressions we substitute σ_u and σ_g given by Eqs 20.63 and 20.62 in Eqs 20.69–20.74. The results are (see Workbook QM20.3):

Workbook

$$^1\Sigma_{g+} = \frac{1s_A(1)1s_A(2) + 1s_B(1)1s_B(2) - 1s_A(2)1s_B(1) - 1s_A(1)1s_B(2)}{2(1-S)}|0,0\rangle$$

$$^1\Sigma_{g-} = \frac{1s_A(1)1s_A(2) + 1s_B(1)1s_B(2) + 1s_A(1)1s_B(2) + 1s_A(2)1s_B(1)}{2(1+S)}|0,0\rangle$$

$$^3\Sigma_{u,m_s} = \frac{1s_A(1)1s_B(2) - 1s_A(2)1s_B(1)}{\sqrt{2(1-S^2)}}|1,m_s\rangle, \quad m_s = -1,0,1$$

$$^1\Sigma_{u,0} = \frac{1s_A(1)1s_B(2) + 1s_A(2)1s_B(1)}{\sqrt{2(1-S^2)}}|0,0\rangle \quad\quad\quad (20.91)$$

The expression for $^1\Sigma_{u+}$ contains the term $1s_A(1)1s_A(2)$ which is a state in which both electrons are on nucleus A, leaving nucleus B with no electrons. A chemist would write this as the ionic state H^-H^+. If we were to assume that the state of the molecule is described correctly by Φ_1 we would run into the following difficulty.

As the distance between nuclei is increased, the molecule evolves into two H atoms. To describe this correctly the ionic terms $1s_A(1)1s_A(2)$ and $1s_B(1)1s_B(2)$ would have to evolve to zero, as R increases. They do not. This tells me that the state Φ_1 cannot alone describe the molecule correctly, for large values of R. As you will see later this difficulty is automatically removed in the CI method but causes serious trouble when we use perturbation theory to calculate the energy of the electrons.

Exercise 20.18

Verify equations Eq. 20.91. I performed this verification in Workbook QM20.3.

Exercise 20.19

(Subtle and difficult.) We have applied the symmetry principle to the electrons in the H_2 molecule. However, the nuclei are also indistinguishable particles. How does this affect the total wave function of the H_2 or D_2 molecules in the electronic ground state? *Hint.* Write the wave function of the molecule as the product of the electronic wave function, the rotational wave function, the vibrational wave function, and the nuclear spin state. The nuclear spin of H is 1/2; that of D is 1. Use the fact that the rotational wave function changes sign upon the interchange of the nuclei if ℓ is odd and remains unchanged if ℓ is even (ℓ is the quantum number of the eigenstates of the angular momentum squared).

The Integrals Required by the Configuration Interaction Method

§34. *Introduction.* We have now constructed six antisymmetrized two-electron wave functions $\Phi_1 \ldots \Phi_6$, that are eigenfunctions of $\hat{\mathbf{S}}^2$ and $\hat{S}_z$. The test wave function to be used in the variational principle is:

$$\Psi(1,2) = \sum_{n=1}^{6} C_n \Phi_n(1,2), \tag{20.92}$$

where $C_1, \ldots, C_6$ are constants to be determined by using the variational principle. For this determination we need to evaluate the matrix elements (see §8)

$$\langle \Phi_i \,|\, \hat{H} \,|\, \Phi_j \rangle$$

and the overlap integrals

$$\langle \Phi_i \mid \Phi_j \rangle$$

In this section we occupy ourselves with these evaluations. The calculations are tedious and require some sophistication. Modern calculations no longer use the basis set given here and the integral evaluation is done differently. Because of this, I will not explain the method of evaluation but give the results.

§35. *The Overlap Matrix is Diagonal.* It can be shown fairly easily that the overlap integrals between the configurations satisfy

$$\langle \Phi_i \mid \Phi_j \rangle \equiv \int d1 \int d2 \Phi_i(1,2)\hat{H}(1,2)\Phi_j(1,2) = \delta_{i,j} \qquad (20.93)$$

The Kroneker delta $\delta_{i,j}$ is equal to 1 when $i = j$, and is zero otherwise. To prove that this is the case, use the fact that the molecular orbitals σ_u and σ_g are orthonormal and the spin states $\mid 0,0 \rangle$, $\mid 1, m_s \rangle$ are also orthonormal.

Exercise 20.20

Prove that Eq. 20.93 is valid for all values of i and j from 1 to 6. *Hint.* Replace the functions Φ_i with the expressions given by Eq. 20.91 and use the orthonormality of the molecular orbitals and of the spin states.

---

§36. *Only the Off-Diagonal Matrix Elements $\langle \Phi_1 \mid \hat{H} \mid \Phi_2 \rangle$ and $\langle \Phi_2 \mid \hat{H} \mid \Phi_1 \rangle$ Differ from Zero.* Because of the way we built the spin states, the singlet configurations are orthogonal to the triplet ones. This means that $\langle \Phi_i \mid \hat{H} \mid \Phi_j \rangle = 0$ if Φ_i is a singlet and Φ_j is a triplet.

The integral $\langle \Phi_1 \mid \hat{H} \mid \Phi_6 \rangle$ is zero because of the symmetry of the two configurations. Φ_1 is a gerade configuration and Φ_6 is ungerade. If we perform a reflection with respect to a plane that is perpendicular to the bond and cuts it in half, then Φ_6 changes sign and Φ_1 does not. The Hamiltonian is not affected by this change, nor is the integral (an integral does not change its value when we perform a symmetry operation). The integral before reflection, $\langle \Phi_1 \mid \hat{H} \mid \Phi_6 \rangle$, must be equal to the integral after the reflection, which is $-\langle \Phi_1 \mid \hat{H} \mid \Phi_6 \rangle$. Therefore, the integral must be zero (if a is a number and we prove that $a = -a$, then a must be equal to zero).

The same argument works for the integral $\langle \Phi_2 \,|\, \hat{H} \,|\, \Phi_6 \rangle$. In addition $\langle \Phi_i \,|\, \hat{H} \,|\, \Phi_j \rangle = \langle \Phi_i \,|\, \hat{H} \,|\, \Phi_j \rangle^*$, therefore $\langle \Phi_6 \,|\, \hat{H} \,|\, \Phi_3 \rangle$ and $\langle \Phi_6 \,|\, \hat{H} \,|\, \Phi_1 \rangle$ are also zero.

§37. *The Hamiltonian Matrix.* These results indicate that the Hamiltonian matrix has the simple form

$$
\overset{\leftrightarrow}{H} =
\begin{pmatrix}
H_{11} & H_{12} & 0 & 0 & 0 & 0 \\
H_{21} & H_{22} & 0 & 0 & 0 & 0 \\
0 & 0 & H_{33} & 0 & 0 & 0 \\
0 & 0 & 0 & H_{44} & 0 & 0 \\
0 & 0 & 0 & 0 & H_{55} & 0 \\
0 & 0 & 0 & 0 & 0 & H_{66}
\end{pmatrix}
\tag{20.94}
$$

To make progress we must evaluate those matrix elements $H_{i,j}$ that are not equal to zero. We do this by following the procedure devised by Slater.[a] For reasons that elude me, Slater decided to solve the eigenvalue problem for $2\hat{H}$ rather than $\hat{H}$. If $\hat{H}\psi = E\psi$ then, of course, $2\hat{H}\psi = 2E\psi$ (and vice versa), and Slater's calculation, by a variational procedure using $2\hat{H}$, will give twice the energy of the ground state. I follow his procedure here, because if I did not, I would have to redo most of his calculations.

So, following Slater, we evaluate $\langle \Phi_i \,|\, 2\hat{H} \,|\, \Phi_j \rangle$ rather than $\langle \Phi_i \,|\, \hat{H} \,|\, \Phi_j \rangle$. From now on, $H_{ij} = \langle \Phi_i \,|\, 2\hat{H} \,|\, \Phi_j \rangle$.

§38. *The Hamiltonian in Atomic Units.* Various international organization recommend the exclusive use of SI units. In spite of this, most people who work in the field of quantum chemistry prefer a specialized set of units, called atomic units. There are good reasons for this: the equations look simpler when written in these units and the numerical values of various quantities are more reasonable (e.g. the energy of a molecule in joules is extremely small). The Hamiltonian of a hydrogen molecule (multiplied by 2), in atomic units, is:

$$
2\hat{H} = -\nabla_1^2 - \nabla_2^2 - \frac{2}{r_{1A}} - \frac{2}{r_{2A}} - \frac{2}{r_{1B}} - \frac{2}{r_{2B}} + \frac{2}{r_{12}} + \frac{2}{R}
\tag{20.95}
$$

$-\nabla_1^2$ and $-\nabla_2^2$ represent twice the kinetic energies of electrons 1 and 2. For $i = 1$ or 2 and $\mu = A$ or B, $r_{i\mu}$ is the distance between electron i and nucleus μ, so

---

[a] J.C. Slater, *Quantum Theory of Molecules and Solids*, Vol. 1, McGraw-Hill, New York, 1963.

Table 20.1 Atomic units and SI units.

Physical quantity	Name of the unit	Value of the unit in SI
Atomic unit of mass	Mass of the electron	9.1094×10^{-31} kg
Atomic unit of charge	Charge of the proton	1.6022×10^{-19} C
Atomic unit of angular momentum	$\hbar$	1.0546×10^{-11} J s
Atomic unit of distance or length	bohr	5.2918×10^{-11} m
Atomic unit of energy	hartree	4.3597×10^{-18} J
Atomic unit of permittivity	$4\pi\varepsilon_0$	1.1127×10^{-10} C^2 J^{-1} m^{-1}

that $-2/r_{i\mu}$ is twice their Coulomb interaction. r_{12} is the distance between the electrons and $1/r_{12}$ is the Coulomb interaction between them. $2/R$ is twice the Coulomb interaction between the nuclei, and R is the distance between them.

§39. *Atomic Units.* The atomic units are defined by taking Planck's constant $\hbar$, the electron mass m, and the proton charge e to be equal to 1, and taking the permittivity of vacuum ε_0 to be equal to $1/4\pi$ (so that $4\pi\varepsilon_0 = 1$). From these four equations, one can derive units for other quantities. The unit of energy is 1 hartree and this is equal to 27.212 eV. The unit of length is 1 bohr, and this is $a_0 = 0.52918$ Å. If you use the Hamiltonian given in Eq. 20.95, then the distances are given in bohr and the energy in hartrees.

A few useful atomic units and the conversion factors to the SI system are given in Table 20.1. The energy of the electron in a hydrogen atom is equal to $-\frac{1}{2}$ hartree and for this reason some people prefer to use the rydberg as energy unit, with the value

$$1 \text{ rydberg} = \tfrac{1}{2} \text{ hartree}$$

§40. *The Matrix Elements in Terms of Atomic Orbitals.* To calculate $H_{ij} = \langle \Phi_i \,|\, 2\hat{H} \,|\, \Phi_j \rangle$, Φ_i and Φ_j have to be expressed in terms of atomic orbitals. I show schematically how this is done for H_{11}. We have (use Eq. 20.69):

$$H_{11} = \langle \Phi_1 \,|\, 2\hat{H} \,|\, \Phi_1 \rangle = \langle {}^1\Sigma_{g+} \,|\, 2\hat{H} \,|\, {}^1\Sigma_{g+} \rangle$$

$$= \langle \sigma_u\sigma_u \,|\, 2\hat{H} \,|\, \sigma_u\sigma_u \rangle \langle 0,0 \,|\, 0,0 \rangle$$

$$= \int d1 d2 \, \sigma_u(1)\sigma_u(2) \, 2\hat{H}\sigma_u(1)\sigma_u(2) \tag{20.96}$$

Here 1 and 2 stand for $\mathbf{r}_1$ and $\mathbf{r}_2$, and $d1$ and $d2$ for $d\mathbf{r}_1$ and $d\mathbf{r}_2$. I have used the fact that $\langle 0,0 \,|\, 0,0 \rangle = 1$.

To express H_{11} in terms of atomic orbitals, use Eq. 20.91 for the orbital part of $^1\Sigma_{u+}$:

$$H_{11} = \left\langle \frac{s_As_A + s_Bs_B - s_Bs_A - s_As_B}{\sqrt{2(1+S)}} \,\middle|\, 2\hat{H} \,\middle|\, \frac{s_As_A + s_Bs_B - s_Bs_A - s_As_B}{\sqrt{2(1+S)}} \right\rangle \qquad (20.97)$$

This integral contains only known functions and operators, and this means that we can evaluate it. However, the techniques used exceed the skills of an undergraduate student. Since they are not used in other areas of physical chemistry, I only give the results. The evaluations are explained in Appendix 5 of Slater's book (cited above).

§41. *Expressions for the Matrix Elements.* Slater gives the following results.

$$H_{11} = \langle \Phi_1 \,|\, 2\hat{H} \,|\, \Phi_1 \rangle = \langle {}^1\Sigma_{g+} \,|\, 2\hat{H} \,|\, {}^1\Sigma_{g+} \rangle \qquad (20.98)$$

$$= \alpha^2 \left[\frac{2(1 - S - K)}{1 + S} \right] + \alpha \left[\frac{2(J + 2K - 2)}{1 + S} + \frac{\frac{5}{8} + \frac{1}{2}J' + K' + 2L}{(1+S)^2} + \frac{2}{w} \right] \qquad (20.99)$$

$$H_{22} = \langle \Phi_2 \,|\, 2\hat{H} \,|\, \Phi_2 \rangle = \langle {}^1\Sigma_{g-} \,|\, 2\hat{H} \,|\, {}^1\Sigma_{g-} \rangle$$

$$= \alpha^2 \left[\frac{2(1 + S + K)}{1 - S} \right] + \alpha \left[\frac{2(J - 2K - 2)}{1 - S} + \frac{\frac{5}{8} + \frac{1}{2}J' + K' - 2L}{(1-S)^2} + \frac{2}{w} \right] \qquad (20.100)$$

$$H_{12} = H_{21} = \langle \Phi_1 \,|\, 2\hat{H} \,|\, \Phi_2 \rangle = \langle {}^1\Sigma_{g+} \,|\, 2\hat{H} \,|\, {}^1\Sigma_{g-} \rangle$$

$$= \frac{\alpha}{16} \left[\frac{5 - 4J'}{1 - S^2} \right] \qquad (20.101)$$

In addition,

$$\langle \Phi_3 \,|\, 2\hat{H} \,|\, \Phi_3 \rangle = \langle \Phi_4 \,|\, 2\hat{H} \,|\, \Phi_4 \rangle = \langle \Phi_5 \,|\, 2\hat{H} \,|\, \Phi_5 \rangle$$

$$= \alpha^2 \left[\frac{2(1 + KS + S^2)}{1 - S^2} \right] + \alpha \left[\frac{-4 + 2J + J' - 4KS - K'}{1 - S^2} + \frac{2}{\alpha R} \right] \qquad (20.102)$$

$$\langle \Phi_6 \mid 2\hat{H} \mid \Phi_6 \rangle = \langle {}^1\Sigma_u \mid 2\hat{H} \mid {}^1\Sigma_u \rangle$$

$$= \alpha^2 \left[\frac{2 + KS + S^2}{1 - S^2} \right] + \alpha \left[\frac{2(-2 + J - 2KS) + \frac{5}{4} - K'}{1 - S^2} + \frac{2}{\alpha R} \right] \tag{20.103}$$

Here S, K, K', J, J', and L are integrals over the atomic orbitals, and α is the exponent in the 1s orbital

$$\alpha = \frac{Z}{a_0} \tag{20.104}$$

If the atomic orbitals of the hydrogen atom are given in atomic units, then $\alpha = 1$ and this is the value used in all calculations done here. However, we could use α as a variational parameter and determine the value of α that gives the lowest energy. We are not going to do this here.

It will take quite a while to explain the symbols appearing in these equations.

§42. *The Overlap Integral $S(R)$.* The quantity

$$S(R) = \int d\mathbf{r} s_A(1) s_B(2) = e^{-\alpha R} \left[1 + \alpha R + \frac{\alpha^2 R^2}{3} \right] \tag{20.105}$$

is called the overlap integral of s_A with s_B or the overlap of s_A with s_B. From now on I will drop the '1' from $1s_A$ and $1s_B$, since we are not using other s-orbitals and there is no danger of confusion.

$S(R)$ depends on the distance R between the nuclei; it is 0 if $\alpha R \gg 1$, and it is 1 when $R = 0$. Physics tells us that this is reasonable. If $R = 0$ the two nuclei are on top of each other, the two orbitals belong to the same nucleus and the integrand becomes $s_A(1) s_A(1)$. Since the orbital s_A is normalized, the integral is equal to 1. When R is larger than the distance over which the orbitals decay to zero (which is, roughly, equal to the Bohr radius) the product $s_A(1) s_B(1)$ is zero for all values of $\mathbf{r}_1$ and this makes the integral equal to zero. A plot of $S(R)$ is shown in Fig. 20.3. It was obtained in Cell 1 of Workbook QM20.4.

§43. *The Integral $J(R)$.*

$$J(R) \equiv \frac{1}{\alpha} \int s_A(1)^2 \left(-\frac{2}{r_{1B}} \right) d1 = -\frac{2}{\alpha R} + 2e^{-2\alpha R}(1 + \frac{1}{\alpha R}) \tag{20.106}$$

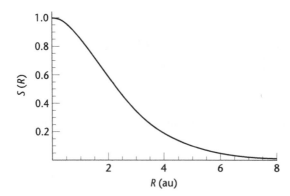

Figure 20.3 The dependence of the overlap integral $S(R) = \int s_A(1)s_B(1)d1$ on the bond length R.

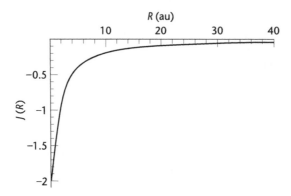

Figure 20.4 The integral $J(R) \equiv \frac{1}{\alpha} \int s_A(r)^2 \left(-\frac{2}{r_B} \right) dr$ as a function of the bond length R.

The integral is twice the average value of the Coulomb interaction of an electron in the s_A orbital of the hydrogen atom A, with the nucleus B. As R increases, the integral becomes equal to $-2/R$; this means that when A and B are far apart, nucleus B "sees" the net charge (equal to 1 au) of the electron (revolving around A).

Fig. 20.4 shows how $J(R)$ depends on the bond length R. This energy is attractive: the electron located on A pulls the nucleus B towards A. This interaction contributes to the formation of the bond between the two atoms. Note the long range, typical of Coulomb interaction.

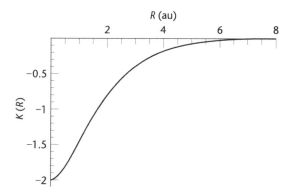

Figure 20.5 The integral $K(R) = \frac{1}{\alpha} \int s_A(1)s_B(1)(-2/r_{12})d1$ as a function of the bond length R.

§44. *The Integral $K(R)$.* The integral K is

$$K(R) = \frac{1}{\alpha} \int s_A(1)s_B(1)\left(-\frac{2}{r_{1B}}\right)d1 = -2e^{-\alpha R}(1 + \alpha R) \qquad (20.107)$$

There is no simple classical electrostatic interpretation of this integral. It decays exponentially with R, it is equal to -2 when $R = 0$ and it generates an attraction between atoms (see Fig. 20.5). This integral appears in the theory because we used the antisymmetry principle.

§45. *The Integral $J'(R)$.*

$$J'(R) = \frac{1}{\alpha} \int \frac{s_A(1)^2 \times 2 \times s_B(2)^2}{r_{12}}d1d2 = \frac{2}{\alpha R} - e^{-2\alpha R}\left[\frac{2}{\alpha R} + \frac{11}{4} + \frac{3\alpha R}{2} + \frac{(\alpha R)^2}{3}\right]$$
$$(20.108)$$

This is twice the average Coulomb interaction between an electron in the $1s_A$ orbital of atom A, with the electron in the $1s_B$ orbital of atom B, divided by α. As $\alpha R \gg 1$, the interaction energy becomes $2/R$, which is twice the Coulomb repulsion of two point charges having the charge $e = 1$ (in atomic units, $e = 1$ and $4\pi\varepsilon_0 = 1$). As R approaches 0, the interaction energy becomes $\frac{5}{4}$ (to see this, expand J' in a power series around $R = 0$ and look at the lowest-order term). This is twice the electrostatic energy of the two electrons in a $1s$ orbital located on the same nucleus. This term causes the atoms to repel each other (see Fig. 20.6).

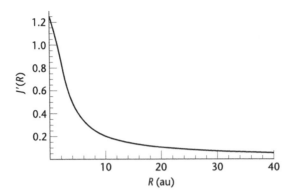

Figure 20.6 The dependence of $J' = \int s_A(1)^2(2/r_{12})s_B(2)^2 d1\,d2$ on the bond length R.

§46. *The Integral $K'(R)$.*

$$K'(R) = \frac{1}{\alpha} \int \frac{s_A(1)s_B(1) \times 2 \times s_B(2)s_A(2)}{r_{12}} d1\,d2$$

$$= \frac{2}{5} \left\{ -e^{-2W} \left(-\frac{25}{8} + \frac{23\alpha R}{4} + 3(\alpha R)^2 + \frac{(\alpha R)^3}{3} \right) \right.$$

$$\left. + \frac{6}{\alpha R} \left[S^2(0.57722 + \ln(\alpha R)) + S'^2 \mathrm{Ei}(4\alpha R) - 2SS' \mathrm{Ei}(-2\alpha R) \right] \right\} \quad (20.109)$$

with

$$S' = e^{\alpha R} \left(1 - \alpha R + \frac{(\alpha R)^3}{3} \right) \quad (20.110)$$

and

$$\mathrm{Ei}(-x) \equiv \int_x^{\infty} \frac{e^{-t}}{t} dt \quad (20.111)$$

$\mathrm{Ei}(x)$ is called the exponential integral of the second kind. **Mathematica** provides the function

$$\mathrm{ExpIntegralEi[\text{-}z]} = -\int_z^{\infty} \frac{e^{-t}}{t} dt$$

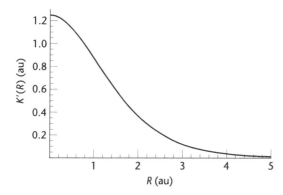

Figure 20.7 The integral $K'(R) = \frac{2}{\alpha} \int (s_A(1)s_B(1)s_A(2)s_B(2)/r_{12})d1\,d2$ as a function of the bond length R.

that calculates the integral. This integral cannot be interpreted in terms of classical electrostatics. It contributes to the energy a term that repels the atoms (see Fig. 20.7).

Since we are going to perform numerical calculations that evaluate K', it is worthwhile to point out some complications. When I evaluated the integral numerically, I found that K' becomes infinite when $R = 0$ (see Cell 5 of Workbook QM20.4). This is not acceptable! As R approaches 0, s_B becomes identical to s_A, and K' becomes equal to J' (compare Eqs 20.109 and 20.108). When $R = 0$, $J' = \frac{5}{8}$ and therefore $K'(R)$ must equal $\frac{5}{8}$, not infinity!

K' cannot be infinite when R is 0, for another reason: when $R = 0$ our system is a He atom and K' is the Coulomb interaction between its electrons. This cannot be infinite!

What is going on? If I use **Mathematica** to calculate $\lim_{R \to 0} K'(R)$, I obtain $\frac{5}{8}$. This is the expected result! Then why do I get infinity when I evaluate the function K' numerically at $R = 0$? The expression for K' contains three terms ($Ei(2\alpha R)$, $Ei(4\alpha R)$, $\ln(\alpha R)$) that become infinite when $R = 0$. To obtain a finite result, these terms must cancel each other when R is small, but before $R = 0$. This cancelation takes place when I perform exact evaluations (i.e. when I take the limit $R \to 0$). However, numerical evaluations on a computer make small round-off errors, and the three terms no longer cancel as they should. You can see a symptom of this in Cell 5 of Workbook QM20.4, where I calculated $K'(R)$ for $R = 10^{-1}, 10^{-2}, 10^{-3}, \ldots$. I get 0.625 (which is $\frac{5}{8}$) for all values of R greater than 10^{-12}. As R becomes smaller than 10^{-12}, $K'(R)$ oscillates wildly with R. Something goes wrong when $R \propto 10^{-12}$, which is the machine precision.

This example warns you that you cannot use computers mindlessly and that you must always be on guard for numerical instabilities caused by the round-off errors (i.e. computers do not work with exact numbers). Using physics and common sense to examine the results is often very helpful in testing that they are reasonable.

§47. *The Integral $L(R)$.* This integral is

$$L = \frac{1}{\alpha} \int s_A(1)^2 s_A(2) s_B(2) \frac{2}{r_{12}} d1 d2$$

$$= e^{-\alpha R} \left(\frac{2}{\alpha R} + \frac{1}{4} + \frac{5}{8\alpha R} \right) + e^{-3\alpha R} \left(-\frac{1}{4} - \frac{5}{8\alpha R} \right) \qquad (20.112)$$

It appears to diverge when $R \to 0$ (since it has terms in $\frac{1}{R}$) but the dangerous terms cancel each other as R becomes smaller and smaller. If $R = 0$, then $s_A = s_B$ and L becomes equal to J' (which is $\frac{5}{8}$). Again, a numerical evaluation of the integral gives $L(0)$ infinite, which is the wrong result. The dependence of $L(R)$ on R is shown in Fig. 20.8.

§48. *Summary.* To find the wave functions and the energies of the H$_2$ molecule, by using the CI method, we have to evaluate the integrals $\langle \Phi_i | 2\hat{H} | \Phi_j \rangle$ and $\langle \Phi_i | \Phi_j \rangle$. The overlap integrals are given by $\langle \Phi_i | \Phi_j \rangle = \delta_{i,j}$. We obtain this result readily by using the properties of the spin states and the symmetry of the orbitals. To evaluate $\langle \Phi_i | 2\hat{H} | \Phi_j \rangle$ we have replaced the molecular orbitals σ_u and σ_g with

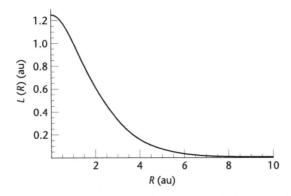

Figure 20.8 The integral $L(R) = \frac{1}{\alpha} \int (s_A(1)^2 s_A(2) s_B(2) \frac{2}{R_{12}}) d1\, d2$.

their expression in terms of s_A and s_B. Once this is done we obtain $\langle \Phi_i \mid 2\hat{H} \mid \Phi_j \rangle$ in terms of integrals over the atomic orbitals. The problem has been thus reduced to pure mathematics and I gave the expression of all the integrals. If you want to see how these expressions are obtained consult Slater's book.

The Ground and Excited State Energies given by Perturbation Theory

§49. *Introduction.* We have constructed the configurations $\Phi_1 \ldots \Phi_6$ with the intention of using them in a configuration interaction calculation. We will do that in "The configuration interaction method" on p. 409. In the present section I explain an approximation that has been used since the late 1920s and is commonly used in introductory textbooks. For us, this is instructive because it dramatizes the need for using the CI method.

Consider the orbital part $\sigma_g(1)\sigma_g(2)$ of the wave function (the configuration)

$$\Phi_1(1,2) \equiv {}^1\Sigma_{g+}(1,2) = \sigma_g(1)\sigma_g(2) \mid 0,0 \rangle$$

One can prove that the wave function of a molecule is a product of single-electron molecular orbitals only if the electrons do not interact with each other. Such a wave function does not describe the physical situation properly. In spite of this, perturbation theory assumes that this gives a reasonable descripion of the ground state of the molecule and uses it to calculate the energy of the system from

$$E_1^0(R) = \langle \Phi_1 \mid \hat{H} \mid \Phi_1 \rangle = \langle {}^1\Sigma_{g+} \mid \hat{H} \mid {}^1\Sigma_{g+} \rangle = H_{11}(R) \tag{20.113}$$

The upperscript 0 is a reminder that the energy E_1^0 is calculated by using perturbation theory. The assumption is that the electron–electron interaction does not modify radically the wave function of the system and we can use it to calculate the energy in Eq. 20.113, rather than use the exact wave function.

Note an inherent contradiction in perturbation theory: the Hamiltonian contains the electrostatic repulsion between electrons but the wave function does not take this effect into account. In a consistent theory, the wave function and the Hamiltonian should be treated at the same level.

Perturbation theory also assumes that the energy of the state ${}^1\Sigma_{g-}$ is

$$E_2^0(R) = \langle \Phi_2 \mid \hat{H} \mid \Phi_2 \rangle = \langle {}^1\Sigma_{g-} \mid \hat{H} \mid {}^1\Sigma_{g-} \rangle \equiv H_{22}(R) \tag{20.114}$$

In contrast to perturbation theory, the CI treatment uses a linear combination of these, and other, configurations.

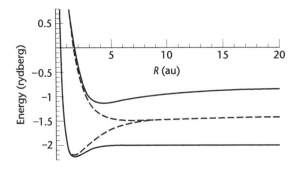

Figure 20.9 The dashed lines show the energies $E_1^0 = H_{11}(R)$ and $E_2^0 = H_{22}(R)$ obtained by perturbation theory; the upper curve is E_2^0. The solid lines show the energies $E_1(R)$ and $E_2(R)$ given by the configuration interaction theory; the upper curve is E_2. The graph was made in Cell 9, of Workbook QM20.7.

We have already developed, in §41, the mathematics needed for calculating E_1^0 and E_2^0. The matrix elements H_{11} and H_{22} are given by Eq. 20.99 and Eq. 20.100, respectively, in terms of the integrals S, J, K, J', K' and L. Formulae for these can be found in §42–§47. Using this information I have calculated E_1^0 and E_2^0 in Workbook QM20.5. The results are plotted, as the dashed lines, in Fig. 20.9.

The form of the ground state potential energy surface E_1^0 is typical of a diatomic molecule. To understand it you must remember several things. When we studied the Born–Oppenheimer approximation in §2, I mentioned that the electronic energy $E(R)$ (the eigenvalue of the BO Hamiltonian) is the potential energy for the nuclear motion. Thus, when the electrons are in the ground state, the nuclei interact with the force (consult your classical mechanics book)

$$F(R) = -\frac{\partial E_1^0(R)}{\partial R} \tag{20.115}$$

This force is directed along the line joining the nuclei. If the force is positive the nuclei repel each other; if it is negative, they attract each other. This information helps us interpret the shape of the curve $E_1^0(R)$. The function has a minimum at a value of R which I denote R_0. You know from calculus that the derivative of a function at the minimum is equal to zero. Therefore, when $R = R_0$, no force acts between the atoms. If $R < R_0$ the derivative is negative and the atoms repel each other. When $R > R_0$ the derivative is positive and the atoms attract each other. A molecule in contact with a medium tends to lose energy and the bond length

R will perform small oscillations around the value R_0. This is why R_0 is the bond length of a molecule in the ground state.

In Cell 3 of Workbook QM20.5 I calculated that $E_1^0(R)$ has a minimum at the distance $R_0 = 1.603$ au $= 0.8484$ Å. The experimental value of the bond length is 0.7416 Å. The energy $E_1^0(R_0) = H_{11}(R_0)$ is equal to -2.1982 rydberg (a rydberg is twice the atomic unit of energy; 1 rydberg is 27.21212 eV). This energy is not measurable, but we can measure the dissociation energy of the molecule. This is defined as $E_1^0(\infty) - E_1^0(R_0)$, the energy of the two atoms minus the minimum energy of the molecule. $E_1^0(\infty) = -1.375$ rydberg (see Cells 5 of Workbook QM20.5). Therefore, the binding energy of the H_2 molecule, at this level of approximation, is

$$-1.375 - (-2.1982) = 0.823 \text{ rydberg } = 11.1998 \text{ eV} \qquad (20.116)$$

(1 rydberg $= 27.21212$ eV). The experimental value is 4.476 eV. This large error is caused by the unreasonable behavior of $H_{11}(R)$ at large R, as you will see next.

§50. *The Behavior of $H_{11}(R)$ at Large R.* To calculate the dissociation energy, in perturbation theory we need to obtain correct values for $H_{11}(R = \infty)$ and for the minimum value of $H_{11}(R)$. There are signs that this calculation has difficulties when R is large. First, we know from experiments that an H_2 molecule in the ground electronic state dissociates into two hydrogen atoms. The energy of a hydrogen atom is -1 rydberg. Therefore, the energy of the dissociated molecule is -2 rydberg. If we use this value, the binding energy should then be

$$-1.375 - (-2) = 0.625 \text{ rydberg} = 8.504 \text{ eV}$$

This is better, but it is still too large. Nevertheless, this calculation tips us off that one source of the large error is the incorrect behavior of $H_{11}(R)$ when R is large. A second sign that this behavior is problematic is provided by the plot of $H_{11}(R)$. Instead of becoming constant when R is larger than approximately 4 Å (when the molecule should have broken into neutral atoms), $H_{11}(R)$ keeps changing as R increases.

An analysis of the formula for $H_{11}(R)$ shows (see Cell 4 of Workbook QM20.5) that

$$H_{11}(R) = -1.375 - \frac{1}{R} \text{ for } R \gg \frac{1}{\alpha} \qquad (20.117)$$

This equation has two troublesome features. Since the molecule breaks into two neutral fragments, there should be no Coulomb interaction between them

($-1/R$ is a Coulomb interaction between two charges of opposite sign, in atomic units). Furthermore, the energy of the constant term should be -2 rydberg (the energy of two hydrogen atoms), not -1.375.

Why is perturbation theory producing this behavior? Let us look at the wave function (see Eq. 20.91)

$$\Phi_1(R) \equiv {}^1\Sigma_{g+}(R)$$

$$= \frac{s_A(1)s_A(2) + s_B(1)s_B(2) - s_A(1)s_B(2) - s_A(2)s_B(1)}{2(1+S)} \,|\, 0,0\rangle$$

In this expression, $s_A(1)s_B(2)$ and $s_A(2)s_B(1)$ represent a system with one electron on nucleus A and the other electron on nucleus B. These two functions, which we can call neutral or covalent, describe a molecule in which each atom has an electron. This gives the correct behavior when R is large. The other two functions are troublesome. $s_A(1)s_A(2)$ represents a molecule with two electrons on A and no electron on B (this is $H^- \cdots H^+$). $1s_B(1)1s_B(2)$ is no better: there are two electrons on B and none on A. These two ionic functions give an incorrect behavior when R is large. The fact that the coefficients of these four functions (in ${}^1\Sigma_{g+}$) are equal, tells us that the ionic structures are as probable as the covalent ones. According to this wave function, when R is large the molecule has a probability of 1/2 of consisting of the ions H^+ and H^- and a probability of 1/2 of consisting of two neutral H atoms. Had the molecule dissociated into ions with the probability 1, the energy of the dissociated molecule would be $-2/R$ (remember that Slater multiplied the Hamiltonian by 2 and we have followed him). Because the probability of dissociating into such a state is 1/2, the average energy of the dissociated system in $-1/R$.

This explains why Φ_1 is not a good wave function to use in perturbation theory. One cure of this problem might be to drop the offending terms from the wave function. However, this might hurt the behavior when R is small and such terms may be needed. As you will see soon, the correct solution is to use configuration interaction.

Exercise 20.21

Use the Heitler–London wave functions

$$\chi_1 = \frac{s_A(1)s_B(2) + s_A(2)s_B(1)}{\sqrt{2(1+S)}} \,|\, 0,0\rangle$$

and

$$\chi_2 = \frac{1s_A(1)1s_B(2) - 1s_A(2)1s_B(1)}{\sqrt{2(1-S)}} \mid 1, m_s \rangle \quad m_s = 1, 0, -1$$

to calculate the energies

$$\mathcal{E}_1 = \langle \chi_1 \mid 2\hat{H} \mid \chi_1 \rangle$$

and

$$\mathcal{E}_2 = \langle \chi_2 \mid 2\hat{H} \mid \chi_2 \rangle$$

Plot $\mathcal{E}_1$ and $\mathcal{E}_2$ versus R. Find the bond length and the binding energy. Analyze the behavior of the wave function at large R.

The Configuration Interaction Method

§51. *The Variational Eigenvalue Problem: a Summary.* The configuration interaction method uses the test function

$$\Psi = \sum_{i=1}^{6} c_i \Phi_i \tag{20.118}$$

and determines the coefficients by using the variational principle. The configurations Φ_i are given by Eqs 20.69–20.74. We have only six configurations Φ_i because we have used a minimal basis set with two orbitals.

The variational principle leads to a generalized eigenvalue equation

$$\sum_{j=1}^{6} \langle \Phi_i \mid \hat{H} \mid \Phi_j \rangle c_j = E \sum_{j=1}^{6} \langle \Phi_i \mid \Phi_j \rangle c_j \tag{20.119}$$

for the Hamiltonian matrix having the elements $H_{ij} \equiv \langle \Phi_i \mid \hat{H} \mid \Phi_j \rangle$. We have already shown that the elements of the overlap matrix $\langle \Phi_i \mid \Phi_j \rangle$ are equal to 1 if $i = j$ and to zero otherwise. This reduces the generalized eigenvalue problem to an ordinary one:

$$\sum_{j=1}^{6} \langle \Phi_i \mid \hat{H} \mid \Phi_j \rangle c_j = E c_i \tag{20.120}$$

The eigenvalue problem for a 6×6 matrix has six eigenvalues, labeled $E_1 \ldots E_6$ and six eigenvectors, labelled

$$\vec{c}^i = \{c_1^i, c_2^i, c_3^i, c_4^i, c_5^i, c_6^i\} \tag{20.121}$$

The vector $\vec{c}^i$ is the eigenvector corresponding to the energy E_i, and has six components $\{c_1^i \ldots c_6^i\}$. This means that the CI wave function for this energy is

$$\Psi_i(1,2) = \sum_{n=1}^{6} c_n^i \Phi_n(1,2) \tag{20.122}$$

where Φ_n are the configuration wave functions.

§52. *The Eigenvalues and the Eigenvectors of Matrix $\overset{\leftrightarrow}{H}$.* We have shown earlier that the Hamiltonian matrix has a simple structure:

$$\overset{\leftrightarrow}{H} = \begin{pmatrix} H_{11} & H_{12} & 0 & 0 & 0 & 0 \\ H_{21} & H_{22} & 0 & 0 & 0 & 0 \\ 0 & 0 & H_{33} & 0 & 0 & 0 \\ 0 & 0 & 0 & H_{44} & 0 & 0 \\ 0 & 0 & 0 & 0 & H_{55} & 0 \\ 0 & 0 & 0 & 0 & 0 & H_{66} \end{pmatrix} \tag{20.123}$$

The matrix elements H_{ij} can be calculated by using Eqs 20.99–20.101 and the integrals calculated in §42–§47. Note that $H_{21} = H_{12}$ and that we will not need the integrals H_{33}–H_{66} (as you will see below).

Because of its "block diagonal" structure, the eigenvalue problem for this matrix is easily solved (by a computer or by hand). The first two eigenvalues are (see Cell 3 of Workbook QM20.6):

$$E_1 = \frac{H_{11} + H_{22} - A}{2} \tag{20.124}$$

and

$$E_2 = \frac{H_{11} + H_{22} + A}{2} \tag{20.125}$$

with

$$A = \sqrt{H_{11}^2 + 4H_{12}^2 - 2H_{11}H_{22} + H_{22}^2} \qquad (20.126)$$

The remaining eigenvalues are

$$E_j = H_{jj}, \, j = 3, 4, 5, 6 \qquad (20.127)$$

The eigenvector corresponding to E_1 is (see Cell 3 of Workbook QM20.6)

$$\vec{c}^{\,1} = \left\{ -\frac{H_{11} + H_{22} + A}{2H_{12}}, 1, 0, 0, 0, 0 \right\} \qquad (20.128)$$

which leads to the molecular wave function:

$$\Psi_1(1, 2) = -\frac{H_{11} + H_{22} + A}{2H_{12}} \, {}^1\Sigma_{g,+} + 1 \, {}^1\Sigma_{g-} \qquad (20.129)$$

The eigenvector corresponding to E_2 is

$$\vec{c}^{\,2} = \left\{ -\frac{H_{11} + H_{22} - A}{2H_{1,2}}, 1, 0, 0, 0, 0 \right\} \qquad (20.130)$$

which leads to the wave function:

$$\Psi_2 = -\frac{H_{11} + H_{22} - A}{2} \, {}^1\Sigma_{g,+} + 2 \, {}^1\Sigma_{g-} \qquad (20.131)$$

These eigenvectors are not normalized.

The remaining wave functions are $\Psi_3 = \Phi_3$ (since $\vec{c}^{\,3} = \{0, 0, 1, 0, 0, 0\}$), $\Psi_4 = \Phi_4 = {}^3\Sigma_{u,0}$, $\Psi_5 = \Phi_5 = {}^3\Sigma_{u,-1}$, and $\Psi_6 = \Phi_6 = {}^1\Sigma_{u,0}$.

§53. *The Coupling between Configurations.* Note a general behavior: the fact that $H_{12} \neq 0$ *couples* the configuration Φ_1 to Φ_2, so that Ψ_1 and Ψ_2 are linear combinations of Φ_1 and Φ_2. Two configurations (Φ_1 and Φ_2) are needed for generating the correct wave functions Ψ_1 and Ψ_2. Because $H_{1i} = 0$ for $i = 3, 4, 5$, and 6 the configurations $\Phi_3 \ldots \Phi_6$ do not appear in Ψ_1. Because $H_{2i} = 0$ for $i = 3, 4, 5$, and 6 the configurations $\Phi_3 \ldots \Phi_6$ do not appear in Ψ_2 either. Since no matrix element couples the configurations with $i \geq 3$ to other configurations, the wave functions

$\Psi_3 \ldots \Psi_6$ are equal to the configurations $\Phi_3 \ldots \Phi_6$. In other words, the energies $E_3 \ldots E_6$ are the same as the ones given by perturbation theory. The CI procedure modifies only the energies E_1 and E_2 and the corresponding wave functions. This happens only because we work with a minimal basis set and have a small number of configurations. Had we had a larger basis set there would be more configurations and more of them would be coupled. All energies will then differ from those given by perturbation theory. The small basis set also limits the accuracy of the energies E_1 and E_2. While they are much better than those given by perturbation theory, they are far from being accurate. Had we had a larger AO basis set, we would have generated more molecular orbitals and more configurations. The new configurations would couple to the old ones, the ground state wave function Ψ_1 would contain more than two configurations and be more accurate. An accurate calculation may have to use hundreds of thousands of configurations.

§54. *When is Perturbation Theory Accurate?* It is easy to see that if we take $H_{12} = 0$ the CI energies in Eqs 20.124 and 20.125 become equal to the energies $E_1^0 = H_{11}$ and $E_2^0 = H_{22}$ given by perturbation theory. With a little algebra one can rewrite Eqs 20.124 and 20.125 as:

$$E_1 = H_{11} + |\frac{H_{22} - H_{11}}{2}| \left[1 - \sqrt{1 + \frac{4H_{12}}{|H_{22} - H_{11}|}} \right] \qquad (20.132)$$

Clearly if

$$\frac{4H_{12}}{|H_{22} - H_{11}|} \ll 1 \qquad (20.133)$$

then the CI energy E_1 is very close to $E_1^0 = H_{11}$; in such a case, perturbation theory gives good results.

Fig. 20.10 shows a plot of H_{11}, H_{22}, and H_{12}. It is clear that the condition Eq. 20.133 is satisfied when R is less than about 2.5 au, but not as R becomes larger. A look at Fig. 20.9 shows that the energies given by perturbation theory are close to the CI ones when R is small and becomes dramatically erroneous as R increases and the condition Eq. 20.133 is violated.

§55. *The Configuration Interaction Energies.* We have now all the equations needed for calculating $E_1(r)$ and $E_2(r)$. We use Eqs 20.124–20.126. The matrix

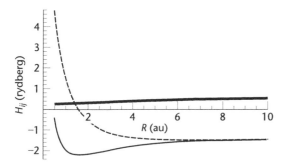

Figure 20.10 H_{11} (thin solid line), H_{22} (dashed line), and H_{12} (thick solid line) as functions of the bond length R. The plot was made in Cell 1 of Workbook QM20.5.

elements H_{11}, H_{22}, and H_{12} are given by Eqs 20.99–20.101. Formulas for the integrals appearing in these expressions are given in §42–§47. Using this information the CI energies are calculated in Cells 1–4 of Workbook QM20.7. The corresponding potential energy surfaces are shown in Fig. 20.9 together with the energies given by perturbation theory. There are substantial differences between the two calculations when R is large.

The binding energy, calculated in Cell 7 of Workbook QM20.7, is

$$0.2373 \text{ rydberg} = 0.2373 \times \frac{27.212}{2} \text{ eV} = 3.203 \text{ eV}$$

This is much closer to the experimental value of 4.476 eV than is the value of 11.2 eV obtained by perturbation theory. The improvement comes mostly from improved values of $E_1(R)$ at large values of R.

The bond length given by CI (Cell 6 of Workbook QM20.7) is 1.668 au = 0.83 Å; this is close to 1.603 au given by perturbation theory and not too far from the experimental value of 0.7416 Å.

Also note, in Fig 20.9, that for large values of R, the ground state energy $E_1(R)$ is constant and equal to -2 rydberg, which is the correct energy of two hydrogen atoms.

§56. *The Configuration Interaction Wave Function of the Ground State.* Here we examine the ground-state wave function

$$\Psi_1(1,2) = -\frac{H_{11} + H_{22} + A}{2H_{12}} \, {}^1\Sigma_{g,+} + 1 \, {}^1\Sigma_{g-} \qquad (20.134)$$

This has the form

$$\Psi(1,2) = c_1^1(R) \, {}^1\Sigma_{g+}(1,2) + c_2^1(R) \, {}^1\Sigma_{g-}(1,2) \qquad (20.135)$$

with the coefficients $c_1^1(R)$ and $c_2^1(R)$ given by

$$c_1^1 = -\frac{-H_{11} + \sqrt{4H_{12}^2 + (H_{11} - H_{22})} + H_{22}}{2H_{12}} \qquad (20.136)$$

$$c_2^1 = 1 \qquad (20.137)$$

This eigenvector is not normalized.

I am most interested in understanding the weight of the ionic functions $s_A(1)s_A(2)$ and $s_B(1)s_B(2)$ and that of the covalent functions $s_A(1)s_B(2)$ and $s_A(2)s_B(1)$ in the ground state wave function $\Psi_1(1,2)$. To find it, I replace $\Phi_1 = {}^1\Sigma_{g+}$ and $\Phi_2 = {}^1\Sigma_{g-}$ with their expression in terms of atomic orbitals, given by the first equation in the group Eq. 20.91. The result is

$$\psi_1(1,2) = a_i(R) \big(s_A(1)s_A(2) + s_B(1)s_B(2) \big) + a_c(R) \big(s_A(1)s_B(2) + s_A(2)s_B(1) \big) \qquad (20.138)$$

The coefficients

$$a_i \equiv c_1^1 + c_2^1 \qquad (20.139)$$

$$a_c \equiv c_1^1 - c_2^1 \qquad (20.140)$$

tell us the extent to which the ground state is covalent (when $a_c \gg a_i$), or ionic (when $a_i \gg a_c$), or mixed (when $a_c \approx a_i$). To describe the character of the wave function, Fig. 20.11 shows the ratio a_i^2/a_c^2 as a function of the bond length R.

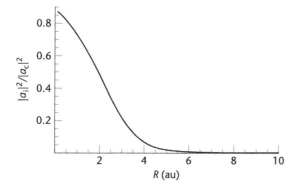

Figure 20.11 Behavior of the ratio a_i^2/a_c^2. a_i and a_c are the ionic and "covalent" coefficients defined by Eqs 20.138–20.140. The plot was made in Cell 12 of Workbook QM20.5 .

You can see that as the bond is stretched, the coefficient a_i becomes much smaller than a_c. For example (see Cell 12 of Workbook QM20.7), if $R = 2$ au then

$$\Psi_1(1,2;R=2) = -4.50\left(s_A(1)s_A(2) + s_B(1)s_B(2)\right)$$
$$- 6.50\left(s_A(1)s_B(2) + s_A(2)s_B(1)\right)$$

However, if $R = 10$ au then

$$\Psi_1(1,2;R=210) = -0.0043\left(s_A(1)s_A(2) + s_B(1)s_B(2)\right)$$
$$- 2.00\left(s_A(1)s_B(2) + s_A(2)s_B(1)\right)$$

Clearly, when R is around equilibrium bond length, the wave function Ψ_1 is a mixture of ionic and covalent contributions. At large R, the ionic term disappears.

The CI method thus cures the problem that perturbation theory had at large values of R. This substantially improves the value of the binding energy.

In this CI calculation, I made a severe approximation: I used a very limited basis set, of only two atomic orbitals, which allowed us to form only two interacting configurations. Had we included more atomic orbitals in the basis set and formed all possible configurations allowed by them, we would have obtained *very* accurate results.

Exercise 20.22

The energy of the triplet states is (in rydberg)

$$E_3 = E_4 = E_5 = \frac{2\alpha^2}{1 - S^2}\left(1 + KS + S^2\right) + \frac{\alpha}{1 - S^2}\left(-4 + 2J + J' - 4KS - K'\right) + \frac{2}{R}$$

(a) Make a plot of E_3 vs R. (b) Write the wave function in terms of products of atomic orbitals and find how the coefficients of these products change with R.

Exercise 20.23

Now consider E_6 and $\Phi_6 = \Psi_6$. You have

$$E_6 = \alpha^2\frac{2(1 + KS + S^2)}{1 - S^2} + \alpha\left[\frac{2(-2 + J - 2KS) + 5/4 - K'}{1 - S^2} + \frac{2}{R\alpha}\right]$$

(a) Plot $E_6(R)$. (b) Analyze the ionic and covalent contributions of the wave function $\Psi_6 = {}^1\Sigma_u$.

Summary

§57. Born and Oppenheimer have shown that to understand the behavior of electrons in a molecule, one should solve the Schrödinger equation for electrons, with fixed nuclear coordinates. The energy eigenvalue of the electrons in this approximation depends on nuclear positions and it is the potential energy of the nuclei. It can be used in the Schrödinger equation for nuclei to study their vibrational and rotational spectra or the strength of the chemical bonds in the molecule.

The Born–Oppenheimer approximation is very accurate for the electronic ground states but it can fail grievously at those nuclear positions for which two electronic excited states are degenerate or almost degenerate.

The most common method for calculating the ground-state electronic wave function uses a variational principle, which states that a function $\Psi(\mathbf{r}_1, \ldots, \mathbf{r}_N)$ that satisfies the condition

$$\int \Psi^*(\mathbf{r}_1, \ldots, \mathbf{r}_N)\,\Psi(\mathbf{r}_1, \ldots, \mathbf{r}_N)d\mathbf{r}_1 \cdots d\mathbf{r}_N = 1 \qquad (20.141)$$

and minimizes the expression

$$E[\Psi] = \int \Psi^*(\mathbf{r}_1, \dots, \mathbf{r}_N) \, \hat{H}_{BO} \, \Psi(\mathbf{r}_1, \dots, \mathbf{r}_N) d\mathbf{r}_1 \cdots d\mathbf{r}_N \qquad (20.142)$$

is the exact solution of the Schrödinger equation

$$\hat{H}_{BO} \Psi = E_0 \Psi \qquad (20.143)$$

and in addition, the minimum value of $E[\psi]$ is equal to the ground-state energy E_1 of the electrons. Here $\hat{H}_{BO}$ is the Born–Oppenheimer Hamiltonian of the electrons.

This statement is used as follows. First, propose a "test function" $\phi(\mathbf{r}_1, \dots, \mathbf{r}_N; a_1, \dots, a_m)$, for the wave function that contains m parameters $a_1, \dots, a_m$. Then use it in Eq. 20.142 (in place of Ψ) to calculate $E[\phi]$ for fixed nuclear coordinates. You obtain an expression for $E[\phi]$ that depends on $a_1, \dots, a_m$. Vary those values until $E[\phi]$ has a minimum. The values of $a_1, \dots, a_m$ you obtain, when introduced into $\phi(\mathbf{r}_1, \dots, \mathbf{r}_N; a_1, \dots, a_m)$, give the best approximation to the ground-state wave function that your test function allows. The corresponding value of $E[\phi]$ is the best approximation to the ground-state energy. Do a minimization for every value of the nuclear coordinates $\mathbf{R}$ to find out how ϕ and $E[\phi]$ depend on $\mathbf{R}$.

It is impossible to perform the high-dimensional integral in Eq. 20.142 numerically or analytically. Because of this, we are forced to represent the wave function by a product of one-dimensional functions $\phi_1(\mathbf{r})$, $\phi_2(\mathbf{r}), \dots, \phi_M(\mathbf{r})$. For a three-electron problem, this representation is

$$\psi(\mathbf{r}_1, \mathbf{r}_2, \mathbf{r}_3) = C_1 \begin{vmatrix} \phi_1(\mathbf{r}_1) & \phi_1(\mathbf{r}_2) & \phi_1(\mathbf{r}_3) \\ \phi_2(\mathbf{r}_1) & \phi_2(\mathbf{r}_2) & \phi_2(\mathbf{r}_3) \\ \phi_3(\mathbf{r}_1) & \phi_3(\mathbf{r}_2) & \phi_3(\mathbf{r}_3) \end{vmatrix} + C_2 \begin{vmatrix} \phi_1(\mathbf{r}_1) & \phi_1(\mathbf{r}_2) & \phi_1(\mathbf{r}_3) \\ \phi_2(\mathbf{r}_1) & \phi_2(\mathbf{r}_2) & \phi_2(\mathbf{r}_3) \\ \phi_4(\mathbf{r}_1) & \phi_4(\mathbf{r}_2) & \phi_4(\mathbf{r}_3) \end{vmatrix}$$

$$+ C_3 \begin{vmatrix} \phi_1(\mathbf{r}_1) & \phi_1(\mathbf{r}_2) & \phi_1(\mathbf{r}_3) \\ \phi_3(\mathbf{r}_1) & \phi_3(\mathbf{r}_2) & \phi_3(\mathbf{r}_3) \\ \phi_4(\mathbf{r}_1) & \phi_4(\mathbf{r}_2) & \phi_4(\mathbf{r}_3) \end{vmatrix} + \cdots$$

The expression is made up of determinants formed with the functions ϕ_1, $\phi_2, \dots$. Using such determinants guarantees that $\psi(\mathbf{r}_1, \mathbf{r}_2, \mathbf{r}_3)$ changes sign when we permute the coordinates of any two particles, thus satisfying a fundamental principle of quantum mechanics.

The coefficients C_1, C_2, $C_3, \dots$ are calculated by using the variational principle.

There are several ways of choosing ϕ_1, ϕ_2,.... A common one is to force them to satisfy an approximate Schrödinger equation with the Hamiltonian H_a for the molecule. This is achieved by representing them as

$$\phi_i(\mathbf{r}) = \sum_j d_{ij} \chi_j(\mathbf{r})$$

where $\chi_j(\mathbf{r})$ are atomic orbitals located on the nuclei of the molecule. The coefficients d_{ij} are obtained by using the variational principle with the Hamiltonian H_a.

For the hydrogen molecule, we have used two atomic orbitals, $\chi_1 = 1s_A$ and $\chi_2 = 1s_B$. Because of this, we can determine ϕ_1 and ϕ_2 by using the symmetry of the molecule. We obtained

$$\phi_1 = \sigma_g = \frac{1s_A + 1s_B}{\sqrt{1 + S^2}} \tag{20.144}$$

$$\phi_2 = \sigma_u = \frac{1s_A - 1s_B}{\sqrt{1 - S^2}} \tag{20.145}$$

We have shown that with these orbitals and the spin states only six configurations (determinants) can be made and we use the configuration interaction (CI) method to calculate the energy of the six electronic states. We studied only the two lowest energies in detail.

It is interesting that including the spin of the electrons turned out to be extremely important even though the spin is not taken into account in the Hamiltonian since all energies involving spin are negligibly small. The spin plays a powerful role through the antisymmetrization principle. If the spin state is antisymmetric (i.e. it is $\frac{1}{\sqrt{2}} \{|\alpha, 1\rangle |\beta, 2\rangle - |\alpha, 2\rangle |\beta, 1\rangle\}$) then the orbital part must be symmetric, to make the product of the orbital and the spin part antisymmetric. This affects the energy of the state dramatically even though we have ignored in the Hamiltonian all interactions involving spin.

We found that as long as the Hamiltonian is independent of spin, the configurations segregate according to their spin states: singlet states couple only to singlets, and triplets only to triplets. One can also show that light absorption involving a transition from a singlet to a triplet state is forbidden, if we ignore the presence of spin terms in the Hamiltonian. Such terms exist but, for most systems they are very small. Nevertheless, because of them the triplet-to-singlet transitions take place, but, for most systems very slowly. This is why a phosphor emits light for a long time after absorbing it: the emission rate is very low and it takes a long time before the population of all the excited states decays.

The calculation of the electronic states and of potential energy surface is central to all of chemistry. Because of lack of time we have not done justice to this very complex and important subject.

21

NUCLEAR MAGNETIC RESONANCE AND ELECTRON SPIN RESONANCE

Introduction

§1. A one-electron atom has several magnetic moments. One is caused by the motion of the "rotating" electron and is proportional to the orbital angular momentum $\hat{\mathbf{L}}$. Another is the intrinsic magnetic moment of the electron and is proportional to the electron spin operator $\hat{\mathbf{S}}$. The third is an intrinsic nuclear magnetic moment, proportional to the nuclear spin operator $\hat{\mathbf{I}}$.

In what follows I will use the symbol $\hat{\mathbf{S}}$ when describing general properties of the spin (be it that of the nucleus or of the electron) and reserve $\hat{\mathbf{I}}$ for the nuclear spin operator. In calculations that involve both the electron and the nuclear spin, I use $\hat{\mathbf{S}}$ for the spin of the electron and $\hat{\mathbf{I}}$ for the spin of the nucleus.

If a system has a magnetic moment $\hat{\mathbf{m}}$ of any kind, it behaves like a magnet. If it is exposed to magnetic field $\mathbf{B}$, it will interact with it. The energy of this interaction is

$$-\hat{\mathbf{m}} \cdot \mathbf{B} \qquad (21.1)$$

In the absence of the magnetic field the spin states are degenerate; an electron or a nucleus with spin up has the same energy as one with spin down. The interaction with a magnetic field splits the degenerate spin-energy levels. For the magnetic fields available, the magnitude of this splitting is proportional to the magnitude B of $\mathbf{B}$; we can make it larger or smaller by varying B.

When we discussed spectroscopy, I explained that the absorption of electromagnetic radiation takes place because the radiation interacts with the charged particles in the molecule. The energy of the interaction between the time-dependent electric field $\mathbf{E}(t)$ of the radiation and the charged particles in the molecule is

$$-\sum_{i=1}^{n} q_i \hat{\mathbf{r}}_i \cdot \mathbf{E}(t) \equiv -\hat{\boldsymbol{\mu}}_e \cdot \mathbf{E}(t) \tag{21.2}$$

where q_i is the charge and $\mathbf{r}_i$ is the position of particle i. $\hat{\boldsymbol{\mu}}_e$ is the electric dipole moment of the molecule.

The interaction energy, Eq. 21.1, between a magnetic field and the spins of the particles in the molecule is very similar to that shown in Eq. 21.2. The only differences are that the magnetic dipole $\hat{\boldsymbol{\mu}}$ replaces the electric dipole $\hat{\boldsymbol{\mu}}_e$ and $\mathbf{B}(t)$ replaces $\mathbf{E}(t)$.

Imagine now that we expose the molecule to two magnetic fields: $\mathbf{B}_0 = \{0, 0, B_0\}$ is static (i.e. independent of time) and oriented along the OZ-axis; $\mathbf{B}(t) = \{B(t), 0, 0\}$ oscillates in time and is oriented along the OX-axis. The energy of the interaction between the magnetic moment $\hat{\mathbf{m}}$ and these two fields is

$$-\hat{m}_z B_0 - \hat{m}_x B(t) \tag{21.3}$$

Here $\hat{m}_x$ and $\hat{m}_z$ are the components of the magnetic moment $\hat{\mathbf{m}}$ along the OX and OZ-axes. The reason why we took the oscillating field perpendicular to the static one will become apparent later.

These two fields have distinct roles. The static field is used to lift the degeneracy of the spin states. Because of it, the energy E_α of the state with spin up differs from the energy E_β of the state with spin down (see Fig. 21.1).

By analogy to what happens when the molecule is exposed to an oscillating electric field, we expect that by exposing a system to an oscillating magnetic field having the frequency

$$\Omega = \frac{E_\alpha - E_\beta}{\hbar}$$

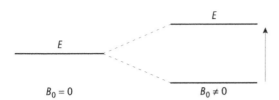

Figure 21.1 This figure shows that in the presence of a static magnetic field $\mathbf{B}_0$, the energy E_α of the state $|\alpha\rangle$ increases and the energy E_β of the state $|\beta\rangle$ decreased; the magnetic field "lifts the degeneracy" of the spin state. A magnetic field oscillating with the frequency $\Omega = (E_\alpha - E_\beta)/\hbar$ will excite the system from state $|\beta\rangle$ to state $|\alpha\rangle$.

we can cause transitions from the state $|\beta\rangle$ to the state $|\alpha\rangle$, just as an oscillating electric field of light causes transitions between the states of a molecule.

A new kind of absorption spectroscopy is thus possible. When it involves the states of the electron spin, this spectroscopy is called electron spin resonance (ESR). The spectroscopy that studies transitions between the states of the nuclear spin is called nuclear magnetic resonance (NMR).

ESR is very useful for studying the properties of radicals or of any system that has "unpaired" electrons. NMR is more generally applicable and it is so useful in studying molecular structure, rate processes, and various molecular motions (e.g. diffusion) that it has invaded all areas of chemistry. It would take more than a thousand book pages to review all of its applications. Here I can only give an outline of the main ideas, in their simplest implementation.

§2. *An Outline of the Chapter.* On pp. 424–427, I begin by adding to the information that you already have about spin states and spin operators. The presentation in the remainder of the chapter follows the pattern used when discussing optical spectroscopy. First we find the energy eigenstates of the system, then derive the selection rules for transitions between these states. For magnetic resonance spectroscopy, the energies of the spin states are determined by the strength of the static magnetic field and by the nature of the interactions between the spins of the nuclei of interest with the other spins.

Sections §4–§9 on pp. 428–434 examines the simplest possible system, in which each spin interacts with the static field $\mathbf{B}_0$ but not with other spins in the molecule or in its environment. The predictions made by this simple (and oversimplified) model are in agreement with observations made with low-resolution instruments.

They are pedagogically useful because we seem to understand better hierarchical arguments, in which each new physical effect modifies (preferably slightly) the situation obtained in its absence.

The first modification (see §10–§12 on pp. 434–435) of this simple theory is needed because the magnetic field acting on a given nucleus differs slightly from the one we apply externally. This happens because the external field affects the state of the electrons, which causes them to exert a very slight magnetic field on the nuclei. This small field must be added to the external one. Because of this, the same nuclei (e.g. protons) having different electronic (i.e. chemical) environments are exposed to slightly different magnetic fields and have slightly different energies. Therefore they generate different peaks in the NMR spectrum. This *chemical shift* is the single most important reason for the usefulness of NMR spectroscopy in determining molecular structure.

The following sections §13–§17 on pp. 436–440, examines the NMR spectrum of a molecule that has two protons that do not interact with each other. The main purpose of this material is to explain how quantum mechanics describes the states of systems with many spins.

The assumption that the nuclear spins do not interact with each other is abandoned in the remainder of the chapter. Each nucleus that has non-zero spin is a small magnet, which creates a magnetic field that acts on the other spins. This interaction causes an additional splitting of the spin energy levels and modifies the NMR spectrum. These modifications offer important clues about molecular structure and they need to be understood.

Sections §20–§33 on pp. 443–456 examines the spectrum of two distinguishable, interacting spins. The word 'distinguishable' is used to indicate that we need not symmetrize or antisymmetrize the spin states because, although pertaining to the same nuclei (i.e. two protons), the two spins have different environments.

The profound modifications of the NMR spectrum when the two protons are indistinguishable are discussed in the final sections, §34–§37 on pp. 457–461.

This extremely abbreviated introduction to NMR gives you some of the tools you need for further studies, but does not do justice to this very interesting, useful, and rich field of physical chemistry.

More Information about Spin Operators and Spin States

§3. Our discussion of spin operators in Chapter 19 is incomplete. We have discussed $\hat{\mathbf{S}}^2$ and $\hat{S}_z$, but nothing was said about the components $\hat{S}_x$ and $\hat{S}_y$. (In this section, I use the letter S for both nuclear and electronic spin operators.)

To understand NMR spectroscopy, we must fill this gap. I will do this for a particle having spin $\frac{1}{2}$, which is the simplest possible system.

As already explained, a particle with spin $\frac{1}{2}$ has two states, which are denoted by $|\alpha\rangle$ and $|\beta\rangle$. They are eigenfunctions of $\hat{S}_z$:

$$\hat{S}_z \,|\,\alpha\rangle = \frac{\hbar}{2}\,|\,\alpha\rangle \tag{21.4}$$

$$\hat{S}_z \,|\,\beta\rangle = -\frac{\hbar}{2}\,|\,\beta\rangle \tag{21.5}$$

and eigenfunctions of $\hat{\mathbf{S}}^2$:

$$\hat{\mathbf{S}}^2 \,|\,\alpha\rangle = \hbar^2 \frac{1}{2}\left(\frac{1}{2}+1\right)\,|\,\alpha\rangle \tag{21.6}$$

$$\hat{\mathbf{S}}^2 \,|\,\beta\rangle = \hbar^2 \frac{1}{2}\left(\frac{1}{2}+1\right)\,|\,\beta\rangle \tag{21.7}$$

If we want to analyze the behavior of the spin during NMR or ESR spectroscopy, we must also know how $\hat{S}_x$ and $\hat{S}_y$ act on $|\alpha\rangle$ and $|\beta\rangle$.

The spin operators are *defined by* the commutation relations

$$\left[\hat{S}_x, \hat{S}_y\right] = i\hbar\hat{S}_z \tag{21.8}$$

$$\left[\hat{S}_y, \hat{S}_z\right] = i\hbar\hat{S}_x \tag{21.9}$$

$$\left[\hat{S}_z, \hat{S}_x\right] = i\hbar\hat{S}_y \tag{21.10}$$

From these relations, we can derive everything we need to know about spin. It is possible to start from Eqs 21.8– 21.10 and show that

$$\hat{S}_x \,|\,\alpha\rangle = \frac{\hbar}{2}\,|\,\beta\rangle \tag{21.11}$$

$$\hat{S}_x \,|\,\beta\rangle = \frac{\hbar}{2}\,|\,\alpha\rangle \tag{21.12}$$

$$\hat{S}_y \mid \alpha \rangle = \frac{i\hbar}{2} \mid \beta \rangle \tag{21.13}$$

$$\hat{S}_y \mid \beta \rangle = -\frac{i\hbar}{2} \mid \alpha \rangle \tag{21.14}$$

Note that the states $\mid \alpha \rangle$ and $\mid \beta \rangle$ are not eigenstates of $\hat{S}_x$ or $\hat{S}_y$. When those operators act on a spin state, they "flip" the spin. For example $\hat{S}_x \mid \alpha \rangle = (\hbar/2) \mid \beta \rangle$ means that $\hat{S}_x$ acts acts on a state with "spin up" to produce a state with "spin down."

Exercise 21.1

Use Eqs 21.11–21.14 to show that

$$\hat{S}_+ \equiv \hat{S}_x + i\hat{S}_y \tag{21.15}$$

and

$$\hat{S}_- \equiv \hat{S}_x - i\hat{S}_y \tag{21.16}$$

have the following properties:

$$\hat{S}_+ \mid \alpha \rangle = 0 \tag{21.17}$$

$$\hat{S}_- \mid \beta \rangle = 0 \tag{21.18}$$

$$\hat{S}_+ \mid \beta \rangle = \hbar \mid \alpha \rangle \tag{21.19}$$

$$\hat{S}_- \mid \alpha \rangle = \hbar \mid \beta \rangle \tag{21.20}$$

Exercise 21.2

Use the commutation relations (Eqs 21.8–21.10) to show that

$$[\hat{S}_z, \hat{S}_+] = \hbar \hat{S}_+ \tag{21.21}$$

$$[\hat{S}_z, \hat{S}_-] = -\hbar \hat{S}_- \tag{21.22}$$

$$[\hat{S}_+, \hat{S}_-] = 2\hbar \hat{S}_z \tag{21.23}$$

$$\left[\hat{\mathbf{S}}^2, \hat{S}_+ \right] = \left[\hat{\mathbf{S}}^2, \hat{S}_- \right] = \left[\hat{\mathbf{S}}^2, \hat{S}_z \right] = 0 \tag{21.24}$$

Note that $\hat{\mathbf{S}}^2 = \hat{S}_x^2 + \hat{S}_y^2 + \hat{S}_z^2$.

Exercise 21.3

Use the commutation relations to show that

$$\hat{S}_+\hat{S}_- = \hat{\mathbf{S}}^2 - \hat{S}_z^2 + \hbar\hat{S}_z \tag{21.25}$$

$$\hat{S}_-\hat{S}_+ = \hat{\mathbf{S}}^2 - \hat{S}_z^2 - \hbar\hat{S}_z \tag{21.26}$$

$$\hat{\mathbf{S}}^2 = \frac{1}{2}\left(\hat{S}_+\hat{S}_- + \hat{S}_-\hat{S}_+\right) + \hat{S}_z^2 \tag{21.27}$$

Exercise 21.4

Suppose we define $|\alpha\rangle$ and $|\beta\rangle$ to be the two-dimensional vectors

$$|\alpha\rangle = \begin{pmatrix} 1 \\ 0 \end{pmatrix}$$

$$|\beta\rangle = \begin{pmatrix} 0 \\ 1 \end{pmatrix}$$

Define the 2×2 matrices

$$\hat{S}_a \equiv \frac{\hbar}{2}\hat{\sigma}_a, \qquad a = x, y, z$$

with the Pauli matrices $\hat{\sigma}_a$ given by

$$\hat{\sigma}_x = \begin{pmatrix} 0 & 1 \\ 1 & 0 \end{pmatrix}$$

$$\hat{\sigma}_y = \begin{pmatrix} 0 & -i \\ i & 0 \end{pmatrix}$$

$$\hat{\sigma}_z = \begin{pmatrix} 1 & 0 \\ 0 & -1 \end{pmatrix}$$

Verify that $|\alpha\rangle$, $|\beta\rangle$, $\hat{S}_x$, $\hat{S}_y$, $\hat{S}_z$ defined in this way satisfy Eqs 21.4–21.27. The necessary algebra is readily performed in **Mathematica** or **Mathcad**. *Note.* This is an example of the isomorphism typical of quantum mechanics. Any calculation performed with the operators $\hat{S}_a$ and the abstract symbols $|\alpha\rangle$ and $|\beta\rangle$ can be performed with these matrices and vectors.

The NMR Spectrum of a System with One Independent Spin

§4. *The Energy of the Spin States for Non-Interacting Spins.* We encounter the simplest possible situation when we study a liquid consisting of molecules that have one nucleus having spin $\frac{1}{2}$ that does not interact with the other spins in the system. In this case the interaction energy between the static magnetic field $\mathbf{B} = \{0, 0, B_0\}$ and the spin is

$$\hat{H} = -\hat{\mathbf{m}} \cdot \mathbf{B}_0 = -\gamma \mathbf{B}_0 \cdot \hat{\mathbf{I}} = -\gamma B_0 I_z \qquad (21.28)$$

The first equality is obtained because the magnetic moment $\hat{\mathbf{m}}$ is proportional to $\hat{\mathbf{I}}$:

$$\hat{\mathbf{m}} = \gamma \hat{\mathbf{I}} \qquad (21.29)$$

The proportionality constant γ is called the *magnetogyric ratio* of the nucleus and it is sometimes written as

$$\gamma = \frac{g_N \mu_N}{\hbar} \qquad (21.30)$$

where g_N is the dimensionless *nuclear g factor* and

$$\mu_N = \frac{e\hbar}{2m_p} \qquad (21.31)$$

is the *nuclear magneton* (m_p is the mass of the proton and e is its charge). From Eqs 21.30 and 21.31 we have

$$\gamma = \frac{g_N e}{2m_p}. \qquad (21.32)$$

γ (or g_N) is determined experimentally and changes from nucleus to nucleus.

The OZ-axis is taken along the direction of the static magnetic field $\mathbf{B}$ and this is why the dot product in the first equation is equal to $B_0 \hat{I}_z$.

We are interested here only in the nuclear spin states of the molecule and because of this $\hat{H}$ given by Eq. 21.31 is the total Hamiltonian of the system in the magnetic field. The time-dependent magnetic field used in NMR spectroscopy does not excite the nuclear or electronic motion; the electronic and nuclear states (e.g. rotation, vibration) do not participate in NMR spectroscopy.

A similar equation gives the energy of interaction of the spin with the oscillating magnetic field $\mathbf{B}(t) = \{B(t), 0, 0\}$:

$$-\hat{\mathbf{m}} \cdot \mathbf{B}(t) = -\gamma \hat{I}_x B_x(t) \tag{21.33}$$

This follows from the fact that the oscillating field $\mathbf{B}(t)$ is oriented along the OX-axis.

§5. *The Energy Levels.* As we have already discussed in Chapter 19, the energy levels of the nuclear spin are

$$E_{m_s} = -\gamma \hbar m_s B_0 \tag{21.34}$$

For a nucleus with spin s, $m_s = -s, -s+1, \ldots, s-1$, s. For a proton, or any other nucleus with spin $\frac{1}{2}$, s is $\frac{1}{2}$ and m_s is either $-\frac{1}{2}$ or $\frac{1}{2}$. Remember that m_s tells us how large is the projection of the spin along the OZ-direction (which is the same as the direction of $\mathbf{B}_0$).

A nucleus with spin $s = \frac{1}{2}$ has two spin states: $|\alpha\rangle$, which has $m_s = \frac{1}{2}$ (spin up), and $|\beta\rangle$, which has $m_s = -\frac{1}{2}$ (spin down). In the state $|\alpha\rangle$ the projection of the spin is along $\mathbf{B}_0$ and in $|\beta\rangle$ it is in the opposite direction. We could have used the notation $|\frac{1}{2}\rangle$ and $|-\frac{1}{2}\rangle$, or $|\uparrow\rangle$ and $|\downarrow\rangle$, but $|\alpha\rangle$ and $|\beta\rangle$ are used more frequently in the chemistry literature.

Because $\gamma > 0$, the energy of the state $|\beta\rangle$ is positive and that of state $|\alpha\rangle$ is negative (see Eq. 21.34). The state with the magnetic moment along the field has lower energy. The zero of the energy scale is the energy of the spin in the absence of the static magnetic field.

In the absence of the static magnetic field, the two spin states are degenerate; the degeneracy is "lifted" by the interaction with the static field. Sometimes we say that the field "splits" the spin levels. The energy difference between the two states is

$$E_\beta - E_\alpha = -\gamma \left(-\frac{\hbar}{2}\right) B_0 - \left(-\gamma \frac{\hbar}{2} B_0\right) = \gamma \hbar B_0 \tag{21.35}$$

and it is proportional to the applied static field.

According to quantum mechanics, exposing the molecule to an oscillating magnetic field (in addition to the static field) having the frequency

$$\Omega = \frac{E_\beta - E_\alpha}{\hbar} = \gamma B_0 \tag{21.36}$$

will cause a transition from the low-energy state $|\alpha\rangle$ to the high-energy state $|\beta\rangle$. This will result in energy absorption from the field and a magnetization of the sample, due to the re-orientation of the spins (hence of the magnetic moments associated with them). NMR spectroscopy detects absorption by monitoring the magnetization of the sample.

The measurements can be conducted in one of two ways: keep the static field $\mathbf{B}_0$ constant and vary the frequency Ω of the oscillating field; or, keep Ω constant and vary $\mathbf{B}_0$. In both cases the magnetization of the sample increases sharply when the condition Eq. 21.36 is satisfied. When this happens the plot of the magnetization versus either B_0 or Ω (depending which one is varied) has a peak. There are as many peaks in an NMR spectrum as transitions.

§6. *The Rate of Energy Absorption.* By analogy with optical spectroscopy, the absorption rate is controlled matrix element of the operator representing the energy of interaction between the system and the oscillating field. In our case this energy is given by Eq. 21.33 and the absorption rate is proportional to

$$-|\langle \psi_f \,|\, \hat{I}_x \,|\, \psi_i \rangle|^2 \tag{21.37}$$

Here, $|\psi_i\rangle$ and $|\psi_f\rangle$ are the initial (before absorption) and the final (after absorption) spin states.

§7. *NMR Notation and Units.* So far in this book the frequency Ω of the radiation absorbed during a transition $E_i \rightarrow E_f$ has been defined as

$$\Omega = \frac{E_f - E_i}{\hbar}$$

where E_i and E_f are the initial and the final states. NMR spectroscopists prefer to use the definition

$$\nu = \frac{E_f - E_i}{h} \tag{21.38}$$

and so do older books on optical spectroscopy. Since the "old" Planck constant h is related to the "new" Planck constant $\hbar$ through

$$\hbar = \frac{h}{2\pi}$$

we have

$$\Omega = 2\pi \nu \tag{21.39}$$

Ω is called the angular frequency and ν, the frequency. Both have units of s^{-1}, which is the hertz (Hz). The order of magnitude of the frequencies at which NMR is performed is 10^6 H or 1 megahertz (MHz). In some books the unit of ν is Hz and that of Ω is radian/second. This is a mistake that leads to inconsistencies; if you encounter such units, just remove the radian everywhere.

In this chapter, I follow NMR practice and use ν for the frequency of absorbed radiation (we have used Ω in all previous chapters). Eq. 21.36 gives

$$\nu = \frac{\Omega}{2\pi} = \frac{\gamma B_0}{2\pi} \tag{21.40}$$

The units of γ are then $s^{-1}T^{-1}$ (not rad $\times$ $s^{-1}T^{-1}$), and those of the nuclear magneton μ_N are $J\,T^{-1}$ (from the relation: energy = $\mu_N \times$ magnetic flux density). Here T stands for tesla, the unit of B.

Table 21.1 gives the spin, the nuclear g factor, and the magnetogyric ratio for a few nuclei used frequently in NMR spectroscopy.

§8. *The Order of Magnitude of Various Quantities.* Let us calculate the magnitude of the nuclear magneton μ_N, the magnetogyric ratio γ, and the excitation frequency when the static field is $B_0 = 1$ T, for the ^{1}H nucleus (the superscript at the left of a chemical symbol is the mass in atomic units). From Eq. 21.31, we have

$$\mu_N = \frac{e\hbar}{2m_p} = \frac{\left(1.602 \times 10^{-19} c\right)\left(1.054 \times 10^{-34}\,\text{Js}\right)}{2 \times 1.673 \times 10^{-27}\,\text{kg}}$$

$$= 5.046 \times 10^{-27}\,\text{J}\,\text{T}^{-1}$$

From Eq. 21.30,

$$\gamma = \frac{g_N \mu_N}{\hbar} = \frac{5.58 \times 5.05 \times 10^{-27}\,\text{J}\,\text{T}^{-1}}{1.054 \times 10^{-34}\,\text{Js}}$$

$$= 26.74 \times 10^7\,\text{T}^{-1}\,\text{s}^{-1}$$

Table 21.1 The nuclear g factor g_N, the gyromagnetic ratio γ, and the transition frequency ν corresponding to $B_0 = 1$ T. Natural abundance gives the percent content of a particular isotope in nature.

Atom	Mass	Natural abundance	g_N	γ ($10^7 s^{-1} T^{-1}$)	ν (MHz)
H	1	99.99	5.5850	26.7399	42.5579
H	2	0.01	0.8570	4.1032	6.5304
Li	7	92.50	2.1710	10.3943	16.5431
C	13	1.11	1.4050	6.7269	10.7061
N	14	99.60	0.4030	1.9295	3.0709
N	15	0.40	−0.5670	−2.7147	−4.3206
O	17	0.04	−0.7570	−3.6244	−5.7684
F	19	100.00	5.2570	25.1695	40.0585
Na	23	100.00	1.4780	7.0764	11.2324
P	31	100.00	2.2634	10.8367	17.2472
S	33	0.74	0.4289	2.0535	3.2682

Then, from Eq. 21.40:

$$\nu = \frac{\gamma B_0}{2\pi} = \frac{26.74 \times 10^7 \ s^{-1} \text{tesla}^{-1} \times 1 \ \text{tesla}}{2\pi}$$

$$= 22.56 \ \text{MHz}$$

§9. *Hot Bands.* When we studied the spectrum of a diatomic molecule, we introduced the concept of hot bands, and I need to remind you what these are. Let us assume that a system has a number of energy eigenstates, having the energies $E_0, E_1, E_2, \ldots$. Statistical mechanics tells us that, if the temperature of the system is T, then the probability p_n that the system has the energy E_n is

$$p_n = \frac{g_n \exp\left[-E_n/k_B T\right]}{\sum_{k \geq 0} \exp\left[-E_k/k_B T\right]} \tag{21.41}$$

Here $k_B = 1.38 \times 10^{-23}$ J/K is Boltzmann's constant and g_n is the degeneracy of the energy E_n. This formula tells me that at 0 K all molecules are in the ground state.

Exercise 21.5

Use Eq. 21.41 to show that as the temperature approaches zero kelvin, p_0 tends towards 1 and p_1, p_2 ...tend to zero.

The only transitions possible at this temperature are $E_0 \rightarrow E_1$, or $E_0 \rightarrow E_2$, etc. There are no transitions $E_1 \rightarrow E_2$ or $E_1 \rightarrow E_3$, etc., because there are no molecules in the state having energy E_1. The same is true for transitions from E_2 to E_3, etc. As T increases, some molecules in the system have the energies E_1, E_2, ..., and they can absorb photons and undergo the transitions $E_1 \rightarrow E_2$, ..., $E_2 \rightarrow E_3$, ... Because of this, as T increases the number of peaks in the spectrum increases. These peaks, which are present only when the temperature is high, are called *hot bands*.

A second point to remember is the presence of stimulated emission. If the number of molecules having the energy E_1 is high, and if the system interacts with radiation of frequency $\Omega = (E_1 - E_0)/\hbar$, the presence of the radiation stimulates the transition $E_1 \rightarrow E_0$ (stimulated emission). An analysis of the kinetics of radiation absorption and stimulated emission, in a two-level system, led us to conclude (see Chapter 10) that the population of the upper level cannot be increased, by photon absorption, beyond $\frac{1}{2}$ (when half the molecules have the energy E_1 and half have the energy E_2).

The same phenomena take place in NMR spectroscopy, where these effects are stronger than in optical spectroscopy. This happens because the energy difference ΔE_{spin} between the spin states is much smaller than the energy difference between the states involved in light absorption. Moreover, ΔE_{spin} is smaller than $k_B T$ and because of this the population of the upper state is very high (see Eq. 21.41).

Exercise 21.6

Calculate the magnitude of $k_B T$ at room temperature in eV.

At room temperature, $k_B T$ is of order 0.025 eV. The energy difference between the two states of a proton, in a field of 1 T, is

$$\hbar\Omega = \hbar\gamma B_0$$

$$= 1.054 \times 10^{-34} \, \text{J s} \, \times 26.73 \times 10^7 \, \text{T}^{-1}\,\text{s}^{-1} \, \times 1 \, \text{T} \, \times \left(6.241 \times 10^{18} \, \frac{\text{eV}}{\text{J}}\right)$$

$$= 1.76 \times 10^{-7} \, \text{eV}$$

Because $\hbar\Omega$ is so much smaller than k_BT, the population of the $|\beta\rangle$ state is very high. Indeed

$$\frac{p_1}{p_0} = \frac{g_1}{g_0}\frac{\exp[-E_1/k_BT]}{\exp[-E_0/k_BT]} = \frac{g_1}{g_0}\exp\left[-\frac{E_1-E_0}{k_BT}\right]$$

$$= \frac{g_1}{g_0}\exp\left[-\frac{1.76\times10^{-7}}{0.0257}\right] = 0.999993\frac{g_1}{g_0}$$

Since g_1/g_0 is 1, the two populations are nearly equal. That means that almost half the nuclei are in the upper state, which is very close to the saturation limit. Because of this, the rate of stimulated emission is almost as high as the absorption rate, which means that the net absorption is very small.

This is one of several reasons why the sensitivity of NMR is so poor compared to that of optical spectroscopy and why NMR spectroscopy for nuclei with low g_N factor (i.e. levels close to each other) is difficult.

The Chemical Shift

§10. *The Magnetic Field Acting on a Nucleus Depends on Environment.* A molecule contains several "magnets," associated with the orbiting electrons, the electronic spins, and the nuclear spins. When a static magnetic field $\mathbf{B}_0$ is turned on, these magnetic moments respond by changing their orientation, just like a magnet tends to turn in the presence of a magnetic field. Because of this, the magnetic field produced by these moments changes. This change is proportional to B_0 and tends to diminish it. The magnitude of the field felt by a nucleus is therefore not B_0 but

$$(1-\sigma)B_0 \tag{21.42}$$

The *shielding constant* σ is positive and its order of magnitude is 10^{-3} to 10^{-5}.

This small modification of the absorption frequency, caused by shielding, can be detected experimentally. This is enormously useful to chemists, since the magnitude of the shielding depends on the chemical nature of the atoms surrounding the nucleus whose spin is being studied.

Replacing B_0 in Eq. 21.40 with $(1-\sigma)B_0$ gives the transition frequency

$$\nu = \frac{\gamma(1-\sigma)B_0}{2\pi} \tag{21.43}$$

§11. *Chemical Shifts.* To eliminate, as much as possible, effects that depend on apparatus, NMR spectroscopists compare the spectrum of the nuclear spins in the substance of interest to that of a reference compound containing the same nucleus. They determine the *chemical shift* δ:

$$\delta = \frac{\nu - \nu_{\text{ref}}}{\nu_{\text{ref}}} \times 10^6$$

$$= \frac{\frac{\gamma B_0}{2\pi}(1 - \sigma) - \frac{\gamma B_0}{2\pi}(1 - \sigma_{\text{ref}})}{\frac{\gamma B_0}{2\pi}(1 - \sigma_{\text{ref}})} \times 10^6$$

$$= \frac{\sigma_{\text{ref}} - \sigma}{1 - \sigma_{\text{ref}}} \times 10^6 \approx (\sigma_{\text{ref}} - \sigma) \times 10^6$$

In the last step, I used $1 - \sigma_{\text{ref}} \approx 1$, which is correct since the shielding constant varies between 10^{-3} and 10^{-5}. The chemical shift is thus expressed in parts per million (ppm).

In principle, any compound can be used as a reference, but it is best if we all use the same compound. For example, $(CH_3)_4Si$ is used for NMR of protons, carbon-13, and silicon-29.

§12. *The Chemical Shift is Very Useful.* Having different chemical shifts for nuclei placed in different environments is very useful to the chemist. For example, not all protons in CH_3CH_2OH are equivalent. The local magnetic field acting on the three protons in methyl differs from that acting on the two protons in CH_2, or on the proton in OH. These protons have different shielding constants and because of this the NMR spectrum has three peaks, one for each "kind" of proton. The intensities of these peaks are in the ratio 3:2:1, which corresponds to the number of protons in each group. Because protons in the methyl groups in all molecules have the same shielding constant, we can recognize the presence of a methyl group in a molecule from the presence of a peak at the appropriate frequency in the NMR spectrum. The same is true for the ethyl and the hydroxyl. This makes NMR extremely useful in determining whether a chemical synthesis has succeeded in producing the intended compound.

As you will see in the next section, the three peaks mentioned above are observed only in low-resolution spectra. At high resolution, these peaks split into closely spaced multiplets. The multiplet frequencies provide additional information about the structure of the molecule.

The NMR Spectrum of a System of Two Non-Interacting Nuclei Having Spin $\frac{1}{2}$

§13. *Introduction.* So far we have analyzed the spectrum of an ensemble of molecules, each having one nucleus with spin $\frac{1}{2}$. Here I discuss the case when the molecule has two nuclei, each having $\frac{1}{2}$ spin, which do not interact with each other. This section is mostly a prelude to the ones in which we examine systems of two nuclei with spin $\frac{1}{2}$ in which the spin–spin interactions are present. You will learn here the notation used for the state of two spins and how the spin operators act on them.

§14. *The Hamiltonian and the States of a System of Two Non-Interacting Spin $\frac{1}{2}$ Particles.* The Hamiltonian of two non-interacting nuclear spins is

$$\hat{H}_0 = -\gamma B_0 (1 - \sigma_1) \hat{I}_z(1) - \gamma B_0 (1 - \sigma_2) \hat{I}_z(2) \tag{21.44}$$

We call the two spins *distinguishable* if their shielding constants σ_1 and σ_2 are not equal. This does not mean that the spins belong to two chemically different nuclei, only that they are located in different environments (e.g. we have two protons with different binding sites in a molecule).

The spin states of two non-interacting nuclei of spin $\frac{1}{2}$ are denoted by

$$| \psi_1 \rangle = | \alpha \alpha \rangle \tag{21.45}$$

$$| \psi_2 \rangle = | \alpha \beta \rangle \tag{21.46}$$

$$| \psi_3 \rangle = | \beta \alpha \rangle \tag{21.47}$$

$$| \psi_4 \rangle = | \beta \beta \rangle \tag{21.48}$$

These symbols have the following meaning. Since $s = \frac{1}{2}$, the spin of one nucleus can have two states: $| \alpha \rangle$, with $m_s = +\frac{1}{2}$, in which the spin is oriented along the static magnetic field $\mathbf{B}_0$ (along the OZ axis); and $| \beta \rangle$, with $m_s = -\frac{1}{2}$, in which the spin is oriented opposite to the external field. The symbol $| \alpha \beta \rangle$ indicates a state in which the spin of nucleus 1 is oriented along the field and the spin of nucleus 2 is oriented against the field. The first symbol in $| \alpha \beta \rangle$ gives the spin state of nucleus 1, and the second, that of nucleus 2.

Because the two nuclei have spin $\frac{1}{2}$, they are fermions. If they were indistinguishable (same shielding constant), their states would have to be antisymmetric with respect to a permutation of the labels of the nuclei. The consequences of this requirement are examined in the last section (p. 457). Here we deal with distinguishable nuclei and the states defined by Eqs 21.45–21.48 are fine.

The energies of these four states are given by

$$E_i = \langle \psi_i \mid \hat{H}_0 \mid \psi_i \rangle, \quad i = 1, 2, 3, 4 \tag{21.49}$$

Let us calculate E_3, as an example of how these symbols are manipulated.

I need to evaluate

$$
\begin{aligned}
E_3^0 &\equiv \langle \psi_3 \mid \hat{H}_0 \mid \psi_3 \rangle \\
&= -\gamma B_0 (1 - \sigma_1) \langle \psi_3 \mid \hat{I}_z(1) \mid \psi_3 \rangle - \gamma B_0 (1 - \sigma_2) \langle \psi_3 \mid \hat{I}_z(2) \mid \psi_3 \rangle \\
&= -\gamma B_0 (1 - \sigma_1) \langle \alpha \, \beta \mid \hat{I}_z(1) \mid \alpha \, \beta \rangle - \gamma B_0 (1 - \sigma_2) \langle \alpha \, \beta \mid \hat{I}_z(2) \mid \alpha \, \beta \rangle
\end{aligned}
\tag{21.50}
$$

The operators $\hat{I}_z(1)$ and $\hat{I}_z(2)$ act on the state $\mid \alpha \, \beta \rangle$, as expected

$$\hat{I}_z(1) \mid \alpha \, \beta \rangle = \frac{\hbar}{2} \mid \alpha \, \beta \rangle \tag{21.51}$$

This is natural since the symbol $\mid \alpha \, \beta \rangle$ tells us that the spin of the nucleus 1 is in the state α and this is an eigenstate of $\hat{I}_z(1)$ with the eigenvalue $\hbar/2$. Similarly,

$$\hat{I}_z(2) \mid \alpha \, \beta \rangle = -\frac{\hbar}{2} \mid \alpha \, \beta \rangle \tag{21.52}$$

If we use Eqs 21.51 and 21.52 in the formula for E_3^0 and take into account the normalization condition

$$\langle \alpha \, \beta \mid \alpha \, \beta \rangle = 1 \tag{21.53}$$

we obtain

$$E_3^0 = \frac{h \nu_0}{2} (\sigma_1 - \sigma_2) \tag{21.54}$$

where ν_0 denotes the frequency

$$\nu_0 = \frac{\gamma B_0}{2\pi} \tag{21.55}$$

The other energies can be calculated in a similar fashion and are:

$$E_1^0 = -\frac{h\nu_0}{2}\{2 - (\sigma_1 + \sigma_2)\}$$ (21.56)

$$E_2^0 = -\frac{h\nu_0}{2}(\sigma_1 - \sigma_2)$$ (21.57)

$$E_4^0 = \frac{h\nu_0}{2}\{2 - (\sigma_1 + \sigma_2)\}$$ (21.58)

Exercise 21.7

Show that Eqs 21.56–21.58 are correct.

§15. *The Order of Magnitude of These Energies.* As an example, let us calculate the spin energies E_1, ..., E_4 of two non-interacting, distinguishable protons when $\nu_0 = 300$ MHz, $\sigma_1 = 5.3 \times 10^{-5}$, $\sigma_2 = 3.1 \times 10^{-5}$. Using Eqs 21.54–21.58, I obtain (see Workbook QM21.3)

$$E_1/h = -2.999874 \times 10^8 \text{ Hz} = -299.9874 \text{ MHz}$$

$$E_2/h = -3300 \text{ Hz}$$

$$E_3/h = 3300 \text{ Hz}$$

$$E_4/h = 2.999874 \times 10^8 \text{ Hz} = -299.9874 \text{ MHz}$$

The order of magnitude of ν_0 is tens or hundreds of megahertz, while σ_1 and σ_2 are of order 10^{-3} to 10^{-5}. Because of this, the energies E_1, ..., E_4 have very different orders of magnitude. In E_1 and E_4, $\sigma_1 + \sigma_2 \ll 2$ and the absolute values of these energies are of order $h\nu_0 \approx h \times 10^8$ Hz. On the other hand, the absolute values of E_2 and E_3 are of the order $h\nu_0(\sigma_1 - \sigma_2)/2 \approx h \times 10^8 \times 10^{-5} = h \times 10^3$ Hz.

§16. *The Selection Rules.* The transition rate from an initial spin state $|\phi_i\rangle$ to a final spin state $|\phi_f\rangle$ is proportional to the matrix element defined by Eq. 21.37. We need to modify this equation slightly because in the present case the interaction energy is the sum of the interaction of each spin with the oscillating field. Therefore, the rate is proportional to

$$|\langle\phi_f | \hat{I}_x(1) + \hat{I}_x(2) | \phi_i\rangle|^2$$ (21.59)

In principle, we could have a transition from any state $|\psi_n\rangle$ to any state $|\psi_m\rangle$ of higher energy. However, for some of them the matrix element in Eq. 21.59 may turn out to be zero (these transitions are forbidden).

Let us start by examining a transition from $|\psi_1\rangle = |\alpha\,\alpha\rangle$ to $|\psi_2\rangle = |\alpha\,\beta\rangle$. The matrix element in Eq. 21.59 becomes

$$|\langle\alpha\,\beta\,|\,I_x(1) + I_x(2)\,|\,\alpha\,\alpha\rangle|^2 \tag{21.60}$$

We have

$$\hat{I}_x(1)\,|\,\alpha\,\alpha\rangle = \frac{\hbar}{2}\,|\,\beta\,\alpha\rangle \tag{21.61}$$

and

$$\hat{I}_x(2)\,|\,\alpha\,\alpha\rangle = \frac{\hbar}{2}\,|\,\alpha\,\beta\rangle \tag{21.62}$$

In Eq. 21.61 the operator $I_x(1)$ acts on the first state in $|\alpha\,\beta\rangle$, which is $|\alpha\rangle$. According to Eq. 21.11 the result is $(\hbar/2)\,|\,\beta\rangle$, which leads to Eq. 21.61. Similarly, using Eq. 21.12 gives Eq. 21.62.

Inserting Eqs 21.61 and 21.62 in Eq. 21.60 gives

$$\left|\frac{\hbar}{2}\langle\alpha\,\beta\,|\,\beta\,\alpha\rangle + \frac{\hbar}{2}\langle\alpha\,\beta\,|\,\alpha\,\beta\rangle\right|^2 \tag{21.63}$$

Due to the orthonormality of the spin wave functions, we know that

$$\langle\alpha\,\beta\,|\,\alpha\,\beta\rangle = 1$$

and

$$\langle\alpha\,\beta\,|\,\beta\,\alpha\rangle = 0$$

Therefore the matrix element in Eq. 21.63 is not zero, and the transition $|\alpha\,\alpha\rangle \rightarrow |\alpha\,\beta\rangle$ is not forbidden.

I will calculate next the matrix element for the transition from $|\psi_1\rangle = |\alpha\,\alpha\rangle$ to $|\psi_4\rangle = |\beta\,\beta\rangle$. I can tell, without performing any calculation, that this transition is forbidden. My anticipation is based on a simple observation. The operator $\hat{I}_x(1) + \hat{I}_x(2)$ acting on a two-spin state such as $|\alpha\,\alpha\rangle$ will "flip" one spin. Therefore it produces $|\beta\alpha\rangle + |\alpha\beta\rangle$. Since flipping one spin does not make the initial state identical to the final state, the matrix element $\langle\beta\beta|\hat{I}_x(1) + \hat{I}_x(2)|\alpha\alpha\rangle$ is equal to zero (for the transition $|\alpha\alpha\rangle \rightarrow |\beta\beta\rangle$, one spin flip can create $|\beta\alpha\rangle$ or $|\alpha\beta\rangle$ but not $|\beta\beta\rangle$). As a result, $\langle\beta\beta|\hat{I}_x(1) + \hat{I}_x(2)|\alpha\alpha\rangle = 0$.

We have found the following selection rules for the NMR spectrum of two non-interacting, distinguishable particles having spin $\frac{1}{2}$: transitions in which the final state differs from the initial state by more than one spin flip are forbidden. Therefore, of all possible transitions, only

$$
\left.
\begin{aligned}
|\alpha\alpha\rangle &\rightarrow |\beta\alpha\rangle & (1 \rightarrow 2) \\
|\alpha\alpha\rangle &\rightarrow |\alpha\beta\rangle & (1 \rightarrow 3) \\
|\beta\alpha\rangle &\rightarrow |\beta\beta\rangle & (2 \rightarrow 4) \\
|\alpha\beta\rangle &\rightarrow |\beta\beta\rangle & (3 \rightarrow 4)
\end{aligned}
\right\} \tag{21.64}
$$

are allowed.

§17. *The Frequencies of the Allowed Transitions.* The frequency of a transition from state $|\psi_i\rangle$ to state $|\psi_f\rangle$ is given by

$$
\nu_{if} = \frac{E_f - E_i}{h} \tag{21.65}
$$

Using Eqs 21.54–21.58 leads to

$$
\nu_{12} = \nu_0(1 - \sigma_1) = \nu_{34} \tag{21.66}
$$

$$
\nu_{13} = \nu_0(1 - \sigma_2) = \nu_{24} \tag{21.67}
$$

Even though there are four allowed transitions, we have only two peaks in the spectrum, because two of these transitions have the same frequencies as the other two.

The following example is calculated in Workbook QM21.3. For $\nu_0 = 300$ MHz, $\sigma_1 = 5.3 \times 10^{-5}$, and $\sigma_2 = 3.1 \times 10^{-5}$, the transition frequencies are

$$
\nu_{12} = \nu_{34} = 300 \times 10^6 \times (1 - 5.3 \times 10^{-5}) = 299.9841 \text{ MHz}
$$

$$
\nu_{13} = \nu_{24} = 300 \times 10^6 \times (1 - 3.1 \times 10^{-5}) = 299.9907 \text{ MHz}
$$

The two peaks are separated by $\nu_{13} - \nu_{12} = 6600$ Hz.

The Spin–Spin Interaction

§18. *Spin–Spin Coupling.* The model described in the previous section (pp. 436–440) assumes that the two spins do not interact with each other. Its results were in agreement with early, low-resolution experiments. As the instruments were improved, several lines were observed where the model predicted only one.

Here are a few examples, taken from a book by Davis.[a] The HD molecule consists of a proton, ^{1}H, having spin $\frac{1}{2}$, and a deuteron, ^{2}H, having spin 1. If the nuclei do not interact with each other, the spectrum will be a superposition of the proton and the deuteron spectra. The proton has spin $\frac{1}{2}$, which means it has two energy levels. The transition from one of these levels to the other produces one peak in the NMR spectrum. The deuteron has spin 1 and three energy levels, corresponding to $m_s = -1, 0, +1$. These levels produce only one peak in the spectrum, because the transition from $m_s = -1$ to $m_s = +1$ is forbidden, and the transition from $m_s = -1$ to $m_s = 0$ has the same frequency as that from $m_s = 0$ to $m_s = +1$. Therefore, if the spins do not interact, the NMR spectrum of HD would have two peaks, one for ^{1}H and one for ^{2}H. This is what is observed if the resolution of the spectrometer is poor.

The high-resolution spectrum tells a different story: it has three peaks, close to each other, located near the frequency where a single ^{1}H peak is expected. In addition, two closely spaced peaks appear near the frequency where the single ^{2}H peak is expected. Note that the spectrum of ^{1}H has three peaks and ^{2}H has three spin states, while the spectrum of ^{2}H has two peaks and ^{1}H has two spin states. The number of peaks for one nucleus equals the number of states of the other nucleus.

This suggests the following explanation for this peak proliferation. The ^{2}H nucleus is exposed to the external magnetic field $\mathbf{B}_0$ and also to the field created by the magnetic dipole $\hat{\mathbf{m}}_1$ of ^{1}H. The ^{1}H protons can be in state $|\alpha\rangle$ or in $|\beta\rangle$. The magnetic moment of the proton in the state $|\alpha\rangle$ has a different orientation than the magnetic moment of one in the state $|\beta\rangle$. Differently oriented magnetic moments of ^{1}H produce different magnetic fields at the position of the deuteron. Therefore, the deuteron in an HD molecule in which the proton is in the state $|\alpha\rangle$ is exposed to a slightly different magnetic field than the deuteron in an HD molecule in which the proton is in the state $|\beta\rangle$. Because of this, the energy of the deuteron spin in a molecule in which the proton spin is in state $|\alpha\rangle$ differs from that of the deuteron in a molecule in which the proton spin is in state $|\beta\rangle$. This is why the deuteron peak in the NMR spectrum splits into two peaks located close to each other.

---

[a] J.C. Davis, *Advanced Physical Chemistry*, Ronald Press Co., New York, 1965.

A similar explanation works for the spectrum of the proton. The deuterons can be in one of the three states, $m_s = -1, 0$ or 1, hence the protons in the sample are exposed to three different magnetic fields and have three different energy levels.

The key point here is that the energies of one nucleus depend on the state of the other nucleus. This is what we mean when we say that the spins interact with each other.

Another example is provided by CH_3CH_2OH. If the protons do not interact with each other the NMR spectrum has three peaks: one due to the protons in CH_3, one due to the protons in CH_2, and one due to the proton in OH. These protons produce different peaks because they have different shielding constants σ.

In the high-resolution spectrum, the peak due to the protons in CH_3 splits into three peaks. Let us assume that this splitting is caused by the magnetic field generated by the protons in the CH_2 group (H in –OH is too far away to affect the protons in CH_3). The states of the two protons in CH_2 are $|\alpha\,\alpha\rangle$, $|\alpha\,\beta\rangle$, $|\beta\,\alpha\rangle$, and $|\beta\,\beta\rangle$. The state $|\alpha\,\alpha\rangle$ has the spins of both protons oriented along B_0. In the state $|\beta\,\beta\rangle$, the spins of both protons are oriented the direction opposite to the field. Because of these different orientations of the spin, the state $|\alpha\,\alpha\rangle$ generates a different magnetic field, acting on the protons in CH_3, than does the state $|\beta\,\beta\rangle$. Yet another field is generated when the protons in CH_2 are in the state $|\alpha\,\beta\rangle$. However, the state $|\beta\,\alpha\rangle$ generates the same field as does $|\alpha\,\beta\rangle$ since it does not matter if proton 1 has spin up and proton 2 has spin down or vice versa. There are, therefore, three different fields acting on the nuclei in CH_3, caused by the four possible states of the two protons in CH_2. This is why the high-resolution spectrum of the protons in CH_3 has three peaks.

The final example is provided by the NMR spectrum of PO_3F^{2-}. The phosphorus and the fluorine nuclei have spin $\frac{1}{2}$ and the high-resolution NMR spectrum has two peaks for the P spectrum and two for the F; the theory that assumes no spin–spin interaction predicts one peak for each nucleus.

§19. *The Interaction Between Nuclear Spins.* The analysis presented above suggests that multiplets, observed in the high-resolution NMR spectrum, appear because of the interaction between the nuclear spins. An obvious candidate for this interaction is the direct interaction between the magnetic moments of the nuclei. It turns out, however, that the tumbling motion of the molecule in a liquid or a gas averages this interaction to zero. Unless the molecules form an ordered solid, the effect of this interaction is negligible.

There is, however, a more subtle interaction between the nuclear spins, which is mediated by the electrons in the molecule. Recall that two magnetic moments are associated with the electrons in a molecule: one is due to the spin of the electron and the other to its orbital motion. The magnetic field of nucleus 1 acts on the magnetic moments of the electrons causing a slight change of the magnetic field the electron produce at the position of the nuclear spin 2. In this way the spin of nucleus 1 affects the magnetic field acting on nucleus 2. The nuclear spins act on each other through the agency of the electrons. The result is an interaction energy of the form

$$H_{\text{int}} = \frac{h}{\hbar^2} J \, \hat{\mathbf{I}}(1) \cdot \hat{\mathbf{I}}(2) \tag{21.68}$$

where $\hat{\mathbf{I}}(1)$ is the nuclear spin operator of nucleus 1 and $\hat{\mathbf{I}}(2)$ is that of nucleus 2. J is the coupling constant between the two spins: it is defined by Eq. 21.68 and it is determined experimentally. The factor $h/\hbar^2$ was introduced to make J have frequency units (s^{-1}).

As you have seen in the previous section some of the states of a system consisting of two non-interacting spins are degenerate. It is very likely that the interaction between the spins will "lift" the degeneracy increasing the number of distinct energy levels. This, in turn, can result in more peaks in the NMR spectrum. The next three sections show that this is what happens.

The Spectrum of Two Distinguishable, Interacting Nuclei

§20. *The Energies of the Spin States.* To understand how the spin–spin interaction modifies the NMR spectrum, we calculate the energies of the spin states of a molecule that has two *distinguishable* protons. This is the simplest possible system that illustrates the effect of the spin–spin interaction. Systems with nuclei having higher spins (e.g. $s = \frac{3}{2}$) or more than two interacting nuclei are treated similarly; there are no conceptual complications, but the algebra becomes much more tedious.

In some situations, the interaction between the spins is strong enough that we cannot use perturbation theory to calculate its effect on the energy levels. Fortunately, many problems involving spins can be solved exactly with a relatively small computational effort. This is not the case in most other areas of quantum mechanics, where accurate solutions require massive computations or are impossible regardless of how much computer time is available.

In this section I calculate the energy level exactly; the results are valid regardless of how strong the spin–spin interaction is.

§21. *The Hamiltonian.* The Hamiltonian of two interacting nuclear spins is

$$\hat{H} = \hat{H}_0 + \hat{H}_{\text{int}} \tag{21.69}$$

Here

$$\hat{H}_0 = -\gamma B_0(1 - \sigma_1)\hat{I}_z(1) - \gamma B_0(1 - \sigma_2)\hat{I}_z(2) \tag{21.70}$$

is the Hamiltonian when the spins do not interact. The two spins are distinguishable because their shielding constants σ_1 and σ_2 are not equal. The operator

$$\hat{H}_{\text{int}} \equiv \frac{h}{\hbar^2} J\, \hat{\mathbf{I}}(1) \cdot \hat{\mathbf{I}}(2) \tag{21.71}$$

is the energy of the interaction between the spins.

§22. *The States of the Interacting Spins.* I want to solve the Schrödinger equation

$$\hat{H} \,|\, \phi \rangle = E \,|\, \phi \rangle \tag{21.72}$$

for two interacting, distinguishable spins. The Hamiltonian is given by Eqs 21.69–21.71, and $|\,\phi\rangle$ and E are unknown. I will assume that

$$|\,\phi\rangle = \sum_{i=1}^{4} c_i \,|\, \psi_i \rangle \tag{21.73}$$

with c_1, ..., c_4 unknown numbers and $|\,\psi_1\rangle$ ... $|\,\psi_4\rangle$ given by Eqs 21.45–21.48. The states $|\,\psi_1\rangle$, ..., $|\,\psi_4\rangle$ are the states of the non-interacting spins.

I plan to determine c_1, ..., c_4 and E so that they satisfy the equation obtained when I replace $|\,\phi\rangle$ in Eq. 21.72 with the expression given by Eq. 21.73.

§23. *Why this Choice of State?* Before doing that, I want to explain why I claim that Eq. 21.73 is capable of representing *exactly* the states of the interacting spins, if c_1, ..., c_4 are well chosen.

We follow here a well-established procedure for solving differential equations. We guess a functional form for the solution, which contains some adjustable parameters. Then, we insert this functional form in the equation and adjust the parameters to give the best solution the functional form is capable of. If we made a good guess, the solution is accurate; if the guess is bad, the solution can be terrible. Eq. 21.73 is our guess and $c_1, \ldots, c_4$ are our parameters. Why do I claim that this guess is perfect?

The solution $|\phi\rangle$ must be flexible enough to provide an answer to all sensible physically questions. In our case there are only four questions that we can ask. One of them is the probability that when the system is in the state $|\phi\rangle$, the spins are in the state $|\psi_1\rangle = |\alpha\alpha\rangle$. Quantum mechanics says that this is given by

$$P_1 \equiv |\langle\psi_1 \,|\, \phi\rangle|^2 = |\langle\alpha\alpha \,|\, \phi\rangle|^2$$

If I use Eq. 21.73 for $|\phi\rangle$, I obtain

$$P_1 = |\langle\psi_1 \,|\, \phi\rangle|^2 = |\sum_{i=1}^{4} c_i\langle\psi_1 \,|\, \psi_i\rangle|^2 = |c_1|^2$$

The last equality holds because

$$\langle\psi_i \,|\, \psi_j\rangle = \begin{cases} 1 & \text{if } i = j \\ 0 & \text{if } i \neq j \end{cases} \tag{21.74}$$

In general,

$$P_i \equiv |\langle\psi_i \,|\, \phi\rangle|^2 = |c_i|^2 \tag{21.75}$$

I find that the guess made in Eq. 21.73 can represent all possible situations arising in the experiment. There is more to this issue, but this is as far as I want to go here. The state $|\phi\rangle$ of the system needs to be a linear superposition of all the states that I can detect in an experiment that puts the system in the state $|\phi\rangle$. The four states $|\psi_1\rangle, \ldots, |\psi_4\rangle$ cover all spin states that I can find when I examine the spin of the system.

Exercise 21.8

Write the state of a system (i.e. the analog of Eq. 21.73) consisting of three interacting, distinguishable protons.

Exercise 21.9

Write the state of a system consisting of two interacting, distinguishable nuclei of spin 1.

§24. *The Galerkin Method: how to Turn an Operator Equation into a Matrix Equation.* Let us start with some nomenclature. When using an expression like Eq. 21.73, we say that we have represented $|\phi\rangle$ in the basis set $|\psi_1\rangle$, $|\psi_2\rangle$, $|\psi_3\rangle$, $|\psi_4\rangle$. Because the states $|\psi_i\rangle$ satisfy Eq. 21.74, we say that the basis set $\{|\psi_i\rangle\}_{i=1}^4$ is *orthonormal*. Since the basis set covers all physical situations of interest, we say that the basis set is *complete*. (The concept of completeness is more general than presented here.)

The Galerkin method starts by taking the scalar product between each element $|\psi_n\rangle$ of the basis set and the eigenvalue equation; this gives

$$\langle\psi_n|\hat{H}|\phi\rangle = E\langle\psi_n|\phi\rangle, \quad n = 1, 2, 3, 4 \tag{21.76}$$

Note that this operation generates four equations. The number of equations is equal to the number of coefficients c_n.

Next, I use Eq. 21.73 for $|\phi\rangle$, the properties of the scalar product, and the orthonormality of the basis set to obtain

$$\sum_{i=1}^4 \langle\psi_n|\hat{H}|\psi_i\rangle c_i = E c_n, \quad n = 1, 2, 3, 4 \tag{21.77}$$

Exercise 21.10

Use Eq. 21.74 and the properties of the scalar product to show that Eq. 21.77 is correct.

Now use your knowledge of linear algebra. $\langle\psi_n|\hat{H}|\psi_m\rangle$ is a number labeled by two indices, n and m. The set of these numbers forms a 4×4 matrix, which I denote by $\overset{\leftrightarrow}{H}$. $\{c_1, c_2, c_3, c_4\}$ are the components of a four-dimensional vector, $\vec{c}$.

In linear-algebra notation, Eq. 21.77 can be written as

$$\overset{\leftrightarrow}{H} \cdot \vec{c} = E\vec{c} \tag{21.78}$$

This is an eigenvalue problem in a four-dimensional space. Its solutions are the four eigenvectors $\vec{c}^{\,i}$, $i = 1, 2, 3, 4$; to each of these vectors corresponds an eigenvalue E_i. Each eigenvector has four components, denoted by

$$\vec{c}^{\,i} = \{c_1^i, c_2^i, c_3^i, c_4^i\}$$

For each eigenvector, we have an eigenstate of the Schrödinger equation:

$$|\phi_i\rangle = \sum_{n=1}^{4} c_n^i \,|\psi_n\rangle \tag{21.79}$$

The eigenenergy E_i of this state is the eigenvalue E_i of the 4×4 matrix $\overset{\leftrightarrow}{H}$.

There are many computer programs that find eigenvalues of a matrix when we give them the matrix elements $\langle n \,|\, \hat{H} \,|\, m \rangle \equiv \overset{\leftrightarrow}{H}_{nm}$.

§25. *The Matrix Elements* $\overset{\leftrightarrow}{H}_{nm} = \langle \psi_n \,|\, \hat{H} \,|\, \psi_m \rangle$. To solve the eigenvalue problem we must calculate the matrix elements $\langle \psi_n \,|\, \hat{H} \,|\, \psi_m \rangle$. Here I perform a few such calculations, to exemplify the methodology used.

We have

$$\langle \psi_n \,|\, \hat{H} \,|\, \psi_m \rangle = \langle \psi_n \,|\, \hat{H}_0 \,|\, \psi_m \rangle + \langle \psi_n \,|\, \hat{H}_{\text{int}} \,|\, \psi_m \rangle \tag{21.80}$$

The diagonal matrix elements $\langle \psi_i \,|\, \hat{H}_0 \,|\, \psi_i \rangle$ have already been calculated and are given by Eqs 21.54–21.58. I repeat those results here for easy reference:

$$\langle \psi_1 \,|\, \hat{H}_0 \,|\, \psi_1 \rangle \equiv E_1^0 = -\frac{h\nu_0}{2}\,[2 - (\sigma_1 + \sigma_2)] \tag{21.81}$$

$$\langle \psi_2 \,|\, \hat{H}_0 \,|\, \psi_2 \rangle \equiv E_2^0 = -\frac{h\nu_0}{2}(\sigma_1 - \sigma_2) \tag{21.82}$$

$$\langle \psi_3 \,|\, \hat{H}_0 \,|\, \psi_3 \rangle \equiv E_3^0 = \frac{h\nu_0}{2}(\sigma_1 - \sigma_2) \tag{21.83}$$

$$\langle \psi_4 \,|\, \hat{H}_0 \,|\, \psi_4 \rangle \equiv E_4^0 = \frac{h\nu_0}{2}\,[2 - (\sigma_1 + \sigma_2)] \tag{21.84}$$

It is not hard to see that

$$\langle \psi_i \,|\, \hat{H}_0 \,|\, \psi_j \rangle = 0 \ \text{ for } i \neq j \tag{21.85}$$

For example, let us look at $\langle \psi_1 \,|\, \hat{H}_0 \,|\, \psi_2 \rangle = \langle \alpha\,\alpha \,|\, \hat{H}_0 \,|\, \alpha\,\beta \rangle$. Since $\hat{H}_0$ contains only the $\hat{I}_z$ operator, its action on $|\,\alpha\,\beta \rangle$ will not flip the spin. Therefore $\langle \psi_1 \,|\, \hat{H}_0 \,|\, \psi_2 \rangle$ is proportional to $\langle \alpha\,\alpha \,|\, \alpha\,\beta \rangle$, which is equal to zero. Similar arguments show that all off-diagonal elements of $\hat{H}_0$ are zero.

Next we need to evaluate the matrix elements of the interaction energy. Let us look at a diagonal matrix element of $\hat{H}_{\text{int}}$, say the third one:

$$\langle \psi_3 \,|\, \hat{H}_{\text{int}} \,|\, \psi_3 \rangle = \frac{hJ}{\hbar^2} \langle \psi_3 \,|\, \hat{\mathbf{I}}(1) \cdot \hat{\mathbf{I}}(2) \,|\, \psi_3 \rangle = \frac{hJ}{\hbar^2} \langle \beta\,\alpha \,|\, \hat{\mathbf{I}}(1) \cdot \hat{\mathbf{I}}(2) \,|\, \beta\,\alpha \rangle \tag{21.86}$$

The dot product of the spin operators is

$$\hat{\mathbf{I}}(1) \cdot \hat{\mathbf{I}}(2) = \hat{I}_x(1)\hat{I}_x(2) + \hat{I}_y(1)\hat{I}_y(2) + \hat{I}_z(1)\hat{I}_z(2) \tag{21.87}$$

We know how the spin operators act on $|\,\beta\,\alpha \rangle$ (use Eqs 21.11–21.14)

$$\hat{I}_x(1)\hat{I}_x(2) \,|\, \beta\,\alpha \rangle = \frac{\hbar}{2}\frac{\hbar}{2} \,|\, \alpha\,\beta \rangle = \left(\frac{\hbar}{2}\right)^2 |\,\alpha\,\beta \rangle \tag{21.88}$$

$$\hat{I}_y(1)\hat{I}_y(2) \,|\, \beta\,\alpha \rangle = \frac{i\hbar}{2}\left(-\frac{i\hbar}{2}\right) |\,\alpha\,\beta \rangle = \left(\frac{\hbar}{2}\right)^2 |\,\alpha\,\beta \rangle \tag{21.89}$$

$$\hat{I}_z(1)\hat{I}_z(2) \,|\, \beta\,\alpha \rangle = \left(-\frac{\hbar}{2}\right)\frac{\hbar}{2} \,|\, \beta\,\alpha \rangle = -\left(\frac{\hbar}{2}\right)^2 |\,\beta\,\alpha \rangle \tag{21.90}$$

Using Eqs 21.87–21.90 we calculate that

$$\langle \beta\,\alpha \,|\, \hat{\mathbf{I}}(1) \cdot \hat{\mathbf{I}}(2) \,|\, \beta\,\alpha \rangle = \left(\frac{\hbar}{2}\right)^2 \langle \beta\,\alpha \,|\, \alpha\,\beta \rangle + \left(\frac{\hbar}{2}\right)^2 \langle \beta\,\alpha \,|\, \alpha\,\beta \rangle - \left(\frac{\hbar}{2}\right)^2 \langle \beta\,\alpha \,|\, \beta\,\alpha \rangle$$

$$= -\left(\frac{\hbar}{2}\right)^2 \tag{21.91}$$

To obtain the last equality, I used $\langle \beta\,\alpha \,|\, \alpha\,\beta \rangle = 0$ and $\langle \beta\,\alpha \,|\, \beta\,\alpha \rangle = 1$.

Using Eqs 21.83, 21.86, and 21.91 in Eq. 21.80 gives

$$\langle \psi_3 \,|\, \hat{H} \,|\, \psi_3 \rangle = \frac{h v_0}{2}(\sigma_1 - \sigma_2) - \frac{hJ}{4}$$

Note that only $\hat{I}_z(1)\hat{I}_z(2)$ contributes to the diagonal elements $\langle \psi_i \,|\, \hat{H} \,|\, \psi_i \rangle$.

Exercise 21.11

Explain why the above statement is correct.

The elements $\langle \psi_1 \,|\, \hat{H} \,|\, \psi_1 \rangle$, $\langle \psi_2 \,|\, \hat{H} \,|\, \psi_3 \rangle$, and $\langle \psi_4 \,|\, \hat{H} \,|\, \psi_4 \rangle$ are calculated similarly. The results are

$$\langle \psi_1 \,|\, \hat{H} \,|\, \psi_1 \rangle = E_1^0 + \frac{hJ}{4} = -\frac{h v_0}{2}(2 - [\sigma_1 + \sigma_2]) + \frac{hJ}{4} \qquad (21.92)$$

$$\langle \psi_2 \,|\, \hat{H} \,|\, \psi_2 \rangle = E_2^0 - \frac{hJ}{4} = -\frac{h v_0}{2}(\sigma_1 - \sigma_2) - \frac{hJ}{4} \qquad (21.93)$$

$$\langle \psi_3 \,|\, \hat{H} \,|\, \psi_3 \rangle = E_3^0 - \frac{hJ}{4} = \frac{h v_0}{2}(\sigma_1 - \sigma_2) - \frac{hJ}{4} \qquad (21.94)$$

$$\langle \psi_4 \,|\, \hat{H} \,|\, \psi_4 \rangle = E_4^0 + \frac{hJ}{4} = \frac{h v_0}{2}(2 - [\sigma_1 + \sigma_2]) + \frac{hJ}{4} \qquad (21.95)$$

§26. *Perturbation Theory.* Perturbation theory assumes that these are the energies that the two interacting, distinguishable protons can take. The argument is simple: if the interaction is weak, we can assume that the states $|\psi_1\rangle, \ldots, |\psi_4\rangle$ of the non-interacting spins provide a good approximation for the states of the interacting spins. This means that the energies of the system are approximated by $\langle \psi_1 \,|\, \hat{H} \,|\, \psi_1 \rangle, \ldots, \langle \psi_4 \,|\, \hat{H} \,|\, \psi_4 \rangle$. Perturbation theory takes into account the effect of the interaction between the spins in the Hamiltonian, but not in the wave function. If we are satisfied that the interaction between spins is weak, we can use these energies to calculate the frequencies at which we will observe peaks in the NMR spectrum. Since an exact solution of the problem is given below I will not discuss the results of perturbation theory further.

Exercise 21.12

Use perturbation theory to calculate the transition frequencies that will be observed in the NMR spectrum. Determine which transitions are allowed and which are forbidden. Make an energy level diagram to illustrate your results.

§27. *The Off-Diagonal Elements.* To solve the eigenvalue problem exactly, we need to calculate the off-diagonal elements $\langle \psi_i \mid \hat{H} \mid \psi_j \rangle$ for $i \neq j$. An example of such a calculation is given below.

$$\overset{\leftrightarrow}{H}_{23} \equiv \langle \psi_2 \mid \hat{H} \mid \psi_3 \rangle = \langle \psi_2 \mid \hat{H}_0 \mid \psi_3 \rangle + \langle \psi_2 \mid \hat{H}_{\text{int}} \mid \psi_3 \rangle$$

$$= -\gamma B_0(1 - \sigma_1)\langle \alpha \beta \mid \hat{I}_z(1) \mid \beta \alpha \rangle - \gamma B_0(1 - \sigma_2)\langle \alpha \beta \mid \hat{I}_z(2) \mid \beta \alpha \rangle$$

$$+ \frac{hJ}{\hbar^2} \Big[\langle \alpha \beta \mid \hat{I}_x(1)\hat{I}_x(2) \mid \beta \alpha \rangle + \langle \alpha \beta \mid \hat{I}_y(1)\hat{I}_y(2) \mid \beta \alpha \rangle$$

$$+ \langle \alpha \beta \mid \hat{I}_z(1)\hat{I}_z(2) \mid \beta \alpha \rangle \Big] \tag{21.96}$$

Using the rules for calculating how the spin operators act on the spin states, we obtain

$$\overset{\leftrightarrow}{H}_{23} = -\gamma B_0(1 - \sigma_1)\left(-\frac{\hbar}{2}\langle \alpha \beta \mid \beta \alpha \rangle \right) - \gamma B_0(1 - \sigma_2)\left(\frac{\hbar}{2}\langle \alpha \beta \mid \beta \alpha \rangle \right)$$

$$+ \frac{hJ}{\hbar^2} \Bigg[\left(\frac{\hbar}{2} \right)^2 \langle \alpha \beta \mid \alpha \beta \rangle + \left(\frac{i\hbar}{2} \right)\left(\frac{-i\hbar}{2} \right) \langle \alpha \beta \mid \alpha \beta \rangle$$

$$+ \left(\frac{\hbar}{2} \right)\left(\frac{\hbar}{2} \right) \langle \alpha \beta \mid \beta \alpha \rangle \Bigg] \tag{21.97}$$

Using the orthonormality conditions gives

$$\overset{\leftrightarrow}{H}_{23} = \frac{hJ}{2} \tag{21.98}$$

The other off-diagonal elements are calculated similarly, and they are all zero except for

$$\overset{\leftrightarrow}{H}_{32} = \frac{hJ}{2} \tag{21.99}$$

We now have all the matrix elements and can write down the Hamiltomian matrix:

$$
\overset{\leftrightarrow}{H} =
\begin{pmatrix}
E_1^0 + \frac{hJ}{4} & 0 & 0 & 0 \\
0 & E_2^0 - \frac{hJ}{4} & \frac{hJ}{2} & 0 \\
0 & \frac{hJ}{2} & E_3^0 - \frac{hJ}{4} & 0 \\
0 & 0 & 0 & E_4^0 + \frac{hJ}{4}
\end{pmatrix}
\tag{21.100}
$$

Next we need to find the eigenvalues and the eigenvectors of this matrix and use them to examine the physical properties of the system.

This matrix is fairly simple and you might want to try to diagonalize it by hand. If you have no experience with this sort of thing, the exercise might be a bit frustrating. I prefer to use a computer to solve this problem. I did this in Workbook QM21.5 and obtained the following results.

§28. *The Lowest Energy E_1 and Eigenvector $|\phi_1\rangle$*. The lowest energy eigenvalue is

$$
E_1 = -\frac{h\nu_0}{2}(2 - [\sigma_1 + \sigma_2]) + \frac{hJ}{4} \equiv E_1^0 + \frac{hJ}{4}
\tag{21.101}
$$

This is equal to the energy E_1^0 of the non-interacting spins plus the term $hJ/4$ introduced by the spin–spin interaction.

The corresponding eigenvector is

$$
\{c_1^1, c_2^1, c_3^1, c_4^1\} = \{1, 0, 0, 0\}
\tag{21.102}
$$

Using Eqs 21.79 and 21.102 we obtain the spin state $|\phi_1\rangle$, corresponding to the energy E_1 :

$$
|\phi_1\rangle = \sum_{i=1}^{4} c_i^1 \, |\psi_i\rangle = |\psi_1\rangle = |\alpha\,\alpha\rangle
\tag{21.103}
$$

In this state both spins are oriented along the static field and this gives them the lowest energy.

§29. *The Eigenvalue E_2 and the Spin State $|\phi_2\rangle$.* The second eigenvalue is more interesting:

$$E_2 = -\frac{h\nu_0}{2}|\sigma_1 - \sigma_2|\sqrt{1 + \left(\frac{J}{\nu_0(\sigma_1 - \sigma_2)}\right)^2} - \frac{hJ}{4} \tag{21.104}$$

The corresponding eigenvector has the components

$$\left.\begin{array}{ll}
c_1^2 = & 0 \\[6pt]
c_2^2 = & -\frac{\nu_0(\sigma_1 - \sigma_2)}{J} - \frac{\nu_0|\sigma_1 - \sigma_2|}{J}\sqrt{1 + \left(\frac{J}{\nu_0(\sigma_2 - \sigma_1)}\right)^2} \\[6pt]
c_3^2 = & 1 \\[6pt]
c_4^2 = & 0
\end{array}\right\} \tag{21.105}$$

The spin state is (use Eqs 21.79 and 21.105)

$$|\phi_2\rangle = \sum_{i=1}^{4} c_i^2 |\psi_i\rangle = c_2^2 |\alpha\,\beta\rangle + c_3^2 |\beta\,\alpha\rangle \tag{21.106}$$

$$= \left[-\frac{\nu_0(\sigma_1 - \sigma_2)}{J} - \frac{\nu_0|\sigma_1 - \sigma_2|}{J}\sqrt{1 + \left(\frac{J}{\nu_0(\sigma_2 - \sigma_1)}\right)^2}\right]|\alpha\,\beta\rangle + |\beta\,\alpha\rangle$$

$$\tag{21.107}$$

What is the meaning of this expression? The state $|\phi_2\rangle$ is an eigenstate of the full Hamiltonian and it is the only state that has physical reality. When the two spins in the ground state $|\phi_1\rangle = |\alpha\,\alpha\rangle$ interact with a magnetic field having the frequency

$$\nu_{12} = \frac{E_2 - E_1}{h}$$

they absorb energy and their state becomes $|\phi_2\rangle$. This is what I mean when I say that $|\phi_2\rangle$ is a state that has physical reality: it can be created in the laboratory. What about the states $|\alpha\,\alpha\rangle$, $|\alpha\,\beta\rangle$, $|\beta\,\alpha\rangle$, and $|\beta\,\beta\rangle$? They are crutches that we use to express the real states $|\phi_1\rangle, \ldots, |\phi_4\rangle$ in a form convenient for calculations. The form encountered in Eq. 21.107 is very common in quantum mechanics. We say that ϕ_2 is a *linear superposition* or a *coherent superposition* of the states $|\alpha\beta\rangle$ and $|\beta\alpha\rangle$. This has the following consequences. If I expose an ensemble of molecules, each having two distinguishable nuclear spins that interact with each other, to an oscillating magnetic field, I create molecules in the excited state $|\phi_2\rangle$. How are the spins in a given molecule oriented? Is nucleus 1 in the state $|\alpha\rangle$ and 2 in $|\beta\rangle$ or is 1 in $|\beta\rangle$ and 2 in $|\alpha\rangle$? According to the rules of quantum mechanics, if the state $|\phi_2\rangle$ is a coherent superposition of these two states, this question cannot be answered with certainty. It is only possible to calculate the probability that one of these situations occurs. In §31 I show how to use $|\phi_2\rangle$ to calculate this probability.

§30. *Why are $|\phi_1\rangle$ and $|\phi_2\rangle$ so Different?* The state $|\phi_1\rangle$ is equal to the state $|\alpha\alpha\rangle$, which is a state of the non-interacting spins. The interaction modifies the energy but does not modify the wave function! This is the situation assumed by perturbation theory and we know that, in general, this is true only if the interaction is weak. However, we have not assumed that the interaction is weak: therefore, the equality of $|\phi\rangle$ with $|\alpha\alpha\rangle$ is true no matter how strong the interaction is. Why is this happening? And why does this not happen to $|\phi_2\rangle$?

You can see why, if you look at the structure of the matrix $\overleftrightarrow{H}$ (see Eq. 21.100). The off-diagonal matrix elements $\langle\psi_1|\hat{H}|\psi_j\rangle$, $j = 2, 3, 4$, and $\langle\psi_j|\hat{H}|\psi_1\rangle$, $j = 2, 3, 4$, are all zero. When this happens, we say that the Hamiltonian $\hat{H}$ does not couple the basis-set function $|\psi_1\rangle$ to the other functions in the basis set. Whenever this happens, $|\psi_1\rangle$ is an eigenstate of the Hamiltonian and $\langle\psi_1|\hat{H}|\psi_1\rangle$ is the corresponding eigenvalue. Such situations are uncommon in other quantum mechanics problems.

$|\phi_2\rangle$ is more complicated than $|\phi_1\rangle$ because the off-diagonal elements $\overleftrightarrow{H}_{23}$ and $\overleftrightarrow{H}_{32}$ are not zero. When this happens we say that the Hamiltonian *couples* the basis-set states $|\psi_2\rangle$ and $|\psi_3\rangle$. As a result, $|\phi_2\rangle$ is a linear combination of $|\psi_2\rangle$ and $|\psi_3\rangle$.

The structure of the matrix $\overleftrightarrow{H}$ and the rules just discussed suggest that $|\phi_3\rangle$ will also be a coherent superposition of $|\psi_2\rangle$ and $|\psi_3\rangle$. Moreover, $|\phi_4\rangle$ will be equal to $|\psi_4\rangle$ and E_4 will be equal to $\langle\psi_4|\hat{H}|\psi_4\rangle$, which is the same result as perturbation theory.

§31. *The Physical Meaning of $|\phi_2\rangle$.* You have learned in Chapters 5 and 6 that only normalized states can be used to calculate the probability that a measurement yields a certain value. Therefore, before discussing the physical meaning of the state $|\phi_2\rangle$ we must normalize it. The normalized state is:

$$|\bar{\phi}_2\rangle = \frac{|\phi_2\rangle}{\sqrt{\langle\phi_2|\phi_2\rangle}} = \frac{c_2^2|\alpha\,\beta\rangle + c_3^2|\beta\,\alpha\rangle}{|c_2^2|^2 + |c_3^2|^2} \equiv \bar{c}_2^2|\alpha\,\beta\rangle + \bar{c}_3^2|\beta\,\alpha\rangle \qquad (21.108)$$

with

$$\bar{c}_2^2 \equiv \frac{c_2^2}{|c_2^2|^2 + |c_3^2|^2} \tag{21.109}$$

and

$$\bar{c}_3^2 \equiv \frac{c_3^2}{|c_2^2|^2 + |c_3^2|^2} \tag{21.110}$$

The first equality follows from the general procedure for obtaining a normalized state $|\bar{\phi}_2\rangle$ from the unnormalized state $|\phi_2\rangle$. The second equality in Eq. 21.108 follows from the properties of the scalar product and Eq. 21.106.

Next, let us assume that we have exposed the system in the state $|\phi_1\rangle = |\alpha\,\alpha\rangle$ to an oscillating magnetic field whose frequency is $(E_2 - E_1)/h$. Some of the molecules will absorb energy from the magnetic field and will be excited to the state $|\phi_2\rangle$. We may ask how are the spins of these excited state oriented? Are they both oriented along the field? The probability that this is the case is

$$|\langle \alpha\,\alpha \mid \bar{\phi}_2\rangle|^2 = |\langle \alpha\,\alpha \mid \left(\bar{c}_2^2 \mid \alpha\,\beta\rangle + \bar{c}_3^2 \mid \beta\,\alpha\rangle\right)\rangle|^2$$

$$= |\bar{c}_2^2 \langle \alpha\,\alpha \mid \alpha\,\beta\rangle + \bar{c}_3^2 \langle \alpha\,\alpha \mid \beta\,\alpha\rangle|^2$$

$$= |\bar{c}_2^2 \times 0 + \bar{c}_3^2 \times 0|^2 = 0$$

Note that the rules of quantum mechanics require that when we calculate probabilities, we use normalized wave functions; this is why $|\bar{\phi}_2\rangle$ appears in the first line of this equation, and not $|\phi_2\rangle$.

The probability that, when the system is in the state $|\phi_2\rangle$, nucleus 1 is oriented along the field (is in state $|\alpha\rangle$) and nucleus 2 is oriented in the opposite direction (is in state $|\beta\rangle$) is

$$|\langle \alpha\,\beta \mid \bar{\phi}_2\rangle|^2 = |\langle \alpha\,\beta \mid \left(\bar{c}_2^2 \mid \alpha\,\beta\rangle + \bar{c}_3^2 \mid \beta\,\alpha\rangle\right)\rangle|^2$$

$$= |\bar{c}_2^2 \langle \alpha\,\beta \mid \alpha\,\beta\rangle + \bar{c}_3^2 \langle \alpha\,\beta \mid \beta\,\alpha\rangle|^2 = |\bar{c}_2^2|^2$$

Similarly (do this as an exercise) the probability that nucleus 1 is oriented opposite to the field and nucleus 2 along it is equal to $|c_3^2|^2$.

We must therefore conceive the state $|\phi_2\rangle$ as one in which the spin orientations are not exactly known. Spin 1 is oriented along the field and spin 2, opposite to it, with the probability $|\bar{c}_2^2|^2$. However, there is a chance that spin 1 is opposite to the field and spin 2 is along it, and this has the probability $|\bar{c}_3^2|^2$.

Exercise 21.13

Use $\nu_0 = 300$ MHz, $\sigma_1 = 5.3 \times 10^{-4}$, $\sigma_2 = 0.31 \times 10^{-4}$, and plot $|\bar{c}_2^2|^2$ as a function of the coupling strength J, or, even better, as a function of $J/\nu_0(\sigma_2 - \sigma_1)$.

§32. *The States* $|\phi_3\rangle$ *and* $|\phi_4\rangle$ *and the energies* E_3 *and* E_4. I will give the other two states without further comment (see Workbook QM21.5). The energy E_3 is (see Cell 4 of Workbook QM21.5):

$$E_3 = \frac{h\nu_0}{2}|\sigma_1 - \sigma_2|\sqrt{1 + \left(\frac{J}{\nu_0(\sigma_2 - \sigma_1)}\right)^2} \tag{21.111}$$

The corresponding eigenvector $\vec{c}^{\,3} = \{c_1^3, c_2^3, c_3^3, c_4^3\}$ is:

$$\left.\begin{aligned}
c_1^3 &= 0 \\
c_2^3 &= -\frac{\nu_0(\sigma_2 - \sigma_1)}{J} - \frac{\nu_0|\sigma_2 - \sigma_1|}{J}\sqrt{1 + \left(\frac{J}{\nu_0(\sigma_2 - \sigma_1)}\right)^2} \\
c_3^3 &= 1 \\
c_4^3 &= 0
\end{aligned}\right\} \tag{21.112}$$

This leads to the following spin state $|\phi_3\rangle$:

$$|\phi_3\rangle = c_2^3|\alpha\,\beta\rangle + |\beta\,\alpha\rangle \tag{21.113}$$

The results for the fourth state are as follows:

$$E_4 = \frac{h\nu_0}{2}(2 - [\sigma_1 + \sigma_2]) \tag{21.114}$$

$$\vec{c}^{\,4} = \{0, 0, 0, 1\} \tag{21.115}$$

$$|\phi_4\rangle = |\beta\,\beta\rangle \tag{21.116}$$

Exercise 21.14

Derive a formula for the probability that when the system is in the state $|\phi_3\rangle$ the first spin is oriented along the field and the second is opposite to it.

§33. *The Orders of Magnitude.* I will calculate now the energies and the wave functions for the parameters used in a previous example: $\nu_0 = 300$ MHz, $\sigma_1 = 5.3 \times 10^{-4}$, $\sigma_2 = 0.31 \times 10^{-4}$, and $J = 6$ Hz. The "old" Planck's constant is $h = 6.626 \times 10^{-34}$ J s. The energies of the spin states are (see Workbook QM21.5)

$$\frac{E_1}{h} = -2.99987 \times 10^8 \text{ s}^{-1}$$

$$\frac{E_2}{h} = -3301.5 \text{ s}^{-1}$$

$$\frac{E_3}{h} = 3298.5 \text{ s}^{-1}$$

$$\frac{E_4}{h} = 2.99987 \times 10^8 \text{ s}^{-1}$$

The allowed transitions are the same as in the case of non-interacting particles, and their frequencies are (see Workbook QM21.5)

$$\nu_{12} = 299.98409 \times 10^8 \text{ MHz}$$

$$\nu_{13} = 299.990697 \times 10^8 \text{ MHz}$$

$$\nu_{24} = 299.9907 \times 10^8 \text{ MHz}$$

$$\nu_{34} = 299.984102 \times 10^8 \text{ MHz}$$

There are now four distinct absorption frequencies, while when the spins did not interact there were only two. The difference between these frequencies is small but it is detectable.

Exercise 21.15

The normalized states $|\bar{\phi}_2\rangle = |\phi_2\rangle/\sqrt{\langle \phi_2 | \phi_2 \rangle}$ and $|\bar{\phi}_3\rangle = |\phi_3\rangle/\sqrt{\langle \phi_3 | \phi_3 \rangle}$ are of the form $a | \alpha \beta\rangle + b | \beta \alpha\rangle$. Calculate a and b for both $|\bar{\phi}_2\rangle$ and $|\bar{\phi}_3\rangle$ when $\nu_0 = 300$ MHz, $\sigma_1 = 5.3 \times 10^{-4}$, $\sigma_2 = 0.31 \times 10^{-4}$, and $J = 6$ Hz. Are $|\bar{\phi}_2\rangle$ and $|\bar{\phi}_3\rangle$ very different from $| \alpha \beta\rangle$ and $| \beta \alpha\rangle$?

Exercise 21.16

Show that if

$$\left[\frac{J}{\nu_0(\sigma_2 - \sigma_1)}\right]^2 \ll 1$$

then the spin–spin interaction is negligible.

The NMR Spectrum of Two Weakly Interacting, Indistinguishable Nuclei Having Spin $\frac{1}{2}$

§34. *Introduction.* So far we have examined the NMR spectra of two identical nuclei (e.g. both protons) that were distinguishable because they had different shielding constants. An example is $Cl_2HC–OH$, where the two protons have different shielding constants because they have a different electronic environment. This is not the case for the protons in CCl_2H_2, because the field caused by the magnetic polarization of the electrons is the same at the two proton locations; these make the shielding constants of these protons equal.

We study now the case when the shielding constants are the same and the two particles are indistinguishable. This means that the states of these particles must be symmetric or antisymmetric (see Chapter 20). In the case of distinguishable particles we used the states $\psi_1 = |\alpha\alpha\rangle$, $\psi_2 = |\alpha\beta\rangle$, $\psi_3 = |\beta\alpha\rangle$, and $\psi_4 = |\beta\beta\rangle$, to construct the theory. When we used perturbation theory, these were the eigenstates of the system. If we went beyond perturbation theory, the states of the two spins were linear combinations of $|\psi_1\rangle$, ..., $|\psi_4\rangle$.

If the two nuclei are indistinguishable these function do not satisfy the symmetry requirements. Indeed, if the particles are fermions, a permutation of their labels must cause the state to change sign; if they are bosons, a permutation of the labels must leave the state unchanged. Let us look at the state $|\alpha\beta\rangle$. In this state, the spin 1 is oriented along the external magnetic field and spin 2 is oriented opposite to the field. If I permute the labels, spin 1 becomes spin 2 and vice versa and the state after this permutation is $|\beta\alpha\rangle$. This violates the symmetry rule which requires that the permutation should lead to either $|\alpha\beta\rangle$ or to $-|\alpha\beta\rangle$.

§35. *The States that Satisfy the Symmetry Requirements.* Because of the symmetry requirements, some of the states $|\alpha\alpha\rangle$, $|\alpha\beta\rangle$, $|\beta\alpha\rangle$, and $|\beta\beta\rangle$ are no longer acceptable. If we denote by $\hat{P}$ an operator that permutes the labels of the nuclei, one can easily see that

$$\hat{P}|\alpha\,\alpha\rangle = |\alpha\,\alpha\rangle$$

$$\hat{P}|\alpha\,\beta\rangle = |\beta\,\alpha\rangle$$

$$\hat{P}|\beta\,\alpha\rangle = |\alpha\,\beta\rangle$$

$$\hat{P}|\beta\,\beta\rangle = |\beta\,\beta\rangle$$

The states

$$|\lambda_1\rangle = |\alpha\,\alpha\rangle \tag{21.117}$$

and

$$|\lambda_4\rangle = |\beta\,\beta\rangle \tag{21.118}$$

are symmetric; they do not change when we permute the labels. The states $|\alpha\,\beta\rangle$ and $|\beta\,\alpha\rangle$ are neither symmetric nor antisymmetric. However, we can combine them to form two new states: the symmetric state

$$|\lambda_3\rangle = \frac{1}{\sqrt{2}}(|\alpha\,\beta\rangle + |\beta\,\alpha\rangle) \tag{21.119}$$

and the antisymmetric state

$$|\lambda_2\rangle = \frac{1}{\sqrt{2}}(|\alpha\,\beta\rangle - |\beta\,\alpha\rangle) \tag{21.120}$$

It is easy to check that $|\lambda_3\rangle$ is symmetric. Indeed

$$\hat{P}|\lambda_3\rangle = \frac{1}{\sqrt{2}}(\hat{P}|\alpha\,\beta\rangle + \hat{P}|\beta\,\alpha\rangle) = \frac{1}{\sqrt{2}}(|\beta\,\alpha\rangle + |\alpha\,\beta\rangle) = |\lambda_3\rangle$$

A similar calculation shows that $|\lambda_2\rangle$ is antisymmetric. The factor of $\sqrt{2}$ is included to enforce orthonormality.

Exercise 21.17

Show that the states $|\lambda_i\rangle$ are orthonormal.

I have now four states $|\lambda_1\rangle$, $|\lambda_2\rangle$, $|\lambda_3\rangle$, $|\lambda_4\rangle$ that either symmetric or antisymmetric when the labels of the nuclei are permuted. Which of these are appropriate for dealing with the NMR spectrum of two indistinguishable protons? The protons have spin $\frac{1}{2}$ and are fermions. It would seem that their state must be antisymmetric and this means that only $|\lambda_2\rangle$ qualifies. However, excluding the other functions would be incorrect. Here is why.

The symmetry principle must be applied to the complete wave function of the system, not just to the spin states. Under the Born–Oppenheimer approximation, this total wave function can be written as

$$\psi(1,2) = N_s(1,2)N_c(1,2)\psi_e(r;1,2) \qquad (21.121)$$

The labels 1 and 2 are those of the nuclei. $N_s(1,2)$ is the state of the nuclear spin, $N_c(1,2)$ describes the rotation and the vibration of the nuclei, and $\psi_e(r;1,2)$ is the wave function of the electrons. The latter depends on the electronic coordinates (symbolized by r) and on the nuclear coordinates 1 and 2 (see Chapter 20). If the nuclei are fermions, as in our case, we must have

$$\hat{P}\psi(1,2) = -\psi(1,2)$$

The permutation operator applied to $\psi(1,2)$ permutes the nuclear labels in each wave function in the product and therefore

$$\hat{P}\psi(1,2) = \hat{P}\left[N_s(1,2)N_c(1,2)\psi_e(r;1,2)\right] = [\hat{P}N_s(1,2)]\,[\hat{P}N_c(1,2)]\,[\hat{P}\psi_e(r;1,2)]$$

In almost all cases $\hat{P}\psi_e = \psi_e$. However, this is not the case for the rotational–vibrational states. In an ensemble of molecules at room temperature, a great variety of rotational states are populated. Some of these are symmetric and some are antisymmetric. For example, for a diatomic molecule the states $Y_\ell^m(\theta,\phi)$, which are a component of N_c, are symmetric if ℓ is even and antisymmetric if ℓ is odd. If N_c is symmetric then N_s must be antisymmetric, to make $\psi(1,2)$ antisymmetric. If N_c is antisymmetric then N_s must be symmetric, for the same reason. Because of this, both the symmetric and the antisymmetric spin wave functions need to be taken into account. This sketchy discussion is only intended to explain why we consider the symmetric spin functions when the particles are fermions.

§36. *Perturbation Theory.* $|\lambda_1\rangle, \dots, |\lambda_4\rangle$ are the states of two non-interacting, indistinguishable particles each having spin $\frac{1}{2}$. They can be used as the starting point for the calculation of the energies of the interacting spins. When we know that the interaction is weak, we can use perturbation theory. This assumes that the coupling between the spins is not strong enough to change substantially the eigenstates of the system. Therefore, $|\lambda_1\rangle, \dots, |\lambda_4\rangle$ are also the eigenstates of the system in which the spins interact. Because of this approximation, the energies in various states of the interacting system are

$$E_i = \langle \lambda_i | \hat{H} | \lambda_i \rangle \tag{21.122}$$

where (see Eq. 21.69)

$$\hat{H} = -\gamma B_0(1 - \sigma)\hat{I}_z(1) - \gamma B_0(1 - \sigma)\hat{I}_z(2) + \frac{h}{\hbar^2} J \, \hat{\mathbf{I}}(1) \cdot \hat{\mathbf{I}}(2)$$

The matrix elements $\langle \lambda_i | \hat{H} | \lambda_i \rangle$ are evaluated by the methods used earlier in this chapter. The results are:

$$E_1 = -h\nu_0(1 - \sigma) + \frac{hJ}{4} \tag{21.123}$$

$$E_2 = -\frac{3hJ}{4} \tag{21.124}$$

$$E_3 = \frac{hJ}{4} \tag{21.125}$$

$$E_4 = h\nu_0(1 - \sigma) + \frac{hJ}{4} \tag{21.126}$$

Exercise 21.18

Derive Eqs 21.123–21.126.

Exercise 21.19

Can you derive the energies (within perturbation theory) of the indistinguishable particles from those of the distinguishable particles, by making $\sigma_1 = \sigma_2$?

Compare these values with those obtained for perturbation theory of the distinguishable spins (Eqs 21.92–21.95). Only the energies E_2 and E_3 differ in the two cases. It would seem that the fact that the particles are indistinguishable does not make much difference. This statement is, however, erroneous. As you see below, the selection rules and the NMR spectra of the two systems are rather different.

§37. *The Selection Rules.* Now compare the selection rules for the two situations (distinguishable and indistinguishable spins). A calculation of the matrix element

$$\langle \lambda_i \,|\, \hat{I}_x(1) \cdot \hat{I}_x(2) \,|\, \lambda_j \rangle, \quad i \neq j, \tag{21.127}$$

which controls absorption rate and determines the selection rules, shows that

- only transitions that cause one "spin flip" are possible (i.e. $|\alpha\,\alpha\rangle \rightarrow |\alpha\,\beta\rangle$ or $|\alpha\,\beta\rangle \rightarrow |\beta\,\beta\rangle$ but not $|\alpha\,\alpha\rangle \rightarrow |\beta\,\beta\rangle$), just as in the case of distinguishable particles.

- only transitions between states of the same symmetry are possible (i.e. only symmetric to symmetric, or antisymmetric to antisymmetric); this is an additional restriction, which operates only for indistinguishable particles.

Exercise 21.20

Derive the selection rules: that is, evaluate the matrix elements in Eq. 21.127.

These two rules restrict the allowed transitions to

$$|\lambda_1\rangle \rightarrow |\lambda_3\rangle \tag{21.128}$$

and

$$|\lambda_3\rangle \rightarrow |\lambda_4\rangle \tag{21.129}$$

The frequencies of these transitions are

$$\nu_{13} = \frac{E_3 - E_1}{h} = \nu_0(1 - \sigma) \tag{21.130}$$

$$\nu_{34} = \frac{E_4 - E_3}{h} = \nu_0(1 - \sigma) \tag{21.131}$$

The two frequencies are equal. Therefore, these allowed transitions generate only one peak in the NMR spectrum. When the spins are distinguishable, there are four peaks.

Exercise 21.21

The symmetry requirement has a strange "all or nothing" character. Even if the two shielding constants differ infinitesimally, the character of the wave functions changes dramatically. Examine whether the change in the spectrum is equally dramatic. Is the change small when $|\sigma_1 - \sigma_2|$ is small?

Exercise 21.22

If the spin–spin interaction is strong, the state of the system is of the form

$$|\phi\rangle = \sum_{i=1}^{4} c_i \, |\lambda_i\rangle$$

Find the energies and the states of the strongly interacting spins (follow the procedure used in sections §20–§33 on pp. 443–457).

APPENDICES

Appendix 1. Values of Some Physical Constants

Adapted from D.A. McQuarrie, *Statistical Mechanics*, Harper & Row, New York, 1973, which cites B.N. Taylor, W.H. Parker, and D.N. Langenberg, *Rev. Mod. Phys.* **41**, 375, 1969.

Quantity	Symbol	Value
Avogadro's number	N_A	6.0222×10^{23}
Planck's constant	h	6.6262×10^{-27} erg s
	$\hbar$	1.0546×10^{-27} erg s
Boltzmann constant	k	1.3806×10^{-16} erg/mol K
Gas constant	R	8.3143×10^{7} erg/mol K
		1.9872 cal/mol K
Speed of light	c	2.9979×10^{10} cm/s
Proton charge	e	4.8032×10^{-10} esu
Electron mass	m_e	9.1096×10^{-28} g
Atomic mass unit	amu	1.6605×10^{-24} g
Bohr magneton	μ_B	9.2741×10^{-21} erg/gauss
Nuclear magneton	μ_N	5.0509×10^{-24} erg/gauss
Permitivity of vacuum	ϵ_0	8.854187×10^{-12} C^2/N^2 m^2

Appendix 2. Energy Conversion Factors

Source: D.A. McQuarrie, *Statistical Mechanics*, Harper & Row, New York, 1973.

	ergs	eV	cm^{-1}	K
1 erg	1	6.2420×10^{11}	5.0348×10^{15}	7.2441×10^{15}
1 eV	1.6021×10^{-12}	1	8.0660×10^3	1.1605×10^4
1 cm^{-1}	1.9862×10^{-16}	1.2398×10^{-4}	1	1.4388
1 K	1.3804×10^{-16}	8.6167×10^{-5}	6.9502×10^{-1}	1
1 kcal	4.1840×10^{10}	2.6116×10^{22}	2.1066×10^{26}	3.3009×10^{26}
1 kcal/mol	6.9446×10^{-14}	4.3348×10^{-2}	3.4964×10^2	5.0307×10^2
1 atomic unit	4.360×10^{-11}	27.21	$2.195 \times 10^{.5}$	3.158×10^5

	kcal	kcal/mol	atomic units
1 erg	2.3901×10^{-11}	1.4394×10^{13}	2.294×10^{10}
1 eV	3.8390×10^{-23}	2.3119×10^1	3.675×10^{-2}
1 cm^{-1}	4.7471×10^{-27}	2.8588×10^{-3}	4.556×10^{-6}
1 K	3.2993×10^{-27}	1.9869×10^{-3}	3.116×10^{-6}
1 kcal	1	6.0222×10^{23}	9.597×10^{20}
1 kcal/mol	1.6598×10^{-24}	1	1.594×10^{-3}
1 atomic unit	1.042×10^{-21}	6.275×10^2	1

FURTHER READING

Here is a list of books that you can read with profit if you decide to expand and refine your knowledge of quantum mechanics. I will give the smallest number of books that covers the material. There are two selection criteria: I like the book and you have adequate background to read it without pain.

1. Ira N. Levine, *Molecular Spectroscopy*, John Wiley & Sons, New York, 1975 and Ira N. Levine, *Quantum Chemistry*, 4th edn, Prentice Hall, New Jersey, 1991.

These books are clearly written and have the right amount of detail for the beginning physical chemist.

2. Michael D. Fayer, *Elements of Quantum Mechanics*, Oxford University Press, New York, 2001.

This book offers further insight into quantum mechanics and spectroscopy. It has a fine balance between the fundamentals and the applications and contains some modern material. It has many interesting comments from an outstanding researcher with vast experience in spectroscopy.

Two other books on spectroscopy are J. L. McHale, *Molecular Spectroscopy*, Prentice Hall, New Jersey, 1999 and W. S. Struve, *Fundamentals of Molecular Spectroscopy*, John Wiley & Sons, New York, 1989.

3. To increase your understanding of the theory of chemical bond read Attila Szabo and Neil. S. Ostlund, *Modern Quantum Chemistry*, Dover Publications, Mineola, New York, 1996. This book is well-paced to take the beginner to a point where you can try reading advanced books and monographs.

4. For a better understanding of the principles of quantum mechanics use Eugen Merzbacher, *Quantum Mechanics*, 2nd edn, John Wiley & Sons, New York, 1970. It makes excellent connections between the physical phenomena and the mathematics. Two books that deal with the fundamentals more formally than Merzbacher are Claude Cohen-Tanoudji, Bernard Diu and Franck Laloë, *Quantum Mechanics*, Vols 1 and 2, John Willey & Sons, New York, 1977 and John J. Sakurai, *Modern Quantum Mechanics*, Benjamin/Cummins Publishing Company Inc., Menlo Park, California, 1985. These books provide enough background to allow reading the original research literature.

INDEX